D0103346

CURRENT PROBLEMS IN ANIMAL BEHAVIOUR

CURRENT PROBLEMS IN ANIMAL BEHAVIOUR

EDITED BY

W. H. THORPE & O. L. ZANGWILL

CAMBRIDGE
AT THE UNIVERSITY PRESS
1963

PUBLISHED BY
THE SYNDICS OF THE CAMBRIDGE UNIVERSITY PRESS

Bentley House, 200 Euston Road, London, N.W. 1
American Branch: 32 East 57th Street, New York 22, N.Y.
West African Office: P.O. Box 33, Ibadan, Nigeria

©

CAMBRIDGE UNIVERSITY PRESS

1961

First printed 1961
Reprinted 1963

First printed in Great Britain by Adlard and Son Limited, Dorking
Reprinted by offset-lithography by Bradford & Dickens Ltd, London, W.C. 1

BEHAVIOUR DISCUSSION GROUP
1953–1958

LIST OF MEMBERS

R. J. Andrew	Department of Zoology, Yale University, New Haven, Conn., U.S.A.
H. B. Barlow	Physiological Laboratory, Cambridge.
A. D. Blest	Department of Zoology, University College, London.
D. E. Broadbent	M.R.C. Applied Psychology Research Unit, Cambridge.
V. R. Cane	Newnham College, Cambridge.
J. H. Crook	Sub-Department of Animal Behaviour, Department of Zoology, Cambridge.
B. A. Cross	Department of Veterinary Anatomy, Cambridge.
R. L. Gregory	Department of Experimental Psychology, Cambridge.
C. G. Gross	Department of Experimental Psychology, Cambridge.
A. D. Harris	Littlemoor Hospital, Oxford.
R. A. Hinde	Sub-Department of Animal Behaviour, Department of Zoology, Cambridge.
J. Kear	The Wildfowl Trust, Slimbridge, Gloucestershire.
G. A. Kerkut	Department of Zoology, Southampton University.
P. Klopfer	Duke University, Durham, North Carolina, U.S.A.
R. B. Klopman	Laboratory of Ornithology, Cornell University, Ithaca, N.Y., U.S.A.
A. Leonard	M.R.C. Applied Psychology Research Unit, Cambridge.

v

P. Marler	University of California, Berkeley, California, U.S.A.
G. V. T. Matthews	The Wildfowl Trust, Slimbridge, Gloucestershire.
D. M. Maynard	Biological Laboratories, Harvard University, Cambridge, Mass., U.S.A.
J. M. Oxbury	Department of Experimental Psychology, Cambridge.
C. H. F. Rowell	Department of Zoology, Cambridge.
T. E. Rowell	Sub-Department of Animal Behaviour, Department of Zoology, Cambridge.
J. R. Smythies	The Maudesley Hospital, London.
W. H. Thorpe	Sub-Department of Animal Behaviour, Department of Zoology, Cambridge.
W. G. van der Kloot	Harvard University, Cambridge, Mass., U.S.A.
M. A. Vince	Department of Experimental Psychology, Cambridge.
D. M. Vowles	Institute of Experimental Psychology, Oxford University.
D. I. Wallis	Department of Zoology, Cambridge.
R. P. Warren	Department of Zoology, Howard University, Washington 1, D.C., U.S.A.
A. J. Watson	Department of Experimental Psychology, Cambridge.
L. Weiskrantz	Department of Experimental Psychology, Cambridge.
J. Wells	Department of Zoology, Cambridge.
M. J. Wells	Department of Zoology, Cambridge.
O. L. Zangwill	Department of Experimental Psychology, Cambridge.

CONTENTS

vii

CONTENTS

PART IV

THEORETICAL APPROACHES TO BEHAVIOUR

GENERAL INTRODUCTION

When O. L. Zangwill returned to Cambridge in 1952 as Professor of Experimental Psychology, there had already developed in the Department of Zoology, and particularly in its attached Field Station at Madingley, a considerable interest in the study of animal behaviour. An active group of research workers, now organised as the Sub-Department of Animal Behaviour under W. H. Thorpe as Reader, had been engaged for some years on the analysis of a variety of behaviour patterns in insects and birds. This work had grown out of the earlier field studies of the ethologists and could be regarded as a continuation of their approach. At the same time, it owed much to advances in other fields of biological study, in particular the comparative physiology of the sense organs. Further, attempts were made wherever possible to submit the data of field observation to more precise experimental analysis. It was felt that the experimental study of animal behaviour, besides being attractive in its own right, had much to contribute to the solution of fundamental problems in biology.

An interest in animal behaviour had existed in the Cambridge Psychological Laboratory before the war and had given rise to some valuable studies. But work in this field was discontinued in 1940, after which the main effort of the Laboratory was concentrated upon the experimental analysis of human skills and its application to practical problems. Although work along these lines continued to flourish in Cambridge, Zangwill believed that the time was ripe for a revival of interest in comparative psychology. Indeed he ventured to hope that this field might well become a major long-term interest of the Psychological Laboratory.

It must be borne in mind that, in the past, the psychologist's approach to problems of behaviour differed in important respects from that of the ethologist. In general, the ethologist has stressed the importance of relating the problems he studies

ix

to behaviour under natural conditions of life. For reasons which are outlined below, he has limited himself very largely to the analysis of the simpler and more highly stereotyped behaviour patterns of animals less highly evolved than the mammals. The psychologist, on the other hand, has found his principal interest in the more general aspects of behaviour, in particular the nature of learning. Accordingly, he has worked mainly with mammals—above all the white rat—and has devoted but little attention to interspecies differences or to the significance of his findings for behaviour in its natural setting. Indeed one may surmise that the psychologist has chosen to work with animals rather than men largely on account of their lesser complexity and greater tolerance of the indignities of experiment! In his defence, however, it may be said that the psychologist is less interested in how particular animals behave than in the explanation of behaviour in general. His ultimate goal is a systematic body of theory applicable at all levels of the evolutionary hierarchy.

To the ethologist, the psychologist obsessed with the rat in the maze was apt to appear so oblivious of the realities of animal behaviour as to be almost a figure of fun. Yet he sometimes forgets that all animals learn something, be it only a simple habituation, and that many animals spend large parts of their lives in learning. On his side, the psychologist—in so far as he knows anything at all about the work of the ethologist—was apt to regard it as anecdotal and as lacking in scientific method. He often failed to realise that the study of insects, birds and fishes, in the wild and in captivity, can be made quite as scientific as that of the white rat in the laboratory and that field observations can often be quantified with at least as much precision as the results of a maze-running experiment.

The view that each party tended to have of the other, while containing elements of truth, was of course ludicrously inadequate. To anyone who knows a certain amount about the work of both, it is obvious that advance in behaviour study has been gravely handicapped by this lack of understanding between two groups who are really working on the same subject. Here, then, appeared a splendid opportunity to build a bridge, and it seemed to us that the best way of doing so was to organise a small group of research workers in zoology and psychology actively engaged in the study of behaviour. By meeting together regularly in informal discussion, it was hoped that workers in each discipline might come to learn and understand the problems, attitudes of mind and methods of approach of the other. In the long run, we hoped, a more unified approach to the study of animal behaviour would emerge.

A group of this character was organised by us in 1953 and consisted at first of approximately six research students or junior staff from each side. Meetings were held three or four times a term in Jesus College to discuss any subject which seemed of mutual interest. The expected difficulties did indeed materialise: the outlook of the two 'sides' was sometimes so different, and the language so technical and specialised, that it was very easy to argue at cross purposes and the arguments sometimes seemed rather sterile as the two sides missed each other's points. None the less, some progress was made towards mutual understanding and the group at least maintained its integrity.

During the first year, the discussions centred around various chapters of Hebb's *Organization of Behaviour* which had appeared in 1949. Then distinguished visitors and, later, individual members of the Group were asked to give an account of their particular lines of investigation, stressing the theoretical implications and the prospects for advance which they felt might emerge from their work. During this time it became clear to all that there was indeed an immense area of common ground over which mutual discussion and argument was extremely profitable. It transpired that what had at first appeared to be very different problems had in fact common bases and that psychologists and zoologists were often working from different directions towards a common solution.

There was, however, one main difference in approach which

is rather fundamental and has to be borne in mind in all general discussions of this kind. As we have said, the psychologist is primarily interested in mechanism and motivation in the individual and its significance for general theories of behaviour. The zoologist, on the other hand, is also interested in comparison of species with a view to understanding how behaviour evolved, how it became stereotyped here and flexible there. In short, biology is still very largely dominated by the concept of evolution and by the desire to trace origins and phylogenies. This has made itself felt in the study of behaviour no less than in that of structure. For instance, no student of the behaviour of related animals can fail to be struck by the uncanny similarity in the fixed or 'instinctive' components of behaviour which a species exhibits. For him, the study of behaviour takes on additional significance within the context of evolutionary biology.

In contrast to the ethologist, the psychologist—even though he may call himself comparative—has seldom concerned himself with the problems of origins or with behaviour within a range of natural species. His comparisons are only too often between rat, monkey and man, and are even then largely limited to differences in 'intelligence' or learning capacity. And whenever psychologists and zoologists seem to get at cross purposes in discussion, it is always worth while to look again at the fundamental method of approach. In our experience, disagreement over the interpretation of some set of facts or some point of terminology only too often resolves itself into one of difference of intellectual attitude.

In the course of two or three years, the Group had organised itself into some sort of unity and cohesion, and had incidentally become a good deal larger. The subject as a whole was creating so much interest that to maintain our number at twelve seemed unduly restrictive and would indeed have resulted in keeping out some who had obviously much to give. The question then arose as to whether the now quite closely-knit body had not achieved some unity of purpose and interest which was worth making available to a wider circle. And so the idea arose of creating a book. There was a good deal of discussion as to how much organisation, guiding and editing such a book would need, what it should be about, and what limits should be set to it. It was felt that for one or two to attempt to organise the rest into a team for book-writing would savour too much of authoritarian methods and almost certainly not yield a result representative of the Group's spirit of free and vigorous discussion.

And so another course was decided upon. Every member was asked to write an essay on an aspect of his own work or discipline which appeared to him interesting and vital for the study of behaviour as a whole. The choice of subject was left entirely to the author. As each essay became available, it was duplicated and copies given to all the other members of the Group. The essayist was then called upon to defend his thesis or yield as gracefully as he could in the face of their combined onslaughts. After each contribution had been thus dismembered and masticated, the author was asked to recreate it in the light of the comments and criticisms received and of the general discussion of his paper.

It soon became clear to the two editors of the present book that the unity of the Group was expressing itself to some extent in the choice of subject and attitude of mind of the contributors, and that the essays, taken together, went some way to demonstrate what such a group as this felt to be important, and why. We had also the feeling that this discussion group did in some sense represent a large body of interest the world over and that its activities might serve, in rather haphazard manner perhaps, to draw attention to some of the most interesting growing points in the study of behaviour at the present time. For these reasons, it seemed to us that the essays which the Group had read and discussed together might, in revised form, prove worthy of publication.

The task of the editors has been to assemble the revised essays in appropriate sections and bring out wherever possible their relations both to one another and to the larger field of study within which they fall. We have also tried to indicate the different approaches to common themes and the common approaches to different themes among the various contributions. As will appear, these similarities and differences not infrequently cut across departmental affiliations and traditional disciplines. Indeed, there is a sense in which the modern study of behaviour is truly inter-disciplinary, and the extent to which this is so can perhaps be judged from a scrutiny of this book.

W. H. THORPE

O. L. ZANGWILL

ACKNOWLEDGEMENT

We are much indebted to Miss E. M. Barraud who has been responsible for typing various drafts, for much proof reading, and for preparing the indexes.

PART I

NEURAL MECHANISMS
AND BEHAVIOUR

INTRODUCTION

It is nowadays accepted in biology that the central nervous system is to be envisaged as the instrument of behaviour. Its functions comprise not only the co-ordination of the different organs and systems of the body but the adaptation of the organism as a whole to its environment. Implicit in Darwin's outlook, this view of behaviour gained strength through the writings of Lloyd Morgan (1894), Loeb (1901) and McDougall (1905), and achieved definite expression in Sherrington's *Integrative Action of the Nervous System* (1906). For Sherrington, it will be remembered, integration is the central feature of all nervous activity from the simplest reflex arc to the most complex achievements of the cerebral cortex. At the higher levels at least, its most significant property lies in the welding of the discrete, specific and segmental activities of the organism— studied with such brilliance by Sherrington himself at the spinal reflex level—into a coherent pattern of adaptive action. Its final outcome is that sustained, plastic and delicate adaptation of the whole organism to its environment which constitutes its behaviour.

If we accept this standpoint, it is plain that the science of behaviour which we hope to build should be firmly grounded in the comparative study of the nervous system. This point was certainly taken not only by Pavlov but by the generation of psychologists who formed the Behaviourist movement in America. Although Behaviourism suffered from a surfeit of conditioned reflexes, and was disfigured by some ugly polemics,

I

it has made possible an approach to psychology which brings it much more fully within the scope of the biological sciences. Indeed, the work of such men as Pavlov, Lashley, Coghill and Weiss may be said to have laid the foundations of a psychology decisively linked to general biological issues.

Although the gain to psychology of an evolutionary outlook can hardly be contested, it has not been altogether without its limitations. In the first place, it has sometimes led to the erection of elaborate theories of behaviour on the most tenuous foundations of physiological fact. And in the second place, it has given rise to broad assumptions about the organisation of nervous activity which have not always been submitted to critical scrutiny. In Part I of this book we shall be concerned essentially with three such assumptions which have acquired wide currency in contemporary thought.

It has been commonly held that, however diverse the structure of the nervous system may be in different animals, its basic functional properties are surprisingly uniform. As Vowles points out, comparative neurophysiology has proceeded largely on the assumption that nervous integration *wherever it may be found* is governed by a few, basically similar, physiological mechanisms. In the nervous system of insects, he suggests, the possibilities of integration are far more restricted than in the vertebrate nervous system owing to the inherent limitations of the insect neuron as an integrating unit. It is therefore probable that insects have evolved types of nervous mechanism very different from those of vertebrates in order to solve very similar problems of adaptation. As Vowles rightly concludes, complexity of behaviour cannot necessarily be equated with equivalent complexity of nervous mechanism and the similarity of behaviour at different phylogenetic levels is in consequence more often apparent than real.

Secondly, it has been widely held by neurologists that increasing complexity of behaviour—in the vertebrate series at least—is correlated with an increasing dominance of the forebrain. This hypothesis, which goes back at least to Sherrington, is certainly true in a very broad sense. But it does not necessarily follow—as is commonly believed—that functions which, in less highly evolved forms, are organised at lower levels of the nervous system become increasingly 'displaced' to the fore-brain with advancing phylogenetic status. Weiskrantz offers a careful analysis of this concept of 'encephalisation', with special reference to the central representation of vision in mammals. He makes it clear that much of the evidence which has been adduced in support of the view that the relative importance of the visual cortex increases with phylogenetic status is open to serious objection. As he rightly points out, experimental methods have been far from uniform and the results are commonly open to more than one interpretation. Further, changes in function with evolutionary advance may well be more complex than is implied by a mere shift in the level of control. Although Weiskrantz's views may well excite controversy, his chapter represents a fresh and much-needed revaluation of an important neurological principle.

Thirdly, it has long been contended that the complexity of behaviour, in mammals at least, is in some ways determined by considerations of brain size. This view has found many protagonists, from Goltz and Loeb at the end of the last century to Lord Adrian in the middle of this. Although arguments in its support have been drawn from many disciplines, it has gained its main impetus from Lashley's long and admirable series of studies on brain mechanisms and intelligence in the rat and monkey. According to Lashley, the efficiency of behaviour is governed in an important sense by the total mass of functional brain tissue. In the case of complex tasks, at least, he found that impairment of performance following lesions of the brain cortex is closely related to the extent of the lesion and within large limits independent of its location. This led Lashley to put forward a principle of mass action governing the integration of behaviour. In his review of Lashley's work, Zangwill points out that this principle is altogether less well-founded than is commonly supposed. Although a relation of some kind may exist between brain mass and performance, he suggests that many of Lashley's findings can be interpreted within a more conservative theoretical framework.

The chapters which form the first part of this book are admittedly critical in intention. None the less, it is hoped that they will serve a constructive purpose. Although intended originally for discussion within a small group, it is believed that their presentation here will make possible wider discussion of the issues which, separately and together, they raise. Apart from stimulating fresh experiment in their respective fields, these chapters perhaps represent yet a further advance towards that 'coalescence' of neurology and psychology for which Lashley has offered so eloquent a plea.

I. NEURAL MECHANISMS IN
INSECT BEHAVIOUR

Any student of animals and plants sometimes finds himself adopting two incompatible points of view. He marvels at the diverse structure of living organisms, with all their complexity and adaptiveness, and yet beneath their superficial variety of structure he seems to see only a few, basically similar, physiological mechanisms. This biological schizophrenia can be allayed, if not cured, by evolutionary diagnosis, with its stress upon homology and common origins. The attractiveness of this treatment has, however, sometimes led to the over-stressing of similarities and the neglect of differences. This is particularly true of comparative neurophysiological studies, and in many text-books one can find the implicit assumption that although the general anatomy of the invertebrate central nervous system (C.N.S.) differs from that of the vertebrate, the properties of their individual neurons and their functional organisation are the same. There is a naïve, but unstated, belief that the invertebrate C.N.S. is really just a simpler version of the vertebrate type. The object of this chapter is to describe some of the properties of the insect nervous system and to suggest what influence these properties may have upon insect behaviour. The properties of individual neurons and groups of neurons will first be discussed; this will be followed by an analysis of receptor and motor systems and simple sensori-motor co-ordinations; and finally the evolution of the brain will be considered.

1. *The properties of nerve cells*

The insect nervous system differs from the vertebrate at all functional levels, not least in the structure of its individual cells. An important feature of the insect neuron is its small cell body, which is almost entirely filled with nucleus. The cell bodies all

5

lie around the periphery of the C.N.S. and as yet they have not been shown to play any part in integration: they may be comparable with the dorsal root ganglion cells in vertebrates. From the cell body the 'axon' makes its way into the central neuropile where all synaptic activity and conduction occur. The axon is typically Y-shaped, a short stalk from the cell body dividing into two or three branches which run to different regions of the

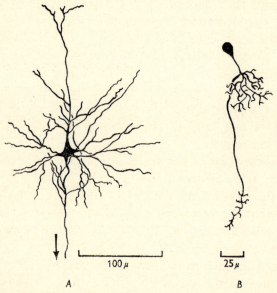

$100\,\mu$ $25\,\mu$

A   B

Fig. 1. Tracings of, A, a pyramidal neuron from the sensori-motor cortex of a cat; and, B, a bipolar neuron from the optic ganglion of a honey bee. Fig. A is from Sholl (1956).

C.N.S. where they subdivide into dense aborescences of fine twig-like processes, giving the cell the form of a piece of string, frayed at both ends. The terminal arborisations from different cells are closely entangled to form synaptic junctions. Synaptic knobs have recently been detected by electron-microscopy (Wigglesworth, 1959), but no details are known of their functional arrangement. Where branches from different cells are intermingled, this raises the possibility of temporal and spatial summation of excitatory and inhibitory processes—the two mechanisms upon which all forms of nervous integration depend.

Nervous integration may be regarded basically as a process whereby there is a change in pattern from input to output of a neural mechanism, rather than a simple relaying of the input pattern with only some form of amplification. The change of pattern must then depend upon the actual pattern of inputs while the output pattern depends not on any simple additive effect of the input components but on the interaction between them—a process perhaps more analogous to chemical than physical reactions.

Nervous integration in this sense must depend upon the patterns of interconnections between neurons. The possibilities of integration are, however, less in insects than in vertebrates for at least three reasons:

(i) In some parts of the vertebrate nervous system, such as the sensory relay centres of the thalamus or the motor neuron pools in the spinal cord, the surface of the cell body provides an important receptive and integrating surface for incoming nerve impulses. Where such a surface is present, synaptic knobs seem to have been especially developed. In other parts of the nervous system, such as the cortex, the cell body provides only a small part of the receptive surface of the neuron (Sholl, 1956). In insects even this small surface area is absent, and the neuron receives its input entirely over its dendritic surfaces and not at all over its cell body (see, for example, the histological appearances shown by Cajal and Sanchez, 1915).

(ii) In insects the receptive surface area of the dendrites is very much smaller than in vertebrates (Vowles, in preparation). This is because there are fewer dendritic branches from each cell in insects, and these branches extend for shorter distances. Both these effects act to reduce the surface area of the dendrites.

(iii) The dendritic arborisations of the insect neuron are very compact structures, very densely fibrillated. Within these arborisations the branches of different presynaptic fibres are

7

very densely and intricately entangled. It seems possible that whether synaptic transmission occurs or not will depend upon how many, rather than which, presynaptic fibres are active. Indeed, it seems significant that in regions where specificity of connection is required this seems to be achieved by reducing the degree of branching of the fibre. The latter then has a single central stalk from which short side branches protrude only slightly, like the teeth in the saw of a sword-fish.

Although future histological and physiological work may revise this picture, it seems probable at present that the insect neuron has fewer potentialities as an integrating unit than has the vertebrate. If the potentialities for interaction between the inputs on a single neuron are so reduced, this will have several important influences upon behaviour. Before discussing this, another property of the insect neuron will be described.

A characteristic of the insect neuron is the slow speed at which it conducts a nerve impulse along its surface. The individual neuron conveys information by means of the frequency of its discharge, and this is *not* influenced by the speed at which an impulse travels. But if a complex activity is performed, which involves a sequence of successive integrations in different parts of the C.N.S., then the speed at which information is carried from place to place will obviously influence the total time taken by the activity, and perhaps its latent period. In fact insect neurons (even the giant fibres) conduct far more slowly than their vertebrate counterparts—about ten times slower—partly because of the thinness of their myelin sheath and their lack of medullation.

Another disadvantage of lengthy integration processes might be that the 'line busy' effect would unduly delay other activities, or if such postponement were impossible a confusion might arise from a simultaneous use of one neuron for two different purposes. This defect could in theory be remedied by increasing the number of neurons available, thus providing alternative routes for carrying the same information, freeing individual neurons to take part in different activities at different times and allowing different integrations to be performed simultaneously in parallel (see, for example, the 'cell assembly' concept of Hebb, 1949).

An increase in the number of neurons in the nervous system would also remedy the deficiency of the neuron as an integrating unit. The possibilities of interaction between numerous inputs can be increased by increasing the number of cells on to which they can fall (either directly or via intermediate cells). However, the nature of the insect vascular and respiratory systems is such that they are unable to cope with bulky nervous tissue. In bees, for example, where the brain is large, special blood vessels and air sacs have had to be evolved to deal with its metabolic needs but even in such an extreme case the volume of the brain is still very small when compared with that of a vertebrate. The fact that insect neurons are much smaller than a vertebrate's (Vowles, in preparation) suggests that the density of insect neurons in the C.N.S. is much greater, and by reducing the size of individual cells the insects have to some extent compensated for the restriction on total brain volume imposed by other physiological systems.

It seems, however, that in the insect nervous system we are dealing with nerve cells which are potentially less efficient as integrating units than are vertebrate neurons, and whose deficiencies cannot wholly be remedied by increasing the number of neurons in the nervous system as a whole. The implications of this will now be discussed.

2. *Receptor systems*

If the neurons of the insect C.N.S. are unable to deal adequately with numerous inputs, this ought to be correlated with the evolution of simple sense organs; for where the number of cells in the C.N.S. is small *and* the receptive surface of each neuron is limited, the first factor will necessitate convergence and the second militate against it. One would therefore expect the insect sense organ to be neurologically simple, with few receptor cells and little complexity of interconnection in the peripheral

9

layers—unlike, for example, the vertebrate retina. When one considers the size and complexity of insect sense organs, however, they seem to deride this expectation. The compound eyes are enormous, the antennae are covered with a thick felt of sensory hairs, as are the mouth parts and tarsi, the exoskeleton and its joints are well supplied with proprioceptors, and there are complex auditory organs in the legs, thorax and antennae. It is, however, important not to confuse physical elaboration with neural complexity, for the two may not occur together—indeed, in the human eye, where an optically poor image (Hartridge, 1947) is moving continuously over the retina (Ditchburn and Ginsborg, 1953), the nervous system compensates for the optical defects, suggesting that physical and neural factors involved in sense organs may be inversely related. Analysis of the actual structure of insect sense organs and of their mode of function does in fact suggest that their neural organisation is simple and that the information which they can convey to the C.N.S. is more restricted than one might expect from their superficial structure. This point will now be discussed for the different sense organs.

In the blowfly (Dethier, 1955; Grabowski and Dethier, 1954; Hodgson, Lettvin and Roeder, 1955) the chemosensory hairs on the feet and the mouth parts each contain only two cells, one sensitive to sugars and the other to acids. In the feeding response, whether or not the proboscis is extended following stimulation of these hairs seems to depend on the difference between the total levels of excitation in all the cells of the two types stimulated. If contact chemoreception is as simple as this, it would be interesting to know the 'taste spectrum' of the fly, and, if this were complex, how the receptors convey information indicating subtleties of taste.

The sensitivity of insects (or indeed of any other animals) to odours has never been investigated as systematically or as precisely as have their other sensory systems. The sense of smell is still the most mysterious of all, in part because of the difficulties of technique in smell experiments, and in part because of the lack of any precise and fruitful hypothesis about the mechanism of olfactory reception. It has, however, been suggested that in insects the axons form the primary sense cells in the antennal nerve, and histological studies support this (Wigglesworth, 1959). A similar situation has also been demonstrated in the proprioceptive nerve from the campaniform sensillae (Pringle, 1938) and in the nerve from the ocelli (Parry, 1947 ; and personal observation), although different species may differ in this respect (Hoyle, 1955a; Ruck, 1957). Where afferent neurons do fuse in this way a high degree of convergence is obtained, but the possibility of integration by synaptic interaction is lost, and in a population of fused fibres information can no longer be conveyed in terms of *which* fibre is active. Even in the cercal receptors of the cockroach, where true synaptic transmission occurs between the sensory neurons and the giant fibres of the ventral nerve cord (Pumphrey and Rawdon-Smith, 1937) the system is so arranged that, providing a minimal number of any of the cercal receptors are stimulated (as by a puff of air), the giant fibres will discharge, and the system has become a simple relay, the receptors communicating only the presence and strength of the stimulus and not its quality or position. Even after this it is the brain rather than the stimulus which determines the speed and time for which the insect will run following the original stimulus (Roeder, 1948).

Similarly, although the optical system of the insect compound eye is complex and elaborate, the neural organisation of the retina is very simple—complexities of interconnection only arise within the optic ganglia of the brain. The compound eye does not seem to have evolved to give an image in the same way as a box camera, or the vertebrate or cephalopod eye, from which the visual world as we know it can be organised and perceived. Rather, with its many radially pointed ommatidia, the insect eye seems to have evolved to give the insect a system very sensitive to movements in the visual field; displacement of a stimulus causing a change in the levels of excitation of the overlapping receptive fields of adjacent ommatidia (Burtt and Catton, 1954). Visual orientation can be interpreted in terms of responses to patterns of movement in the visual field, and even the form perception of bees seems to depend upon the pattern of flicker which the bee sees when it moves relatively to a flower (Hertz, 1929–1935, 1937; Wolf and Zerrahn-Wolf, 1935). Such sensitivity is enhanced by the failure of a flashing light to fuse into a continuous stimulus at high frequencies of flicker, and Autrum and Stoecker (1950) have shown that in some insects the flicker fusion frequency (f.f.f.) may rise to as high as 300 c.p.s. However, this seems to be a function of the optic ganglia reacting upon the retina, for if they are removed the f.f.f. of the isolated retina (as measured by the electro-retino-gram) falls to a very low value (Autrum and Gallwitz, 1951). Further, Hassenstein (1951) has shown that the perception of movement is achieved by correlating activity only in adjacent pairs (or sometimes triplets) of ommatidia, and the insect does not react to stroboscopic movement if the successive stimuli are separated in space by more than two ommatidial angles. Thus analysis of visual function shows that the compound eye performs neurologically simple activities.

A sensory system where greater complexity of behaviour is found is the auditory mechanism, for the single receptor cell and its axon can convey more information than usual, thus compensating for the simplicity of the chordotonal organs as neural wholes. This has been accomplished by coding the information in terms of the rate of change of frequency of discharge in single neurons (Pumphrey, 1940; Pringle, 1954; Haskell, 1956). In crickets, locusts, cicadas and some other insects, singing plays an important part in courtship and threat behaviour. Each species has a variety of songs, and different songs have different functions (Jacobs, 1953). The insects have therefore to respond selectively to the different songs of their own species, and to ignore those of other species, which must have a different pattern. In these insects the song consists of a sound of high but variable frequency which is varied systematically in intensity (loudness). The frequency of this basic carrier note is unimportant, and the pattern of the song is the temporal pattern of changes in loudness. The chordotonal sensillae of insects are sensitive to a wide range of frequencies, but if stimulated by a tone modulated in intensity the frequency of their discharge follows the pattern of modulation. This means that in the song the information is coded in terms of temporal variations in intensity (amplitude modulation), and in the receptor this is translated into temporal variations in frequency of discharge (frequency modulation). In such a system the potential number of different messages is very great, and would seem to be limited by the characteristics of the sound-producing system, and the ability of the receptor system to give separable bursts of impulses.

The chordotonal organs also have a directional sensitivity to sound (Pumphrey, 1940). The efficiency of stimulation of the receptors depends on the direction from which the incident sound comes, relative to the tympanic membrane. The nerves can convey information about this in terms of how many fibres are active and the average frequency of discharge as the insect turns relative to the source of sound. It seems possible that by comparing the stimulation on the two sides of the body the insect might get a directional 'fix' for the source of sound at least over short distances. In the ear of Noctuids there are only two sensory cells, and the direction of the sound might be represented by the relative latencies of the discharges from the two cells on the two sides of the body (Roeder and Treat, 1957; Roeder, 1959). This method is obviously far less complicated than that used by mammals (Stevens and Davis, 1938).

It will be seen therefore that in the auditory receptors a potentially large amount of information can be received and conveyed by utilising a system of coding not employed elsewhere; and this can be correlated with the methods of sound production available to animals without lungs.

In the light of these considerations, it seems reasonable to conclude that the neural structure of insect receptors is more simple than might be anticipated from their physical structure, and this is in accordance with the hypothesis proposed in this chapter.

3. *Motor systems*

Since the sensory systems of insects show a simplicity of neural organisation it might be expected that their motor systems would be correspondingly uncomplicated, for the properties of neurons which limit integration in one system should also limit it in the other. An animal's behaviour can be described in terms of movements, attitudes and postures, and these can all be analysed into patterns of contractions of single muscles. A discussion of the motor system of an insect must therefore start with the control of contractions in single muscles.

Three factors have to be considered in this connection:

  (i) How a particular tension is maintained.

  (ii) How the tension can be changed.

  (iii) How the rate of change of tension can be controlled.

In vertebrates the amount of tension developed depends upon the total number of muscle fibres contracting: since each fibre is innervated by only a single motor neuron, the tension is therefore determined by the total number of motor neurons discharging. Where a wide range of tensions and a fine degree of control are needed, this can be accomplished by having a large number of motor neurons, each supplying only a few muscle fibres. In the eye muscles, for example, a motor neuron may supply as few as two muscle fibres, while in the biceps, where less fine control is needed, a single neuron may control as many as 150 fibres (Young, 1957).

Different tensions are produced by varying the number of motor neurons active, by means of central excitation or inhibition within the spinal cord. Similarly, a constant tension is maintained by a feed-back system from the proprioceptors, which respond to variations in tension and recruit or inhibit motor neurons to counteract these changes. Different types of muscles have been developed for different purposes, rapidly acting and fatiguing types being used for voluntary activity and rapid postural adjustments, and slow acting and slow fatiguing types for maintaining long tonic contractions (see, for example, Creed, Denny Brown, Eccles, Liddell and Sherrington, 1932; Walsh, 1957).

In insects, where both the number of motor neurons and muscle fibres are small, one might expect a muscle to show a rather restricted range of coarsely controlled tensions. In fact, the number of motor neurons controlling a single muscle is very small indeed, and may be as few as four or two (Hoyle, 1957). In spite of this, the muscle tension is controlled over a wide range with a high degree of sensitivity. This has been achieved by evolving a different type of control of muscle contraction from that found in vertebrates: every muscle fibre is innervated by most of the motor neurons, and most of the motor neurons innervate every fibre. Each neuron has several endings in each muscle fibre and the muscle, which has no effective action potential, contracts only around the regions of the active endings. The extent and character of the contraction therefore depend upon the relative activity of the different motor neurons: these have different functional properties, some producing rapid, phasic contractions and others a slow, tonic activity. Inhibitory fibres have also been described, which depress a muscle's excitability and can therefore reduce the effect of other neurons. It is clear, therefore, that the integration of excitatory and inhibitory processes is decentralised on to the muscle itself, and this could perhaps be taken to support the idea that the neurons themselves are incapable of sufficiently adequate integrations. It seems probable that the primitive type of control may be peripheral, and that insects have refined this to an extremely efficient system. The advantages to the vertebrate in developing central control may lie in the greater possibilities for rapid complex co-ordination and variability of response. This will be discussed in the section dealing with the brain.

The maintenance of a particular tension again seems to be

controlled by a proprioceptive feed-back system, probably from proprioceptors which act as strain gauges in the chitinous exo-skeleton (although note Finlayson and Lowenstein, 1955 and 1958). The limitations of these receptors have already been discussed. In vertebrates it is well known that peripherally the rate of change of tension is controlled by a feed-back system from the proprioceptors. This acts to damp the contraction, which might otherwise be damagingly rapid (Ruch, 1951). This control is a complex activity involving not only simple inhibitory processes but also a system whereby the intrafusal fibres can be contracted relatively independently of the general muscle fibres, thus making them independent of the tension developed in these general fibres. The intrafusal system can also come under direct central control, and the brain can thus activate or depress contractions by under- or over-facilitating the existing reflex arcs (Granit, 1955).

In insects such a complex neural system does not seem to exist, and the rates of change of tension can be mediated by the motor neurons already described. Hoyle (1955b) has shown that in the locust the sudden sharp rise in tension necessary for jumping is produced by timing the impulses in the 'inhibitory' and 'excitatory' neurons, so that the first volley hyperpolarises the muscle membrane and the second can then cause a more rapid and greater depolarisation (and hence contraction) than from the normal resting potential. Again the peripheral factor controlling rate of change of tension operates on the muscle fibre and not on the neuron.

4. *Co-ordination of sensory and motor systems*

Since the motor neurons provide the final common paths for the excitations involved in many different activities, and since the capacities of the individual neurons to deal with numerous inputs are limited, there must be some restriction on both the number of different behaviour patterns in which a muscle can be active and the number of different sensory activities with which it can be co-ordinated. These restrictions could be over-come by performing the necessary integrations elsewhere in the C.N.S. but, as will be seen later, there is reason to suppose that this cannot be done. Alternatively, it is possible either to have many of the same components shared by different behaviour patterns, or to have rather simple forms of behaviour in which each act involves only a part of the general musculature. It might prove difficult to distinguish between these alternatives: if, for example, the legs showed the same standing postures during feeding, courting and egg-laying, should one consider the 'standing' as being shared by all three patterns of behaviour, or merely that the legs, when not in other use, have to do something and just continue to stand? The difference would seem to be that in the first case certain groups of muscles are actively drawn into the different acts, whereas in the second they are either not affected, or have some of their potential activities inhibited.

Lack of information on this point makes it difficult to decide whether insects have adopted either or both of these alter-natives. In general it seems that where behaviour is complex, involving the activity of many different muscles, this complexity is one of the temporal sequence of different movements, rather than of simultaneous, spatial patterns. For example, when a female insect is searching for a suitable substrate on which to lay her eggs, this may involve a flight controlled by eyes and antennae, but once the substrate is found the isolated abdomen itself can lay the eggs satisfactorily (unless a feat such as wood-boring is necessary), particularly as the ovipositor may be supplied with its own muscles and chemoreceptors to guide it (Dethier, 1947a). Similarly, during the mating of insects such as locusts, the female may wander about and eat while her abdomen copulates with that of the male on her back. This would suggest that any single act consists of a fairly simple pattern of muscular contractions, and that other muscular activity which does not interfere with it is not affected.

Some activities, of course, by their nature involve numbers of different muscles. Walking is a good example, but Hughes (1957) has shown that the co-ordinations involved are largely intrasegmental, each leg responding to the strains on the exoskeleton in its own neighbourhood. This would explain the immediate adaptive change in stepping sequence following leg amputations (see Roeder, 1953). Another example is the cleaning of the antenna in crickets, which involves co-ordination between antenna, legs and mouth parts. Huber (1955a) has shown that in the cricket the antenna is cleaned by the following sequence of movements:

(i) The antenna is depressed.
(ii) The ipsilateral foreleg is brought forward and scraped over the head and eye, then drawing the antenna down between the maxillae.
(iii) The antenna is elevated and drawn up through the maxillae.

The motor centres for the control of the mouth parts, antennae and legs lie in the suboesophageal ganglia, the deutocerebrum (antenno-motor centre) and the prothoracic ganglia respectively. It might seem reasonable to suppose that the activity of these centres would be co-ordinated, either by direct interconnections or through mediation by some higher centres. In fact Huber has shown that if the higher brain centres are destroyed the co-ordination of cleaning movements is not impaired. Further, if the connections to the motor centres for the legs in the prothoracic ganglia are severed behind the suboesophageal ganglion, stages (i) and (iii) still occur, but are now inadaptive in that in the absence of stage (ii) the antenna is not placed in the maxillae. This would suggest that the cleaning movements are not co-ordinated with each other in a closed-circuit system, but once the stimulus for cleaning has been applied (usually by foreign matter on the antenna) the three activities appear in a stereotyped fashion in the correct temporal sequence. In other words, although the movements of the individual appendages are monitored by a feed-back system from their own proprioceptors, there are no neural circuits which allow the activities of different appendages to influence each other directly.

A similar situation is also found in the cleaning of wings by the Diptera (Heinz, 1949). The wing-cleaning movements normally consist of rubbing the hind leg over the wing's surface. One might have expected this to involve a co-ordination between motor centres of both leg and wing, and also between the sensory centres for these appendages. But this seems not to be the case: Heinz showed that the normal leg movements occurred if flies had their wings amputated at emergence; if the missing wing were replaced by an artificial wing made of toilet paper glued to the thorax, then the leg movements were adjusted (within physical limits) to clean the paper wing as adequately as the normal one. This might suggest that the control of cleaning involves co-ordination of leg movement with leg 'sensation', the wing itself playing little part except to provide a stimulus for the leg receptors.

The apparent simplicity of the control circuits described above might also be correlated with the slow speed of conduction of nerve impulses in the insect. In closed circuits where co-ordination is obtained by the continuous circulation of information (e.g. in a negative feed-back loop), the speed with which a message is carried from one part of the system to another will obviously influence the character of the control. If the conduction is slow and the circuit is long and involves many components, two effects would be expected:

(i) The transient error would be large (thus giving the appearance of over-damping).

(ii) The delays caused by recirculating information after each change of state would be cumulative, and if the system has inertia it would tend to oscillate (imitating an under-damped system).

As already stated, insect neurons do conduct slowly (about 2–3 metres per second), and although this may be fast enough for an open circuit, or for the fast start to an activity, its cumulative effect might be important in closed systems. This seems to be well illustrated by prey-catching in the Praying Mantis, as worked out by Mittelstaedt's (1957) brilliant study.

When hunting, the Mantis takes up its position at a place where prey might be expected (e.g. near a flower head). When ready, the Mantis remains still until an insect such as a bee enters its field of view. The Mantis then turns it head to fixate its prey. By this movement the image of the prey is brought approximately on to the position where it stimulates both eyes

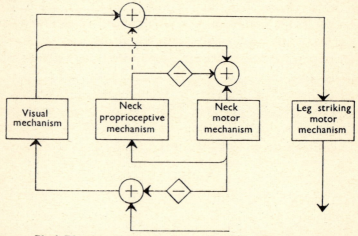

Fig. 2. Block Diagram showing how the different mechanisms are connected in the prey-catching activity of the Praying Mantis. (After Mittelstaedt, 1957.)

nearly equally. This fixating movement is relatively slow and may take up 100 milliseconds. After fixation has occurred, the Mantis usually lets fly with its raptorial forelegs and catches the prey accurately and precisely. The legs have to move for the right distance in the right direction, and this in spite of the fact that the action may be over in less than 30 milliseconds. Mittelstaedt has analysed the organisation of the prey-catching mechanisms and expressed this in terms of an information flow diagram (Figure 2). From this diagram it will be seen that the fixation movement is controlled by a circuit consisting of two closed loops, each loop having only two components. One loop connects the visual mechanisms with the mechanisms controlling movement of the head on the thorax, and the latter is controlled by a subsidiary loop through the neck proprioceptors, which respond to displacements of the head on the thorax. Since these are closed loops, one would expect the fixation movement to be slow; and since it is slow, it will have little inertia and not show much oscillation. This in fact is found. On the other hand, the grasping movement is controlled only by a single motor mechanism, and this control is arranged in an open chain system, no feed-back from the proprioceptors in the legs being employed to monitor the grasping. This means that once the visual mechanism has computed the correct information to activate the motor mechanism correctly, the grasping can be performed rapidly in an all-or-none manner; and, by omission of the proprioceptive feed-back, potential lag and oscillatory error are eliminated. The accuracy is also improved by having only a single input to the leg motor mechanism, a situation which keeps the amplification of the system low.

5. *The evolution and function of the brain*

When complex actions occur in vertebrates these normally involve the brain, which controls their initiation, co-ordination and regulation. One of the important features of vertebrate evolution has, of course, been the increasing dominance of the brain over activities which were originally governed by the spinal cord: the nervous control of behaviour has been increasingly centralised. This process of encephalisation presumably started when the primitive chordates developed a bilateral symmetry and habitually moved with the front end forward (de Beer, 1944). When an animal moves habitually with one end leading, it is obviously useful for sense organs to be developed at the front end in order to gain information about the environment through which the animal is about to move. Correlated with this, one finds the brain developed primitively as a series of sensory ganglia, associated with the anterior sense organs: the brain starts as the 'common sen-

'sorium' of Aristotle. The remainder of the primitive brain seems to consist largely of motor centres for control of the head's appendages (jaws, etc.) and relay centres for passing information back to the spinal motor mechanisms where the actual activities are co-ordinated and executed (Herrick, 1931).

From this primitive arrangement, typical of both vertebrates and invertebrates, the brain seems to have evolved by the combination of three different trends. These trends will now be discussed and compared for vertebrates and insects. The vertebrate brain will be considered first.

(i) Higher centres are developed in the brain itself for co-ordinating and controlling complex patterns of behaviour, such as those normally considered to be instinctive. The reasons for this are diverse. First, the time factor may be important: when different motor activities have to be co-ordinated with each other and with sensory information, it will reduce delays in transmission to have the centres close to each other—the sensory centres already being present in the brain. Secondly, by grouping the different centres together instead of having them scattered about the central nervous system, the actual volume of conducting tissue is reduced. Thirdly, if space has to be found to accommodate the increase in volume of the nervous system (space which has to be protected from the external environment and isolated from movements of the viscera, the body musculature and its appendages, which might affect a delicate nervous mechanism), then—given the basic ground plan of the vertebrate body—it is difficult to see where else except in the skull these requirements could be found. Further, by having the cranium at the front end, it does not interfere, by its clumsy bulk, with the general physical abilities of the body.

(ii) The central nervous mechanisms concerned with sensory and motor activities become increasingly refined. This has happened in two ways, which can perhaps be symbolised by the two models shown in Figure 3. In these models, each of the small circles represents a single 'neural unit'. This clumsy and imprecise term is used here in a functional sense to refer to a population of neurons which act together in a single function and only in that function: e.g. a group of motor neurons controlling the contraction of a single muscle, or the group of cells in the cortex which receive impulses from the fovea rather than from other retinal regions. Co-ordinated patterns of behaviour are then achieved by the interaction of the neural units, and for this it is necessary to have the different neural units potentially able to interact with each other. This could be done in two

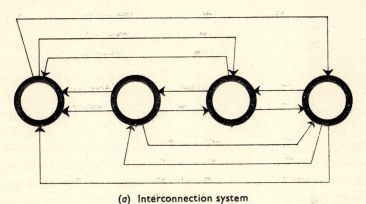

(a) Interconnection system

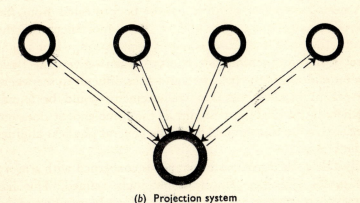

(b) Projection system

Fig. 3. A diagram to illustrate the two ways in which integration might be achieved in a set of nervous centres.

different ways, corresponding to the *interconnection* and *projection* models respectively.

In the interconnection system (Figure 3*a*) the neural units are directly connected with each other. In the projection system (Figure 3*b*) they are connected via a separate centre, which might serve both as an integrating and as a relay centre, and which must be able to deal with a large number of different inputs and outputs. The first type of system appears to represent the histological structure of the corpus striatum of birds, while the second is more representative of the thalamo-cortical system of mammals. Refinement of the sensory and motor mechanisms can be considered in terms of increasing the specificity of function of a single neuron unit and by increasing the number of neurons involved in each unit (thus increasing its potential states). In the first case, it seems probable that primitively the dominance of the brain over reflex activity was an affair of influencing the course of whole groups of reflexes, but that it later became so refined that separate neural units were developed for influencing individual limbs and segments, while later still, in primates, neural units representing single muscles (probably as part of simple movements) have developed in the cerebral cortex. This refinement of function of the neural units must obviously lead to an increase in the number of units, either by addition of new units or by splitting of the original ones. This will have important consequences according to whether the interconnection or projection system is used. In the interconnection type, an increase in the number of units means a great increase in the number of connections (the addition of a single unit in the model shown needs an extra eight connections); but in the projection system only two extra connections are needed—to and from the lower centre—providing that the lower centre can deal adequately with the new connections. The greater the number of units added, the greater the disparity between the number of new connections required in the two systems. Where the potential volume of nervous tissue is limited, either by the actual space available or by the inability of the vascular and respiratory systems to deal adequately with bulky tissue, this limit will be reached sooner if the system is of the interconnection type.

These limiting conditions are, of course, a marked feature of the insect's structure and physiology, and so the development of the brain along the lines of an interconnection system will be very limited. The development of a projection system will also be very restricted, for this arrangement demands a lower centre which will deal with many inputs and outputs. If each of the neurons comprising this centre has the capacity to interact with a large number of other neurons, then any increase in the number of neural units may be accomplished without a corresponding increase in the components of the lower centre. As has already been said, it is precisely in the lack of this capacity that the insect neuron differs from that of the vertebrate. If insects had developed the projection system, every increase in the number of neural units would have necessitated a corresponding increase in the size and number of components of the lower centre. Hence the same factors which limit the development of the interconnection system would have come into play.

As in insects the volume of the brain as a whole is limited, one would expect the number of neural units comprising it to be likewise limited. This should mean that the neural units of insects are less refined than those of vertebrates, which in turn implies that the corresponding behavioural units are more gross. This implies further that on the sensory side the perceptual world will be less rich in detail; and that on the motor side the patterns of activity will be composed of only a few combinations of fairly stereotyped, simple movements—each movement itself being performed by groups of muscles contracting in fixed relationships to one another.

On the sensory side, lack of information makes it difficult to check this prediction. It is known that insects are sensitive to almost any stimulus one likes to name, and to many, such as polarised light, that we cannot ourselves detect. But, in general, it would seem that the configurations of stimuli to which an insect responds are either fairly simple (e.g. the bee's response to the flower as a pattern of flicker) or complex but very specific (e.g. the song of cicadas). Readiness to respond to specific stimuli may in itself lead to neglect of other aspects of the world, and the evidence of specific, complex sensitivities cannot in itself be taken to imply a richness of perceptual detail.

The analysis of the motor aspect of behaviour already given does seem to support the above prediction. If one considers egg-laying and mating movements of the abdomen, toilet movements, nest-building movements and so on, these are indeed characterised by their stereotypy and simplicity. But much more evidence on this point is needed before a dogmatic statement can be made.

(iii) In the vertebrate brain it is probable that complex motor patterns are produced by facilitating or inhibiting the simple reflex mechanisms and their components already present in the spinal cord. In addition, the brain itself controls the details of integration necessary for the efficient steering of behaviour—by complex interactions between the cortex, cerebellum and reticular formation (see Brodal, 1957). In insects, however, the peripheral (or segmental) reflex mechanisms preserve a high degree of autonomy (see Maier and Schneirla, 1935) and the details of their co-ordination do not seem to be regulated by the brain. The absence of co-ordinating centres for motor patterns from the brain is characteristic: centres for the control of locomotor, respiratory, copulatory and ovipository behaviour have been demonstrated in various parts of the central nerve cord and its ganglia, and these activities can be carried on in the absence of the brain (for review, see Roeder, 1953). Indeed, in the Praying Mantis, where the female may start to eat the male's head at an early stage in courtship, this may even be adaptive, for the male's body continues its mating activities (Roeder, 1935). One consequence of the absence of any cerebellar-like activity in the insect brain might be that they could not learn motor skills unless the segmental mechanisms themselves are modifiable—a possibility that no one has investigated.

The study of localisation of function in the insect brain has only just started. At our present state of knowledge, it seems that the supra- and sub-oesophageal ganglia regulate the excitability of the different segmental behaviour mechanisms (in the sense of modifying their threshold, intensity and duration) but do not co-ordinate their components. The appearance (or otherwise) of an activity seems to depend upon the balance of excitation and inhibition in the head ganglia. Huber (1955) has suggested that there may be actual competition by different behaviour patterns for the available excitation. If this is so, then the performance of one activity should influence the 'specific action potential' of a subsequent activity. The situation might be equivalent to the non-specific exhaustion of different activities in spiders (Drees, 1952), and could be correlated with the finding of Bastock and Manning (1955) that different parts of the courtship behaviour of *Drosophila* apparently occur in response to similar stimuli but at different threshold values.

Within the protocerebrum lie some strangely shaped lobes—the corpora pedunculata or mushroom bodies. It appears likely that these lobes regulate the flow of impulses to the sub-oesophageal ganglia, and thus control the release of a particular activity. Huber (1955a and b; Oberholzer and Huber, 1957) has shown that, in crickets, mechanical and electrical stimulation of the α- and β-lobes, and tracts in their vicinity, results in the continuous performance of a specific instinctive activity, such as courtship or threat stridulation. Damage to the calyx results in an inability to sing the correct song, which loses its normal pattern. Vowles (1955) has shown further that the calyx and the α-lobe receive tracts from the sensory centres of the nervous system, while the β-lobe has motor connections. It seems likely, therefore, that the sensory inputs to the calyx and α-lobe regulate activity in the β-lobe and hence determine the behaviour to be performed. In ants (Vowles, in preparation), lesions placed between the corpora pedunculata and the optic ganglia do not blind the insect but produce disruption of guided behaviour in a previously learned T-maze (see Figure 4).

This suggests that the mushroom bodies are important in relation to 'memory'.

The learning of ants is much less complex than that of vertebrates (Schneirla, 1946), and Vowles (1958) has shown that they show no sign of intra-ocular transfer in a T-maze. This may be correlated with the small size of tracts connecting the two sides of the brain; for there is nothing resembling the corpus callosum of vertebrates. It seems possible, therefore, that ants may learn independently with the two sides of the brain!

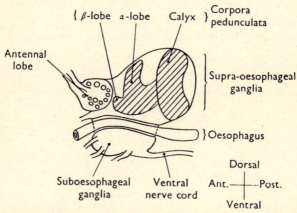

Fig. 4. A diagrammatic representation of a sagittal section through the ant's brain, showing the anatomical relationships of the different lobes and ganglia.

6. *Conclusion*

Biologically, the insects are probably the most successful group of animals, particularly in the tropics. It has always been a wonder to mankind that animals which are so small should have a repertoire of behaviour which is so varied and so adequate. Even recently an eminent zoologist (Pantin, in the Croonian Lecture to the Royal Society, 1952) has suggested that perhaps insects have mechanisms for nervous integration different from those of vertebrates, and as yet undiscovered. If the evidence put forward in this chapter can be accepted, it shows the dangers of equating behavioural complexity—which may itself be only apparent—with complexity of neural mechanism. It shows too the risk of homologising vertebrate and insect behaviour mechanisms, for different animals may use different neural devices to solve the same problems of adaptation. The properties of the insect neuron and the small size of the insect nervous system render necessary a functional organisation of behaviour far simpler than is often supposed. What should arouse our wonder is the success with which simplicity has been crowned.

II. ENCEPHALISATION AND THE SCOTOMA

1. *Introduction*

Within the past twenty or thirty years, two beliefs about the visual nervous system have been expressed with increasing frequency and authority until today they appear to be in no danger whatever of being denied acceptance. These are: (1) partial lesions of the striate cortex of mammals, but especially of man, give rise to clear-cut blind areas, or scotomata, in the field of vision; (2) visual function undergoes encephalisation[1] with increasing phylogenetic development, as a result of which lesions of the striate cortex of highly evolved vertebrates produce much more severe impairment than in lower vertebrates. If acceptance of the second assertion implies that *partial* striate lesions do *not* produce clear-cut field defects in lower forms, then at least some amplification as to what defects, if any, they *do* produce would appear to be an interesting if scarcely discussed question. In this connection, it will be borne in mind that explanation of how such defects can occur in man is often based upon anatomical findings in the monkey, which is, at the same time, declared to be phylogenetically inferior. Any effort to fill in this scotoma in our knowledge, however, eventually forces one to reconsider the two initial assertions.

The mammalian visual pathways have received intensive investigation, especially the course of the optic nerve fibres to the dorsal lateral geniculate nucleus (l.g.n.) of the thalamus, and the course of the fibres originating in the l.g.n. to the striate cortex. The finer details of this organisation will be considered later. At this point I would like to mention briefly two main gaps in our knowledge which may well turn out to be directly relevant to the issues under discussion:

[1] This may be defined as an evolutionary process in which the forebrain progressively takes over functions which, in more primitive forms, are organised at levels below that of the cerebrum.

(i) In what structures other than the l.g.n., the superior colliculus, and the pretectal nuclei do fibres of the optic nerve (or tract as it is called beyond the optic chiasma) terminate?

That there are thalamic connections other than to the l.g.n. seems plausible. Debate has waxed and waned for fifty years on the question of whether the pulvinar receives optic tract fibres. With Marchi and Weigert techniques, fibres go directly into the pulvinar of monkey, but these techniques are not adequate to distinguish between terminal and non-terminal fibres. Other techniques, such as that of transneuronal degeneration used so effectively by Clark and Penman (1934) do not reveal degeneration in the pulvinar, but that might be because of multiple input to the pulvinar (transneuronal degeneration being in any case a rare phenomenon in the central nervous system). Again, O'Leary (1940) using Golgi material, describes optic tract fibres going to the cat's *ventral* l.g.n., which does not undergo degeneration with lesions of the striate cortex. Bard aptly points out that 'scarcely anything is known of the function or connections of the ventral lateral geniculate body' (1956, p. 1149). What is the route of visual fibres to the reticular activating system? Such questions are of interest on two accounts: First because they might represent additional cortical pathways (the pulvinar projects widely to the parietal and temporal lobes) which may be open after removal of the striate cortex. Secondly there is considerable evidence that lesions well outside the striate cortex can affect visually-guided behaviour (Chow, 1952*a*; Mishkin, 1954; Mishkin and Pribram, 1954), but as yet, no clue has been provided which would indicate how visual information reaches these areas. Related, perhaps, to these questions is the further one of what possible significance can be attached to the multiple vision receiving areas which have been uncovered by the electrophysiologists (cf. Bard, 1956).

(ii) From the viewpoint of devil's advocate, one can ask how well established is the conclusion that the l.g.n. projects *only* to the striate cortex?

31

Poliak found complete degeneration in the l.g.n. following striate cortex removal in monkey (1933) and chimpanzee (Poliak and Hayashi, 1936); in addition, with lesion of the monkey optic radiations (using Marchi material) he found fibres going solely to the striate cortex (1932), although in an earlier investigation he reported fibres going outside the limits of the striate cortex of the cat (1927). The striate cortex of man and monkey has sharply defined limits in terms of its histological appearance, from which it derives its name. But such is not the case for the rat, at least along its lateral and antero-lateral border, where the separation between striate and non-striate is gradual and blurred (Lashley, 1934). If an investigator claims that the striate cortex has been removed in the rat, how does he know? Again, in the cat there is some disagreement as to whether striate cortex removal leads to complete generation of the dorsal l.g.n. (Smith, 1937b, 1938) or partial degeneration (Minkowski, cited by Walls, 1953). Indeed, what is meant by 'complete degeneration' of the l.g.n.? That the large ganglion cells disappear after striate removal seems definite. But there are many small cells left. Are these 'glial' cells small and/or shrunken nerve cells? There may well be standards which the anatomists have developed to make these distinctions, but in many studies there is no mention of such standards.

2. *Encephalisation of visual function*

The effects of decortication are commonly said to show increasing severity with ascending phylogenetic status. Frogs without their cerebral cortex are reported to be practically unaltered in their visual responsiveness and decorticate birds can still fly from perch to perch. Rats lacking the visual cortex can perform intensity discriminations in a maze or jumping-stand apparatus and guinea-pigs are not greatly altered after striate cortex removal, at least in making optokinetic responses. Cats lacking the striate cortex also show optokinetic responses and can make intensity discriminations, as can also dogs. Dogs and monkeys can acquire a conditioned eyelid response to a flash of light after removal of the striate cortex and the monkey at least is stated to make discriminations based on total luminous flux, though not on brightness differences. On the other hand, man is totally and permanently blind after extensive injury to the striate cortex. Such is a sketchy but typical summary (cf. Marquis, 1935, 1942, for detailed reviews of the effects of striate removal at different phylogenetic levels).

In considering these conclusions in detail, some points of interest emerge. (The following discussion, for the sake of brevity and because of the availability of source material will be limited to mammals. Also the term 'destriate' will be used to refer both to the case of removal of striate cortex and the case of occipital lobectomy.) It is anticipating the argument somewhat, but a point which will emerge later is that one's results in measuring the limits of visual capacity depend very precisely on the method of measurement. One clinician's scotoma is another's amblyopia. And so one can ask to what extent various mammalian species have been compared on procedures *which are equivalent*. I defer for the moment the both trivial and profound meaning of this question, which would have one enquire to what extent a monkey in the water maze or a jumping-stand is equivalent to a rat reaching for a banana with a long stick or dismantling a puzzle box. A simpler and more specific meaning will be dealt with first.

The discussion can be prefaced by a historical comment: In general, mammals, and especially cats, dogs and monkeys, were for long described as totally and permanently blind (but retaining, as does man, a pupillary reflex) after complete striate removal. Such animals, under normal conditions of illumination, bump into objects, cannot locate their food visually, are apt to fall off raised surfaces, show no visual placing reaction, and do not respond to objects moved in front of them, even when such movements are threatening (cf. Smith, 1937*b*; Smith and Warkentin, 1939, for reviews). All the same, both Goltz and Munk had observed, over fifty years ago, that dogs without their occipital lobes would close their eyelids when presented with very marked changes in illumination (cited by Smith, 1937*a*), and gradually it has become evident that such animals can be taught to make certain discriminations based on light, often just as rapidly as normal animals.

At this point it will be convenient to restrict the discussion for the moment to the destriate cat, since a most illuminating analysis of this preparation has been carried out by K. U. Smith and his co-workers (1937*a* and *b*, 1938, 1939, 1943). If under normal conditions such a cat is placed within a large rotating cylinder on which are painted stripes perpendicular to the direction of rotation, it will respond with following movements of the head and eyes. The essential features of these optokinetic responses appear to be similar to those shown by normal cats, except in that the operated cat is much more stereotyped in its response pattern (Smith, 1937*a*; Smith and Warkentin, 1939). As the separation between the stripes is gradually increased, operated cats stop responding much sooner than normal cats. The slenderest stripes to which they will respond are not quite as slender as those to which a normal animal will respond. And while a destriate cat will generally not respond to isolated moving objects under normal illumination, Smith was able to obtain closure of the eyes when the experimenter's hand interrupted a very intense light (500-watt bulb at 72 cm. from the eyes). Kennedy (1939) studied destriate cats' responses to moving stimuli by training them to approach a moving pattern (placed on a rotating disc) in a discrimination box and to avoid a matched stationary pattern. These animals were severely impaired compared with normals, especially when the pattern was a complex one. However, when a *simple* pattern was rotated at a much higher speed than would be required for a normal cat, the operated cats could make the discrimination.

What about intensity discrimination? Cats trained in a discrimination box showed no change in differential threshold following striate removal, although they did suffer loss of the habit and had to be retrained (Smith, 1937*b*). This finding of

34

no threshold change held only under limited conditions, viz. when there was no general illumination in the apparatus other than that produced by the stimuli to be discriminated. As soon as even a low level of illumination was introduced, the cats were severely impaired, even though pre-operatively they had shown very rapid learning under identical conditions. Mead is quoted (by Morgan and Stellar, 1950, p. 169) as finding a marked increase in $\Delta I/I$ after destriation, but we do not know what conditions of illumination prevailed. On the question of absolute threshold, evidence is again conflicting. Bridgman and Smith claimed (on the basis of a single animal) that the absolute threshold, under dark-adaptation, was raised 'about 500 times' (1942). Gunter, in an unpublished note, found no change in the scotopic luminosity curve, but a severe loss of sensitivity under photopic conditions (1952).

Finally, Smith presented inconclusive evidence that destriate cats might be able to learn a simple pattern discrimination between horizontal and vertical stripes, but if they can learn it at all it is only with difficulty. He could not obtain visual placing responses of the fore-limbs on a striped table top, even with high illumination (1937a).

How can one characterise the visual world of the destriate cat? The popular description of such an animal as capable only of responding to non-patterned light (cf. Marquis and Hilgard, 1936) raises difficulty, for is a moving striation a 'pattern' or not? Smith suggests that the destriate cat remains capable of responding more or less normally to situations presenting marked gradients of retinal illumination but shows itself as much impaired under circumstances in which such gradients are markedly reduced (1937, p. 361). Another way of putting it might be to say that those problems which are simplest for the normal cat to learn (in terms of trials required for acquisition) or, in the case of unlearned behaviour such as the opto-kinetic response, those conditions which are optimal for eliciting a response in the normal animal are the least likely to be affected by striate removal. Thus, the most effective stimulus to optokinetic movement is a moving pattern occupying the greater part of the visual field. The brighter the light source and the nearer to one a moving object, the more likely is palpebral closure of the eyelids to occur. Increasing the general illumination in a simultaneous threshold situation should increase the 'contrast', which should hinder discrimination. But perhaps this characteristic can be made more specific in later discussion.

If it is true that only a limited set of conditions produce visual responses in destriate cats, one would like to know whether different degrees of residual function in other species are due simply to the phylogenetic status of these animals, or to the different set of conditions under which they were studied, or both. We will compare the cat with the dog and primates on the one hand, and with rodents on the other.

A destriate dog or monkey will respond to moving objects occupying 'the greatest part of the visual field' but not to isolated objects (Rademaker and ter Braak, 1948). I can find no reference to any human cases tested with large striped fields, but it should be clear that simple test of response to a single moving object or a threatening gesture will not be adequate for present purposes. (It must be recognised throughout all of the discussion that human cases of extensive bilateral occipital brain injury are rare and invariably involve widespread damage. In fourteen cases of total blindness due to brain injury from cerebral infarction reviewed by Symonds and MacKenzie (1957), four died within three months of the onset of blindness.)

As for intensity discrimination and its break-down under conditions of moderate and high illumination, there are no studies which are exactly comparable. Kluver (1936, 1937, 1941, 1942) was able to show that destriate monkeys could discriminate intensities of light, but only on the basis of their 'total luminous flux' and not on their energy per unit area. (This may have been true for cats of Smith and Mead as well, but this relationship was not reported as having been studied.) Kluver noted that moderate conditions of illumination interfered completely with the discrimination. 'Under daylight conditions the animals could not even respond to a single light . . . although presented under favorable conditions' (1941, p. 42). Similarly, the condition in which Marquis (1934) demonstrated intensity discrimination in destriate dogs was one of very low illumination. On the human level, data are very sparse. Bender and Krieger (1951) describe a single case, however, which is relevant. One of their patients was a man who 'had apparently been completely blind for many years. When he was examined under ordinary room daylight, he could not tell whether it was light or dark. He was then placed in a dark room for 15 minutes, after which a brilliant light . . . was turned on and off. Precautions were taken to eliminate auditory and thermal clues. Nevertheless, the patient could tell the difference between light and dark' (1951, p. 77). Unfortunately, not much is said of the suspected pathology, or even whether the patient had a normal pupillary reflex. And while it seems extraordinary that such a response had not been uncovered for 'many years', it suggests that human cases may not be blind under at least one condition which applies to destriate cats, dogs and monkeys—who were, after all, likewise for many years described as being totally blind.

Finally, there are the observations by Marquis and Hilgard (1936, 1937) showing that a conditioned eyelid response to a flash of light (unconditioned stimulus: puff of air) can be demonstrated in destriate dogs and monkeys, although with a slightly longer latency than pre-operatively. Since we have no observations on other species, we cannot relate it to an argument on encephalisation, although the authors point out that the dogs seem to suffer less severe amnesic effects for the habit than monkeys.

When we turn to rodents, an interesting complication arises. It certainly seems true that a guinea-pig without its striate cortex will respond optokinetically to a large moving striped field, and *not* to a moving, isolated object. But this is also the case with the guinea-pig in its normal state, and so striate cortex removal produces no change in its movement vision, at least as demonstrated optokinetically (Smith and Bridgman, 1943). According to these authors, the lack of following response to isolated moving objects is a characteristic of rodents generally, and this is supported by evidence of Rademaker and ter Braak (1948).

This presents the opportunity to make a slightly digressive and general comment on the question of encephalisation. Most commonly the doctrine is invoked in such a way as to imply that the particular function remains unchanged as it climbs the cerebral ladder. Thus, one speaks of noticing little change in destriate rats, but a great change in destriate monkeys. But the rat is rather limited, visually, in any case. The same comment can be made on the effects of lesions in the sensori-motor cortex—effects likewise often invoked in support of the doctrine. Evidence points to distal musculature and joints being most sensitively affected by such lesions in primates. Lower animals have, at best, only a slight degree of fine manipulative control over such musculature. In both instances, we are not comparing the effect of a single variable (cortical removal) on the *same* dependent variables. While I would not like to go to the opposite extreme and maintain that the sole consequence of increasing cortical development is a corresponding increase in the complexity of behaviour, I do believe that the possibility of progressive *change* in function with advancing phylogenetic status has not received adequate emphasis. Nor has the complication thereby introduced into the interpretation of the effects of cortical damage in relation to phylogenetic level been given the attention it deserves.

And while we are on this general issue, it might be worth while pointing out a few other facts not generally stressed. Thus, even in the case of the decorticate pigeon, generally described as being relatively unimpaired, Dusser de Barenne points out (in discussing a study by Schrader) that 'distinct visual disturbances can be observed. The behaviour is such that one is induced to assume that it does not recognise the object depicted on its retina. It manifests no fear under circumstances in which the normal pigeon does' (1933, p. 891). Furthermore, in the phylogenetic development of the visual system, the progression often takes a few steps backwards. 'It is customary to consider man as the proud and finished product of an evolutionary line starting from rather imperfect forms and gradually approaching human standards by increasing differentiation . . . It needs no more than a glance at the photographs (of the striate cortex) to see that the lamination of the striate area is more elaborate in tarsius than in the monkeys, and more elaborate in the monkeys than in apes or man' (von Bonin, 1942, p. 426). Supporting this is Clarke and Penman's finding (1934) that the macaque l.g.n. shows greater differentiation into laminae than that of man.

While it appears that the destriate rodent is not altered in its response to moving objects, it is not the case, however, that such a lesion is without any effect whatever. Lashley (1931a, 1935) has undertaken a very thorough analysis of these visual changes in rats. A destriate rat can learn a simple intensity discrimination as rapidly as a normal rat, although it suffers complete lack of retention for the habit if acquired pre-operatively. It can discriminate a black card from a black card with a white square on it, although with difficulty. It cannot discriminate patterns such as inverted versus erect triangle, or horizontal versus vertical striations. It can discriminate discs of different sizes, but when the luminosities are equated the discrimination breaks down. It will not respond to a small moving pattern—but we are not told whether a normal rat will respond to it.

On the whole, these results conform to the general pattern already found for the destriate cat, as discussed above. In the case of the simple intensity discrimination (for which the simple Yerkes discrimination box was used), very low illumination was probably employed. (Lashley's description is not very clear on this point but it would appear that in one at least of the situations described, the only illumination was that provided by the positive stimulus (1935).) We do not know whether normal illumination would interfere with this discrimination.

Smith has pointed out that in the condition under which some simple pattern discrimination was demonstrated (using the jumping-stand), high illumination was used (Smith, 1937b).

Again, therefore, there is no evidence to reject the view that in infra-human mammals, and possibly even humans, simple intensities can be discriminated under conditions of low illumination but not otherwise, and that simple gradients of light can be discriminated provided such gradients are very marked. The fact that the guinea-pig normally, in some aspects at least, resembles a cat without its striate cortex, and the additional fact that the guinea-pig is usually described as an almost pure rod animal, yields the rather tempting possibility that cortical removal produces a residual capacity similar to that maintained in a pure rod system, a hypothesis which did not escape Smith (1937b). But this would produce a paradox, for if the guinea-pig has no or almost no cones, can we thereby conclude that its striate cortex is functionless and could be removed with impunity? It is true, as we have seen, that the destriate guinea-pig is virtually unchanged in its movement perception. But it is highly likely that if the destriate guinea-pig were tested extensively in situations comparable to those employed by Lashley for the rat (another rodent generally considered to have a predominantly rod retina) it similarly would be deficient in responding differentially to complexly patterned lights. The paradox might be resolved by stressing not the type of unit receptor as such (i.e. rod or cone), but the extent to which such receptors can be considered as being independent of each other, by virtue of their relations with retinal ganglion cells. It happens, of course, that the receptors of many predominantly rod retinas or part-retinas (such as the human periphery) demonstrate a high degree of non-independence, but this is not a *necessary* relationship and there may well be predominantly rod retinas which contain both highly independent and non-independent systems of receptors. We might then re-phrase the hypothesis about the effects of destriation to state that it produces a preparation in which the input from the highly non-independent system remains relatively intact, while the input from the independent system is impaired. With a highly non-independent system, we might predict no difficulty in discriminations based on total luminous flux but impaired discrimination of *local* differences of brightness (acuity). Further, differential discrimination of the intensity of stimuli presented *simultaneously*—the method used in all the studies cited above—would be less fine than in the normal animal, particularly as the general illumination is increased.

The suggestion that the destriate mammal behaves roughly like a human with only extreme peripheral vision emphasises the need for accurate histological verification of lesions purporting to be 'complete'. In some species, especially monkey and chimpanzee, the region most likely to be left intact in a supposedly complete occipital lobectomy lies along the lips of the internal calcarine fissure, on the medial surface of the brain, and extends quite far rostrally. In view of this, it is noteworthy that Kluver did not publish any histological verifications of his lesions, and his finding that discrimination in the destriate animal is based solely on luminous flux might have been due simply to the retention of extreme peripheral vision. Lashley (1931) claims that such discrimination is not, in fact, a characteristic of the human periphery, but cites no evidence. Gross and Weiskrantz (1958), however, found that the human subject appears to match stimuli of different sizes in the far periphery solely on the basis of flux, at least within the limits studied (2–3 degrees visual angle).

The criticism advanced against Kluver cannot be levelled against Lashley, who pioneered adequate methods of reconstruction of lesions (Lashley, 1922, 1934, 1939, 1942; Lashley and Frank, 1934). His work sufficed to show that lesions in the striate cortex alone interfered with visual discrimination in the rat and that a further cortical lesion following striate removal produced no further decrement in intensity discrimination. Further, the relationship between striate lesions and deficit in pattern discrimination appears to be all-or-none—one animal with just 700 cells (out of a normal total of 35,000) remaining in one l.g.n. still showed pattern discrimination. The fact that Lashley (1931a) found a luminous flux discrimination in his destriate rats is obviously relevant to the preceding argument.

On the other hand, the remarks made in the Introduction, on the projection of the l.g.n., are still pertinent. Lashley (1934) and Smith (1938) refer to degeneration only of the 'ganglion' cells following destriation. Smith does, however, comment that he found 'medium-sized cells, showing pathological character-istics as compared to the ganglion cells of the external geni-culate, distributed at infrequent intervals throughout the nucleus' (1938, p. 258). Perhaps the only crucial test would be one conducted with a completely decorticated animal. Such a preparation is obviously a difficult one, and it is not surprising that there is some lack of agreement about the presence of visual responsiveness. Culler and Metler (1934) reported conditioning of foot-withdrawal to light in a decorticate dog. But the response was not adaptive, and there has been said to have been a suggestion of auditory artifact in the conditioning procedure. Some earlier workers, such as Koudrin in Pavlov's laboratory, claimed to obtain salivation to light in a decorticate dog (cited in Marquis and Hilgard, 1936) and Goltz and Munk reported (cited by Smith, 1937a) reflex closure of the eyelids to bright light. Ten Cate, on the other hand, was unsuccessful in setting up conditioned light responses in decorticate cats, although auditory conditioning was still possible (Smith, 1937b).

3. *The effects of sub-total lesions*

So far, we have been considering only the effects of total ablation of the striate cortex—or at least of lesions purporting to be total. In the case of partial ablations, it is often held—at any rate in the case of man—that the effects are qualitatively similar to those produced by total ablation, though less exten-sive in scope. As Marquis wrote: 'Destructive lesions of the area striata abolish all visual sensitivity, even the crudest perception of light and dark. Small isolated lesions result in corresponding small scotomas; destruction of the calcarine cortex of one hemisphere causes homonymous hemianopia; bilateral destruction, total blindness. But in each instance the blindness is absolute for the region of the visual field concerned' (1935, p. 807).

There is, of course, no good reason—logical or empirical—why a sub-total lesion should produce effects qualitatively similar to those of a total lesion. Indeed, several lines of evidence suggest that, in animals at least, something of an all-or-none relation holds between striate destruction and resulting deficit. As has already been noted, Lashley (1935) found that 1/50th of the striate cortex was sufficient for complex pattern discrimination. Furthermore, he noted that intensity discriminations acquired pre-operatively were not lost following sub-total lesions, but were in general lost following total lesions—the animal retaining the capacity, however, for relearning as rapidly as it had pre-operatively. Smith (1937b) also found an all-or-none relationship between striate lesions and retention of intensity discrimination in the cat. In this connection, Lashley (1935) aptly recalls an earlier comment by Goldstein to the effect that the brain-injured patient is disinclined to shift from one method of performance to another until the old method just cannot be made to work.

But it is obvious that results of this kind might be obtained even if the only effect of sub-total lesions were to produce limited scotomata, or fields severely depressed for the discrimination of fine gradients, since the subject might then learn to use the intact parts of the field. Amnesia of intensity habits might simply imply that the world suddenly looks quite different and undifferentiated, so that not even a flash of light or an illuminated panel would be sufficiently similar for transfer to be obtained. For reasons which will become apparent, this is a difficult claim to evaluate experimentally, but the problem is far more interesting than that of total striate removal. Indeed, the analysis of sub-total effects is much more likely to give us an understanding of the functional organisation of a system than outright destruction of the whole system.

In so far as one can evaluate the animal evidence at all, one is not driven to the conclusion that limited lesions produce limited scotomata. Let me qualify this quickly by rejecting the case of a complete unilateral lobectomy, since this is apparently equivalent, at least in primates, to removing a whole system, i.e. one may consider the homonymous halves of the retinae to constitute a single system in so far as their relationship to each l.g.n. and striate cortex is concerned. It does seem to be the case that hemianopia (blindness of one-half of the field) results from such lesions in monkey (Settlage, 1939; Harlow, 1939) and chimpanzee (Spence and Fulton, 1936; Walker and Fulton, 1938). But what of the case when *part* of either or both striate cortices is destroyed?

Lashley was uncertain whether rats with partial striate lesions could be said to have scotomata. There were tasks on which the partial destriate rats were deficient, but it was not a simple matter to relate these to the presence or absence of field defects. For example, a partial destriate rat could discriminate an upright triangle from an inverted one, but not a triangle enclosed by a circle from one not so enclosed—although separate tests of acuity indicated that the circle should have been discriminated (Lashley, 1931a). In an animal with a very large lesion—49/50ths of its visual cortex—he noted a consistent shift in fixation (Lashley, 1939), but if any shifts were detected in rats with smaller lesions, I can find no mention of them. Minkowski examined dogs with a variety of striate lesions, and concluded that it was impossible to obtain lasting visual defects unless at least half of the visual cortex of one hemisphere was destroyed (cited by Lashley, 1935).

One of the pioneers of experimental neurology, Ferrier, studied the effects of occipital lesions in monkeys. He repeatedly stressed the absence of effect. A typical comment was 'In two cases I removed the greater portion of both occipital lobes at the same time without causing the slightest appreciable impairment of vision. One of these animals within two hours of the operation was able to run about freely, avoiding obstacles, to pick up such a minute object as a raisin without the slightest hesitation or want of precision and to act in accordance with its visual experience in a perfectly normal manner' (Ferrier, 1886, p. 274). From Ferrier's drawings it would appear that his lesions included the entire lateral extent of *area striata*. It is difficult to ascertain how much medial tissue must have remained, but it is fairly certain from what we now know that his lesions could not have included the entire striate cortex.

Both Poliak and Kluver had the occasion to observe monkeys with large bilateral, but incomplete, striate lesions. Poliak comments 'it was remarkable to observe that, in spite of the extensive bilateral damage to the cortex connected with central vision, the animal on the tenth day after the operation picked "lice", though in a somewhat superficial, perfunctory way; while on the twelfth day it picked up peanuts very well, shelled them apparently under the control of its eyes, and picked out the objects in a poorly illuminated room even when these were little distinguishable from the ground by their colour' (1933, p. 561). Poliak, it might be noted, perhaps more than any other single anatomist, has pressed for a point-to-point representation. Kluver's description of a monkey with a bilateral lesion of the 'macular' cortex (i.e. the lateral cortex) was very similar. He noted that the animal picked up bits of food without error 24 hours after operation. 'The general impression gained three weeks after the operation was that the animal was as lively and attended to incidental visual and auditory stimuli as would a normal animal. The performance in pulling in a food box was entirely normal. There were no difficulties in promptly locating the end of a ·3 mm. thick black thread lying on a black cardboard' (1937, p. 398).

Wilson and Mishkin (1955) and Mishkin and Weiskrantz (1958) have observed monkeys with lesions placed similarly to that in Kluver's animal. The entire lateral striate cortex was removed bilaterally, a region which according to electro-physiological data corresponds to a central retinal area of

16–20° in diameter (Marshall and Talbot, 1942). In behavioural testing, such animals were found to be slightly deficient—relative to normals—on size-discrimination, patterned-string problems, shape-discrimination, and visual 'learning set' tasks, in addition to showing a lower critical fusion frequency for flicker, than they had pre-operatively. These behavioural results are chiefly of interest in contexts other than the present one, but it is pertinent to note that in neither study was any behaviour noted that would clearly indicate a discrete area of blindness in the field. It should also be noted that as the monkey retina is similar to the human (Poliak, 1941), a fairly marked alteration in acuity might have been predicted. In fact, the only change apparent to gross observation was a transient 'mispointing'—i.e. the animal would tend, when reaching for a peanut, to 'miss' it by an inch or two. This phase disappeared, as I recall, in about ten days. Otherwise, it was impossible, by gross behavioural observation, to distinguish these animals from normals.

Extreme caution should be exercised in interpreting these observations. Even in humans it is difficult to find and plot scotomata (this may be, of course, not because a scotoma is elusive in the manner of a needle in a haystack, but because it is like a balloon which can be inflated by some methods and punctured by others). The visual world is sufficiently redundant so that, under normal conditions, one can easily fill in and complete the missing bits, probably without any awareness of so doing. Such appears to be the interpretation that Settlage (1939) placed upon the performance of his monkeys, who had sustained complete lobectomy on one side and incomplete and varying lesions on the other side. Immediately post-operatively he noted a surprise reaction as a peanut shifted across the animal's vertical meridian, but this surprise reaction disappeared in time. 'As a matter of fact, it becomes impossible to carry out manipulations in the blind fields of which the monkey will not immediately become aware. ... There is no grossly observable skewing of the eyes, nor is there any easily detectable

increase in the amount of head movement and eye movement'
(1939, p. 127). We shall refer to this study again later.

It is clear that this problem will not be settled until a reliable
method of measuring field defects in animals can be devised.
Such a task is obviously most formidable but Cowey (1958) has
made a most promising start in achieving it. While human cases
are clearly more amenable to measurement of field defects, they
are very difficult to assess for extent of brain damage. Even
when necropsy data are available, it is often difficult to be
certain of the precise extent of the lesion. We shall deal with the
human material as it becomes relevant at various steps of the
argument. For the moment, admitting that we stand in danger
of outdistancing our data, I would like to ask the reader's
indulgence in considering the hypothesis that limited, partial
striate lesions do *not* produce scotomata, or at least not of the
size which would be predicted on the view that there is a strict
point-to-point projection of the retinal surface on the visual
cortex.

If such a hypothesis is accepted, it is clear that it could be true
only if more than a single route were open to retinally elicited
impulses travelling to the striate cortex. Three possibilities for
alternative pathways will be discussed. These possibilities are
not mutually exclusive.

(i) The fibres of the optic tract might not have a one-to-one
relationship with the cells of the l.g.n., but show a certain
degree of overlap between the terminal fields, such as shown in
Figure 5. This sort of notion has been put forward by Lorente
de Nó and others (Fulton, 1949, p. 349), although prevailing
anatomical opinion seems to favour strongly the one-to-one
schema for vision. The anatomical support for a one-to-one
schema stems, for example, from the fact that limited lesions
of the retina give rise to discrete and sharply delimited transneuronal degeneration in the l.g.n.; and striate cortex lesions,
equally, produce sharply defined zones of retrograde degeneration in the l.g.n. Such is the principal way by which the retina
has been 'projected' to the cortex. But such lesions, clearly,

have not been made with the degree of precision required to settle whether zones of overlap equivalent to, say, 5–10° of visual angle, exist. For it would be necessary to show that *adjacent* retinal or cortical lesions produce comparably adjacent zones of degeneration in the l.g.n. Another anatomical approach has been the examination of silver-stained material in the l.g.n. after enucleation of one eye, since this stain permits the identification of both cell bodies and their processes, and the tracing of degeneration in some of its stages. Glees (1941, 1942) and Glees and Clark (1941) have carried out such an examination in rabbit, cat and monkey. In all three species considerable ramification of the optic tract fibres in the region of the l.g.n. was found. In the rabbit and cat large numbers of terminal boutons on each l.g.n. cell as well as on the dendrites were reported, suggesting multiple endings and supporting (but not proving) overlap. In the monkey, however, Glees and Clark found only one bouton per l.g.n. cell, and none attached to the dendrites. Clark (1942) also points out that transneuronal degeneration occurs reliably and clearly in monkey and man but to a much lesser extent in cat, rat and ferret—again supporting greater uniqueness of pathways in primates. Be this as it may, and despite the fact that it is not at all clear what functional significance should be attached to the boutons, it may be noted that there is a paradox in Glees and Clark's findings in the monkey. They report that each optic tract fibre ramifies into five or six branches, but there is only one bouton per cell. But we know (and Clark mentions some of these data in his review paper (1942)) that the l.g.n. of monkey contains somewhere from 1,000,000 to 1,800,000 cells (Chow, Blum and Blum, 1950; Bruesch and Arey, 1942). The optic nerve contains approximately 1,200,000 fibres—about 20% of which are estimated to by-pass the l.g.n. and go to the mid-brain (Chow, Blum and Blum, 1950). Somewhere along the line, then, we are missing 3–5 million boutons. We can only conclude that there are endings not represented by boutons, or that Glees and Clark did not sample widely enough (or that more than one fibre branch ends on the same bouton, which seems rather unlikely). (For further discussion of this point, cf. Walls, 1953.)

On the other hand, there is at least a certain degree of anatomical support for interaction between cells in the l.g.n., within each lamina if not between laminae. (Three layers of each primate l.g.n. receive their input from one retina, the other three from the other retina.) O'Leary found an abundance of short-axon multi-polar cells within each l.g.n. lamina of cat (1940). Walls says the existence of such intercalated neurons in the mammalian l.g.n. is well established. 'In fact, no mammalian l.g.n. is known *not* to contain them,' although he admits they have not been properly studied in many species. They have been seen, in addition to cat, in man, rabbit and monkey (Walls, 1953). O'Leary speculates that such neurons may 'play the role of synchronisers for groups of principal cells' within the l.g.n. (1940, p. 427) and it is of interest that Lindsley (1957) has found that rapid, repetitive rates of firing in the optic nerve are transduced to rates approximating the c.f.f. somewhere between the optic tract and the striate cortex.

The assertion that partial striate lesions lead to *sharp* zones of degeneration in all layers of the l.g.n. would, if true, make it necessary to look for alternative pathways peripheral to the geniculo-striate system. The intercalated fibres might provide such a possibility (if some of their connections, at least, were with the ramifying terminal axons of the tract). Equally, and concomitantly, there would appear to be ample opportunity within the retina itself for alternative pathways. The non-independence of adjacent retinal elements in some parts of the retina, at least, seems well established (Barlow, 1953). At the same time, it is usually claimed, though not yet established electrophysiologically, that the macular region consists of receptor units which have a one-to-one relationship with retinal ganglion cells. None the less, even the possibility of overlap within the geniculo-striate system should not be entirely ruled out. The extent to which this might be so depends upon the extent to which degenerated zones of the l.g.n. are completely functionless and non-degenerated zones are completely normal. Relevant to the latter question is Lashley's comment that 'after partial lesions in the striate area (of rat), ganglion cells disappear completely from a limited area of the lateral geniculate nucleus and those which remain in the other regions are *always* less sharply stained than cells in other nuclei and appear somewhat shrunken' (1939, p. 53, my italics). Elsewhere he noted that there 'sometimes appears to be an

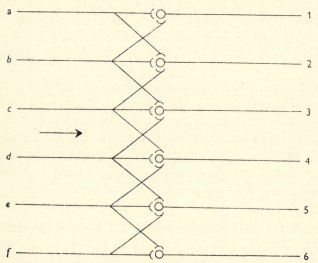

Fig. 5. Schematic representation of relationship between optic nerve fibres (*a, b, c, d, e, f*) and fibres going to striate cortex (1, 2, 3, 4, 5, 6).

increase in the number of glial cells throughout the nucleus' (1934, p. 61). On the former issue, one wants to know to what extent non-ganglionic l.g.n. cells are neurally functional. May some of the small cells in zones of degeneration be O'Leary's intercalated cells?

A network such as that shown in the diagram (Figure 5) would have some interesting properties: (*a*) With very small striate lesions, there would be no blocking of impulses, and hence no scotoma. As the lesion is enlarged, a small scotoma would begin to appear, but it would always remain smaller than the scotoma which would be predicted on a strict point-to-point basis until the limiting case of complete removal was reached. It will be appreciated, therefore, that there is no contradiction between the present hypothesis and the classical neurological correlations between site of occipital damage and location of scotoma in the visual field (Holmes, 1919; Holmes and Lister, 1916; Spalding, 1952a and b). It will also be appreciated that the human material, even with post mortem examination, is not as a rule adequate to settle whether the *quantitative* relationship between amount of damage and size of scotoma is as predicted on a point-to-point basis.

Nor, indeed, will a typical clinical perimetric chart of a scotoma help here, since it can easily be shown to change its size and shape over a considerable range by simple manipulation of relevant parameters—i.e. size of stimulus object (Hughes, 1954). We require, rather, a measure in terms of optimal conditions—so that by a scotoma is meant a region of the visual field in which *no responses to light can be demonstrated even under the most favourable conditions*. Considered from this point of view, the conclusion of Poppelreuter, based on an extensive experience of World War I brain injury cases, is very pertinent. He declared that he could never find an *absolutely blind* scotoma; some rudimentary function was always present. He tried to order the different levels of visual function as follows: (1) amorphous light sensitivity, (2) size perception without definite form, (3) amorphous form perception, (4) perception of discrete objects, (5) mild amblyopia, (6) normal vision (cited by Kluver, 1927). More recently, Bender and Krieger (1951) have reported finding some visual responsiveness in perimetrically blind fields; under conditions of faint illumination, patients could respond to changes in intensity, movement and flicker. They could localise the position in space of such stimuli, although less accurately than in the 'intact' parts of the field. The authors try to show that dispersion of light into the intact parts of the field would not account for these results, although their arguments seem a bit forced.

At this point, also, I would like to suggest the possibility that the recovery in Settlage's monkeys, mentioned above, may have been due not to simple learning based on redundancy within the visual field, but to the scotoma having been in fact smaller than would ordinarily have been predicted on the basis of the locus and extent of the cortical lesion. Spence and Fulton (1936) also found gross observational evidence for a 'macular' scotoma in a chimpanzee. Poliak and Hayashki, who examined the brain of this animal, comment 'we suspect that considerably more than the macular cortex proper was eliminated' (1936, p. 58)—in fact, only the most rostral portion of the l.g.n. was completely normal. They also draw attention to 'the surprisingly large degree of visual function which was preserved in our ape although only a fraction of its total nervous-visual mechanism was left intact' (1936, p. 60).

(b) It might be predicted that, in addition to the possibility of function within the 'blind' region, there might be malfunction in the 'normal' region, since this would now be in a somewhat more complex relation to the periphery. It is significant, I believe, that Bender and Teuber (1947, 1948, 1949) and others have found evidence of malfunction in perimetrically normal areas of the visual field in cases of occipital lobe injury in man.

(c) If it is assumed that the organism learns to use peaked information arriving at the cortex to correlate with a motor co-ordinate system (and that some degree of learning is important has been evident since Stratton's work half a century ago), it can be supposed that with a partial striate lesion the peak will shift to a locus normally correlated with a slightly different position in the visual field. We might therefore predict that 'mispointing' would occur in such cases. As has been noted, the partially destriate monkey 'mispoints' for a brief period after operation and similar phenomena have been observed in patients with occipital injuries (Bender and Tueber, 1947). A lens could no doubt be designed to give similar patterns of displacement in a normal animal, and it is noteworthy that humans wearing simple displacement lenses at first mispoint but rapidly adjust to them very well.

These relations can be displayed in very schematic form in Figure 5. It is supposed that a stimulus exciting fibres c, d and e would normally generate a peak at position 4. Assume now that fibres 4, 5 and 6 are cut. The same stimulus will now give rise to a peak at position 3 instead of position 4. A peak at position 3 would ordinarily be associated with impulses travelling along fibres b, c and d, let us say, or at least with impulses generated by a stimulus centred at position c rather than position d. There has been a 'lateral shift' in the apparent position of the stimulus.

(d) If overlap exists at any level central to the retinal ganglion cell, it would be predicted that a retinal lesion should produce a field defect having properties more akin to the classical all-or-none type than a cortical lesion. I know of no study that has specifically investigated this possibility. It might also be possible to determine the degree of overlap in the system by making a peripheral retinal lesion and a bilateral macular cortical lesion and testing the adequacy of visual response (on classical hypothesis).

A 'many-to-many' rather than a 'one-to-one' network would make the central visual area more akin, functionally, to other sensory and motor cortical areas, in which a high degree of lability has been demonstrated (George Clark, 1948). The approach adopted by Glees, Cole, Liddell and Phillips (1950; cf. also Glees and Cole, 1950, and Cole and Glees, 1954) in measuring shifts of excitability of motor cortex following small lesions might well be applied profitably to the striate cortex (but recording, of course, rather than electrically stimulating). Similarly, the techniques developed by Kennard and her co-workers for studying development of the motor cortex would appear to be useful for the investigation of the striate cortex (Kennard, 1938; Kennard and McCulloch, 1943; Ward and Kennard, 1942).

The question is frequently raised as to why we should expect to find a system in which there is so much possibility for blurring. Aside from the fact that maximal efficiency is not always attained in biological systems, a number of hypotheses have been advanced in which functional significance is attached to overlapping and/or converging systems—such as the compression of information into fewer channels, and especially perhaps the information contained in a moving stimulus (Barlow, 1953, and Ch. XIII, this book). It might even be suggested that certain functions involving binocular vision actually demand a degree of lability. For example, binocular fusion can occur when similarly shaped patterns fall on retinal points which are not exactly 'corresponding'. The degree of non-correspondence that can be tolerated has been determined in the classical work of Blum (cf. Ogle, 1950). Stereoscopic depth perception would be difficult to envisage without such lability. Cases of strabismus are said to demonstrate such lability (Duke-Elder, 1949; Bender and Teuber, 1949).

(ii) While the anatomical evidence does not support interaction between laminae in the l.g.n. (and hence between the input from separate eyes), some electrophysiological data would support such interaction (Bishop and Davis, 1953), although the issue is controversial (Marshall and Talbot, 1940). If there is interaction of this kind, we know nothing of its spatial properties, and such information would clearly be relevant to this general discussion. Even assuming that there is *no* interaction between laminae, one may ask to what extent there is 'correspondence' between points on the separate laminae—i.e. do 'corresponding' points as defined by retinal relationships converge to the same point on the striate cortex? There is no evidence on this point, at least with the degree of precision required. Cortical lesions in man, however, often produce scotomata which are similar in shape but not exactly congruent for the two retinal fields of vision. Thus it may be that vision can be filled along the edges of a scotoma by one of the two eyes—i.e. that the binocular scotoma is smaller than either monocular scotoma.

(iii) It must be considered significant that the neural response to light stimuli—at least from the retinal ganglion cells onwards—shows a certain similarity throughout the vertebrate family. Coding seems to be done mainly, if not entirely, in the terms of 'off', 'on', and 'on-off' responses. Individual ganglion cells have characteristically one of these three types of response (Granit and Tansley, 1948), as do individual fibres of the optic nerve (Hartline, 1938), single cells within the l.g.n. (de Vallois *et al.*, 1957) and, if they have any response at all, single cells within the striate cortex (Jung and Baumgartner, 1955). Recent work has shown that different *laminae* of the monkey l.g.n. are associated with different types of response—the dorsal layers showing on-response, ventral layers off-response, and intermediate layers on-off response (de Vallois *et al.*, 1957). (Incidentally, single layers within the l.g.n. were also shown not to be related to narrow wavebands of light, as suggested by Le Gros Clark, although within any one layer some cells were found which had this property.) This segregation of the three types of responses, first into different fibres and then into different l.g.n. layers, suggests that it is important for coding purposes that they do not mix, at least until they reach a certain cortical stage. Walls suggests that the three layers of the l.g.n. (corresponding to each half-retina) 'are related to each other as are three maps of the same country, one of which is geodetic, a second climatological, and the third agricultural. Just as three maps are required to keep such types of information apart and intelligible, so also a mammal with highly differentiated vision requires a multiplicity of genicular maps if the cortex is to be able to make full use of the classified information the retina sends them' (Walls, 1953, p. 75).

It would scarcely be appropriate to speculate here as to what aspects of vision might correspond to geodesy, climate and agriculture. It might be thought that a mechanism of alignment of the two eyes so as to produce fusion might work better with either an off- or an on-type response (more likely the former, for other reasons). Thus, if alignment is controlled by the convergence on to a single group of cells in the striate cortex of cells representing each eye, as suggested by Delisle Burns (1957), so that such a group is maximally either active or inhibited when corresponding points are fired, an on-off response from either eye would tend to produce confusion—temporal summation being confused with spatial summation. Similarly, for absolute sensitivity to light under conditions of darkness, an on-response would seem to be most suitable. For moving patterns, one might think an on-off response most effective. But regardless of the adequacy of these superficial guesses, the relevance of such an analysis for present purposes is that it is logically possible that the three systems, whatever their 'adequate stimuli', have projections to the cortex which are not topographically or perhaps even topologically, identical.

It might be easier to separate the three systems, cortically, in infra-primate mammals, where the area of overlap of the two visual fields is much less extensive than in primates. For this reason, the observations of Rademaker and ter Braak (1948) are of interest. They reported that lesions in the caudal part of the striate area of the dog were sufficient to eliminate the blinking response and conjugate ocular movement. Rostral lesions were without effect on these responses, but they did destroy the optic 'placing reaction', on which caudal lesions were without effect. It is regrettable that the authors did not attempt to relate these defects to possible scotomata, and to the projection of the retina to the cortex. Lashley (1931a) also found the striate cortex of the rat was not 'equipotential' with respect to pattern vision, for lateral striate lesions have a more deteriorating effect than medial lesions. It has already been mentioned that he found it difficult to relate such a deficit clearly to a scotoma.

Even in the human material, there have been many observations supporting 'dissociation' of different types of visual function following occipital damage. Thus Riddoch (1917) discussed several World War I cases in which there was a hemianopia for pattern vision but not for movement vision.

Recovery for movement vision, when it occurred, typically began in the peripheral part of the field. He discussed a case also, with loss of stereoscopic vision but not pattern vision, as do Bender and Teuber as well (1948, 1949). Holmes also described a case in which 'only large white objects when moving may be recognised ... when they cease to move, he sees them no longer; they disappear' (1919, p. 198). Riddoch suggests that the dissociation of object and movement perception might be similar to the dissociation between pain, temperature and touch in the somatic sensory system. The complementary syndrome has been reported by Pötzl, Gelb and Goldstein, Best and others (cited by Kluver, 1927), in which objects could be perceived only when stationary. Thus, one of Gelb and Goldstein's patients in observing the hand of a stop watch, reported that it could be seen in successive positions, but not in motion.

Directly relevant to this question is a conclusion of Poppelreuter's, summarised by Kluver as follows: 'Poppelreuter arrives "inductively" on the basis of some hundred cases, at "principles of dissociation" as regards the visual system in cerebral lesions. According to him, we have to do, firstly, with a number of part-systems: (1) the brightness system; (2) the color system; (3) the form system; (4) movement; (5) direction. *These different part-systems may be more or less independently disturbed*' (Kluver, 1927, p. 342, my italics).

Summary

I have tried to show that within the mammalian class the striate cortex does not show any considerable alteration in function as a consequence of phylogenetic evolution. Without denying the probability that vision comes to depend increasingly upon the cerebral cortex, I have argued that the effects of removal of the striate cortex appear to be surprisingly similar throughout the mammalian series. Furthermore, it has been suggested that sub-total lesions of the visual cortex produce considerably less visual impairment than might be predicted on the classical theory of 'point-to-point' projection of the retinal surface on the cortex. Some alternative theories of visual representation are suggested.

III. LASHLEY'S CONCEPT OF
CEREBRAL MASS ACTION[1]

1. *Introduction*

In 1824 Pierre Flourens published his classical study of the effects of ablation of various parts of the central nervous system upon the behaviour of vertebrates. His choice of 'guinea-pigs' was catholic, embracing quail, moles, dormice and snakes in addition to the common domestic animals. These experiments led Flourens to the view that 'all sensory, intellectual and volitional faculties' reside in the cerebral hemispheres and must be regarded as occupying concurrently the same seat in these structures. By this he appeared to mean that the various categories of psychological function are not localised in different regions of the brain but are to be viewed as in some sense the product of its total action. 'Feeling, willing and perceiving are but one single and essentially unitary faculty, residing in a single organ' (Flourens, 1824, p. 212).

Flourens also reported that cerebral lesions, if not exceeding certain limits, are followed by virtually complete restitution of function. This would appear to indicate that the parts of the brain remaining undamaged are in some way able to carry out the functions normally exercised by the whole brain.[2] But the present writer has been unable to trace in Flourens any precise statement of the limits within which complete functional

---

[1] Lashley's death in September 1958 was announced just as this chapter was being completed. The present writer would like to dedicate it to the memory of a very distinguished man.

[2] Flourens' work has had great influence upon our modern conceptions of the functions of the brain. It is notable that Boring, in writing of Lashley's work, comments that it seems to represent both a fundamental advance in the psychophysiology of the brain and a return to the position of Flourens a century before (Boring, 1929, p. 560). With modesty unusual in a historian, Boring concludes that the next step is quite unpredictable.

restitution is possible or any systematic comparison of the effects of lesions varying in extent. Although Flourens arrived at something very close to Lashley's idea of equipotentiality, it cannot be said that he anticipated the principle of cerebral mass action.

The first worker to deal at all precisely with the possible relation of behaviour to cerebral mass appears to have been F. Goltz, who reported a long series of decortication experiments in dogs in the closing decades of the last century. His precedence was fully acknowledged by Lashley. On the basis of long—if unsystematic—observation, Goltz concluded that the deterioration in behaviour produced by operations on the brain is roughly proportional to the extent of the ablation (1881, p. 33). While not denying localisation of function—in principle at least—Goltz was clearly of the view that intelligence cannot be localised in any particular region of the brain. This view was in decided contrast to the general opinion of the time.[1]

This relation between extent of cerebral lesion and severity of behaviour defect was taken up early by Lashley in his long and admirable series of studies on brain mechanisms and intelligence. As early as 1920 we find him beginning to cast doubt upon the adequacy of the conventional theories of habit and to speculate about learning in terms of the equipotentiality of large areas of the cerebral cortex.[2] And in the same year he first raised the question of whether there is a significant relation between the *mass* of functionally effective cerebral tissue and the rate of learning. An experiment bearing on this question

[1] The work of Fritsch and Hitzig, Ferrier, Munk and others had given strong impetus to the belief in the cerebral localisation of psychological function which, no doubt because of the extravagancies of phrenology, had fallen into disrepute. None the less, some influence of Flourens remained. In 1901 Loeb, who was a pupil of Goltz, wrote that the associative processes (with which he virtually identified intelligence) are distributed over the entire cerebral cortex and suffer only as the result of extensive and bilateral brain damage (Loeb, 1901, p. 274).

[2] Broadly speaking, it may be said that theories of habit have been traditionally conceived in terms of the establishment and facilitation of specific nervous pathways and connections. This type of theory, which goes back at least to Locke, finds perhaps its clearest expression in William James (1890, 1, pp. 104–20). Hebb (1949) has offered a most sophisticated modern variant.

(Lashley, 1920) was reported in spite of the fact that the results in this particular case were negative. None the less, there can be little doubt that Lashley felt that he stood upon the threshold of an important discovery.

The issue of mass action was brought up in a much more systematic way in a paper published by Lashley in 1926. Here he pointed out that a relation between brain mass and the complexity of behaviour had been suggested by several disciplines, e.g. comparative anatomy, cerebral pathology and experimental physiology. Although somewhat anecdotal, this evidence does perhaps suggest that the greater the size of the brain (or the greater the quantity of brain tissue left intact in an operated animal) the more delicate and highly adapted the behaviour. At the same time, Lashley pointed out that it is difficult to see how brain mass as such can govern behaviour and that proper understanding of this relationship may well entail new conceptions of nervous action. As a first step, he proposed to undertake a programme of research designed to study in quantitative terms the relations between learning, retention and the amount of brain tissue at the disposal of the animal.

Lashley's programme (Lashley, 1926) was envisaged in three main stages: First, he proposed to study the effects of brain lesions of varying extent upon the learning and retention of different types of habit in the rat. In this study, special attention was to be given to the role of the cortical sensory projection areas. Secondly, he undertook to test the validity of his findings in the rat for the monkey and other higher forms. And thirdly, he proposed to review the neurological literature in order to determine the relative importance of extent and locus of injury for the severity and duration of the after-effects of brain injury in man. Apart perhaps from the third stage, this programme was virtually accomplished in studies undertaken by Lashley and his co-workers over a quarter of a century. Our purpose in this chapter is to review their progress and attempt to assess their outcome.

2. *Lashley's early experiments*

The first experiment directly relevant to mass action was reported by Lashley in 1920. This followed on from an earlier experiment (Lashley and Franz, 1917), in which it had been shown that loss of the 'double platform box' habit in rats occurs only after extensive lesions in the anterior parts of the cerebrum. In the present experiment Lashley studied the formation of the same habit after brain injury. Contrary to expectation, he found that amounts of excision varying between 14 and 50% of the cortex produced roughly the same decrement in learning. In the case of this habit, then, there appeared to be '. . . no relation between the absolute quantity of cerebral cortex function and the ability to learn' (Lashley, 1920, p. 101).

Undeterred by this negative outcome, Lashley then turned to the study of less complex habits, in particular those based upon brightness discrimination. Now earlier experiments (Lashley, 1920, 1922) had shown that this type of habit, when acquired pre-operatively, is lost only after bilateral excision of the occipital area (though it might none the less be re-acquired at normal rate). His present problem was to determine whether more limited (sub-total) lesions in the occipital lobes lead to loss of habit proportional to the extent of the excisions. In general, this was found to be the case—at any rate within the range of excisions used in this study, which varied between 2 and 44% of the neopallium (Lashley, 1926). It could therefore be concluded that the degree to which a brightness discrimination habit is retained after operation is a function of the total amount of cortex in the occipital third of the brain remaining intact. This conclusion was reaffirmed in a later study (Lashley, 1929).

The most convincing evidence in support of a mass action principle was adduced by Lashley some years later in his noteworthy monograph, *Brain Mechanisms and Intelligence* (1929). This work is so well known that only a summary account need be given here. Briefly, Lashley was concerned to study the effects on maze performance in the rat of cerebral excisions in various loci and of varying extent. His findings led him to three main conclusions:

(1) The formation of maze habits is impaired by cerebral lesions, the degree of impairment being closely related to the extent of the lesion and independent of its locus.

(2) The retention of maze habits learned prior to operation is likewise impaired by cerebral lesions, the degree of impairment being again proportional to the extent of the lesion and independent of its locus.

(3) The defect in performance produced by any given extent of lesion is a function of the complexity of the task.

These conclusions undoubtedly suggest that the efficiency of maze performance is in some way governed by considerations of quantity and provide strong support for the mass action hypothesis. At the same time, one may note that they did not pass unchallenged. Hunter (1930), in particular, delivered a strong critical attack which produced a sharp rejoinder from Lashley (1931c). The gist of Hunter's argument is that the correlations found by Lashley between extent of brain lesion and relative defect of performance might well be explained in terms of progressive cortical sensory loss. His argument will be considered in greater detail at a later stage. Before embarking on controversy, however, let us take a closer look at the concept of mass action itself.

3. *The concept of mass action*

It is by no means easy to elucidate precisely what Lashley meant by 'mass action' and the present account cannot claim to give an authoritative statement of his views. In the first place, Lashley was an exceedingly cautious theorist and modified his ideas more than once in the light of fresh experimental data. And in the second place, it is not always easy to disentangle his views on the specific issue of mass action from those which he held on related matters, in particular cortical equipotentiality and the nature of the learning process. For these reasons, the present account is to be regarded as no more than a summary, and by no means necessarily accurate, statement of Lashley's position.

Before we can understand what Lashley had in mind by mass action, it is necessary to enquire what he meant by *equipotentiality*.[1] As we have seen, his experiments had suggested that the formation and execution of habits are dependent upon the integrity of the whole cortex (as with maze performance) or upon the appropriate sensory projection area (as with brightness discrimination). In both cases, however, no subordinate area of the cortex appears to have greater importance than any other area and there is no evidence of differential localisation of the component parts of a habit. It can therefore be said that all parts of the cortex which subserve a given habit contribute equally to its efficiency. Equipotentiality, then, is to be defined as '. . . the apparent capacity of the functional area to carry out, with or without reduction in efficiency, the functions which are lost by destruction of the whole' (Lashley, 1929, p. 25). This capacity is assumed to vary from one region of the brain to another and with the character of the functions involved. Indeed Lashley expressly stipulates that it may hold only for the association areas and for functions more complex than simple sensitivity or motor co-ordination. It seems likely that what Lashley is trying to say is that, in the case of complex behaviour, there is complete lack of specialisation within an area (which may be the whole cortex) itself specialised for the control of this behaviour. Contrary to what is often supposed, the idea of equipotentiality is by no means inconsistent with the cortical localisation of function.

Mass action (or mass function) is introduced in the first instance as a factor limiting equipotentiality. This latter '. . . is not absolute but is subject to a law of mass action whereby the efficiency of a whole complex function may be reduced in proportion to the extent of brain injury within an area whose functions are not more specialised for one component than for

[1] Lashley's use of this term clearly derives from embryology, in which it has been taken to mean the capacity of embryonic tissue to produce any or all of the parts of the developed organism. It has a quite different meaning in physics.

another' (loc. cit.). Mass action, then, is to be conceived as a general factor governing the activity of such parts of the cortex as are demonstrably equipotential in the control of a given behaviour pattern.

These ideas were more fully explained in a paper published by Lashley in 1931. Here he begins by stressing that his principles are certainly not inconsistent with the facts of cerebral localisation. 'No one today,' he writes, 'can seriously believe that the different parts of the cerebral cortex all have the same functions or can entertain for a moment the proposition that because the mind is a unit the brain must also act as a unit' (1931*b*, p. 245). At the same time, he reminds us how little we know regarding which functions may be held to be localised or the relative fineness of localisation. Classical localisation theory, moreover, has failed to explain '. . . how the specialised parts of the cortex interact to produce the integration evident in thought and behaviour' (1931*b*, p. 246).

Lashley introduces his concept of mass action in the following way: He asks us to consider the function of a part of the brain within which a precise degree of localisation is known to exist and chooses the visual cortex of the rat as a representative example. He then points out that the results of excision of the visual cortex vary with the performance under examination. In the first place, habits based upon detail or pattern vision are lost following the excision of a small area in the lateral part of the area striata. In the case of detail vision, then, localisation is clearly of the utmost importance. In the second place, habits based upon brightness discrimination are lost only after destruction of the *entire* striate area and sub-total lesions produce a loss of habit proportional to their size and independent of their location. There is still localisation, but '. . . its effects are independent of that finer localisation which is essential to pattern vision' (1931*b*, p. 245). In the third place, habits based upon a number of sensory modalities, e.g. maze performance, are impaired by lesions in the visual cortex to a degree proportional to their relative extent. Lashley reminds us

6

that his own experiments have shown that equal amounts of destruction in the motor, somaesthetic and visual areas are attended by equal amounts of retardation in maze learning, suggesting that these areas contribute in some non-specific way to the efficiency of behaviour. This contribution, which is quite independent of the special function of the region concerned, is held to vary in direct proportion to its relative mass.

It follows from Lashley's argument that one and the same cortical area must be supposed to function at one time as a highly differentiated system and at another as an amorphous mass of nervous tissue. How are these very different modes of function to be envisaged in concrete neurophysiological terms? Lashley is not very specific here but he does suggest that the process is essentially one of *facilitation*. In the case of brightness discrimination, for instance, he suggests that an influence is exerted by the entire visual cortex upon the subcortical visual nuclei, the degree of facilitation being proportional to the amount of visual cortex remaining intact. (This idea was, however, modified later as the result of further experiments.) Here, then, mass action is to be viewed as a process of facilitation exercised by the cortex upon the subcortex—an idea that reappears in the work of Beach (1940) on sexual mechanisms in the male rat. Although of a general nature, this facilitation is specific in so far as it is limited to a particular class of functions (or habits) and is exercised by a particular region of the cerebral cortex.

It is not altogether clear from Lashley's writings whether the mass action of the cortex is always to be envisaged as a process of facilitation. In discussing maze performance, it is true, he seems often to have in mind some process of trans-cortical facilitation, though it is difficult to see how this is to be reconciled with the apparently negative effects of trans-cortical (linear) lesions upon behaviour (Lashley, 1944, 1950). At times indeed, Lashley appears to be thinking less in terms of facilitation, conceived as the direct influence of one part of the brain upon another, than in terms of some intrinsic property or dimension of brain function, akin perhaps to Head's (1926) concept of *vigilance*. Head, it will be borne in mind, postulated a general factor governing the overall efficiency of the nervous system, or of its parts, to which he gave the name of vigilance. Broadly, the lower the level of vigilance, the less capable is the organism of precise and adapted response. Applied to Lashley's findings, this might mean that brain mass is one of the factors governing the vigilance, and thereby the efficiency level, of an equipotential cerebral system.[1] None the less, it is only fair to add that Lashley expressly denied that vigilance or similar concepts could be helpfully applied to the problem of mass function (1929, p. 168).

Whatever the precise nature of mass action, Lashley appears to have been convinced—at all events in 1931—that the concept was a valid one and that his experimental findings could be explained in no other way. 'The limiting condition for efficiency,' he writes, 'is the surface area or mass of cortical tissue and not the specific anatomical relations of its parts' (1931*b*, p. 251). Although Lashley was later prepared to consider other interpretations of his findings (cf. Section 8, below), one may surmise that he was never really shaken in his belief in the principle of mass action.

4. *Some implications of mass action*

There can be no doubt that Lashley's thinking about a variety of problems in psychology was greatly influenced by the mass action hypothesis. In the first place, it fitted well with his views on the nature of habit. Although Lashley lacked a positive theory, he was convinced that the learning process does not depend upon specific anatomical structures or fixed systems of neural connections (cf. 1929, pp. 122–5, 127–31; 1950, pp. 459–62). He adduced over and over again the negative effects

---

[1] This interpretation of mass action might appear to gain support from Lashley's own reference to phenomena of *regeneration*. In the regeneration of hydroids, he points out, the number of tentacles regenerated is correlated with the size of the original mass of tissue (1931*b*, pp. 253–4).

of linear (trans-cortical) lesions on behaviour as evidence that the correlations between brain mass and performance level cannot be explained in terms of the reduplication of equivalent neuronal arcs (1929, 1944, 1950). As a basis for habit, he was led to seek some mechanism widely distributed throughout the cortex and able to facilitate a variety of activities by its total mass rather than by 'specific integrations'.

Secondly, Lashley was clearly of the opinion that mass facilitation in the cortex '. . . in some way underlies the intelligent activities of the organism' (1929, p. 250). Although averse to psychological disputation, he was plainly drawn to theories which conceive of intelligence as a general capacity rather than an assemblage or mosaic of abilities.[1] In keeping with this outlook, Lashley showed scant sympathy for attempts to ascribe different aspects of intelligent activity to cerebral mechanisms localised in different areas of the brain and took other authors severely to task for attempting to do so. In spite of the fact that Lashley's principles are certainly not incompatible with concepts of localisation, there can be no doubt that his work served to create a climate of opinion distinctly hostile to the whole idea.

5. *Visual discrimination in the rat*

It has been pointed out that the original conception of mass action was largely based on the results of experiments on brightness discrimination in the rat (Lashley, 1926). These results, and the interpretations placed upon them by Lashley at the time, are still quoted in the text-books (cf. Davis, 1957) in spite of the fact that they were largely vitiated by Lashley's own later work. A short account of this work may therefore be given.

The original finding was that habits based upon reaction to brightness differences are impaired only by lesions within the occipital third of the brain and that, within this region, the

[1] Lashley's work was warmly welcomed by Spearman as providing powerful, if indirect, support for his conception of a 'general factor' (*g*) in human intelligence (Spearman, 1937, II, p. 237).

extent of damage determines the degree of habit loss. Now Lashley showed in his later experiments on pattern vision (Lashley, 1931*b*; Lashley and Frank, 1934) that destruction of a small region in the lateral part of the area striata abolishes the animal's capacity for pattern discrimination. He thought that this region is almost certainly the projection area for the binocular visual field and is presumably of great importance for fine discrimination. In the light of this unexpected finding, Lashley was led to re-examine his earlier data on brightness discrimination. He concluded that the correlations between extent of lesion and amount of habit loss reported earlier were almost certainly due to encroachment on this same small region. That is to say, the correlations were due not to overall mass action but to the involvement of a small cortical area critical to the performance of visual discrimination habits (Lashley, 1932).

Even this is not the end of the story. Three years later, Lashley reported that habits of brightness discrimination are perfectly retained after occipital lesion *provided only that a small part of the geniculo-striate system remains intact* (Lashley, 1935). This conclusion was later extended to pattern vision (Lashley, 1939). It was therefore admitted by Lashley that post-operative loss of visual habits appears to obey an 'all-or-none' law, the essential condition being the preservation of a minimal amount of striate cortex in connection with the subcortical visual nuclei. Although this may still be conceived to mediate some essential facilitation, it is clear that the effect cannot be held to depend on the factor of total mass.

6. *The 'non-sensory function' of the visual cortex*

We have seen that Lashley supposed that the projection areas of the cortex contribute to the efficiency of behaviour in some way other than by virtue of their sensory functions. This conception of a 'non-sensory function' of the sensory areas is crucial to the whole idea of mass action. It arose from the findings of Lashley (1929) and Lashley and Wiley (1933) that lesions of comparable size within the different projection fields produce comparable impairment in maze performance. In view of the fact that the different senses are known to contribute in very different degree to maze running, Lashley felt able to conclude that the different projection areas contribute in a non-specific way to maze performance, and that this contribution is proportional to their relative mass.

In the 1929 monograph Lashley briefly cited some findings which appeared further to support his claim. He pointed out that the maze performance of rats which had sustained enucleation of the eyes is superior to that of animals with excision of the visual cortex (1929, pp. 110–2). Unfortunately, few animals were used and the individual differences were large. More satisfactory were the experiments reported a few years later by Tsang (1934, 1936). This worker showed first, that removal of the striate areas in rats leads to a deficit in maze learning very much greater than that produced by enucleation of the eyes; and secondly, that removal of the striate cortex together with enucleation leads to a defect very much greater than that produced by the latter operation alone. These results were very generally taken to constitute strong *prima facie* evidence in support of the mass action theory.

The whole problem is obviously a most important one and it is a great pity that later work on it has produced inconsistent and somewhat controversial results. In 1941 Finley reported a study which appeared to show that excisions of the visual cortex give rise to no greater decrement in maze learning than is produced by peripheral blinding or total elimination of visual stimuli. Lashley himself took up the challenge. He criticised Finley's work strongly on both anatomical and methodological grounds and reported some fresh experiments together with a particularly close analysis of the results (Lashley, 1943). The conclusion was that total destruction of the striate area does in fact produce deterioration in maze performance in blind rats and that this deterioration is about ten times as great as that found to follow enucleation of the eyes or section of the posterior radiations. This work of Lashley's has been widely accepted as reliable; none the less, a more recent study by Pickett (1952) has failed to provide confirmation.

It has been pointed out in several quarters that the experiments upon which this controversy hinges cannot be regarded as strictly comparable (cf., e.g., Morgan and Stellar, 1950, pp. 488–9). In the first place, there has been a good deal of dispute as to the precise extent of the lesions in the different studies. Lashley, in particular, has been taken to task for exceeding the strict limits of the striate cortex. In the second place, the apparatus, technique and training methods have varied a good deal. Thus Finley (1941)—and also Pickett (1952)—used elevated mazes, which might be thought to make greater demands upon vision than the alley mazes used by Lashley and Tsang. It may be hoped that future experiments will be conducted under conditions agreed in advance by all parties to the dispute.

Although it is always invidious to attempt to arbitrate between the contradictory results of different workers, it is at any rate clear that the supposed non-sensory role of the visual cortex (or indeed of any other sensory area) in maze performance has not yet been definitely established. This is not of course to say that the sensory areas are necessarily without functions other than those imposed upon them by the anatomical lay-out of the sensory projection systems. They may indeed have other functions, either of a relatively specific kind or in relation to the total integrative activity of the brain. But it is not at present possible to give a definite answer to this crucial question in neurological theory.

7. *Mass action and maze performance*

The strongest evidence for mass action is undoubtedly Lashley's work on the effects of cerebral lesions on the learning and retention of maze habits. Apart from the 1929 monograph to which brief reference has already been made, there is an important study which Lashley published (with L. J. Wiley) in 1933. This was on a much larger scale than is usual in ablation work and was designed to throw light on a number of problems, not least that which the authors designated as that of the 'continuity of mass relations'.

Broadly, the question is whether the law of mass action holds over the whole range of cerebral excisions—from the smallest to the largest—or whether the lesion must exceed a certain critical size before there is any appreciable impairment. The authors point out that the latter alternative has been suggested by the results of certain earlier studies, e.g. those of Lashley and Franz (1917), Lashley (1920, 1929), Loucks (1931) and Maier (1932a). If sustained, it would indicate that the law of mass action holds only between certain limits of lesion size, which might well vary from one activity to another. But Lashley and Wiley were unable to find evidence of a limitation of this kind with the series of mazes used in their study. In no case was there any suggestion of a critical amount of destruction resulting in a sharp rise in error scores. The relation between extent of lesion and retardation in learning was found to be curvilinear, error scores appearing as a logarithmic function of the extent of lesion. This was taken as establishing the 'continuity of mass relations' from the smallest to the largest lesions.[1]

It would be easy to cite many other experiments reported during the last thirty years which give apparent confirmation to the mass action hypothesis. Tsang (1937), for example, reported an interesting study of rats hemidecorticated in infancy and trained in the maze at maturity. Although the detrimental effects were found to be far less severe than in the case of animals operated at adult age, no clear influence of the locus of the excision could be established. Tsang therefore postulated a factor common to all four cerebral quadrants

---

[1] None the less, it is important to note that an important conclusion reached by Lashley in his earlier study was not borne out in these experiments. No certain influence of the size of the lesion upon the proportional difficulty of the various mazes could be established. Lashley and Wiley were therefore unable to confirm that the degree of retardation in learning after brain operation is a function of the complexity of the task.

which contributes in a non-specific way to the integration of behaviour. This is presumably to be envisaged as mass action in the sense of trans-cortical facilitation.

Taking the evidence as a whole, it is beyond reasonable doubt that the deterioration in maze learning—or in the execution of maze habits acquired before operation—correlates highly with the extent of the lesion, and shows no obvious dependence on its location. This basic finding, which we owe above all to Lashley, seems reasonably secure. But its interpretation is a much more difficult matter. Lashley was himself well aware of many possible interpretations of his results and indeed referred on one occasion (Lashley, 1943) to mass action as a 'residual hypothesis' to be entertained, presumably, only if other and more obvious explanations fail. Although his belief in this particular hypothesis appears never to have been seriously shaken, it is only fair to make reference to some alternative explanations which have been put forward.

8. *Other explanations of Lashley's findings*

In 1943 Lashley adduced six possible ways of explaining the correlations between size of lesion and degree of impairment in behaviour. These explanations (which are by no means mutually exclusive) comprise (*a*) indirect effects of operation (e.g. diaschisis and pressure effects); (*b*) summated cortical sensory defects; (*c*) progressive invasion of some area critical to performance; (*d*) progressive invasion of the association areas; (*e*) interference with the interaction of diverse functional areas; and (*f*) dynamic mass action of the cortex. It is probable that yet other explanations might be devised but these seem sufficient for a preliminary survey.

Lashley had little difficulty in disposing of the first of these possibilities, which neurological opinion generally would regard as remote, except during the acute post-operative stages. The sixth—that favoured by Lashley himself—has already been sufficiently discussed. We may therefore confine ourselves to the four explanations that remain.

Summated sensory defects: It is obviously possible that the correlation between size of lesion and degree of defect might be due to progressive encroachment upon the cortical sensory fields by the larger lesions. As we have seen, this was the view of Hunter (1930) and indeed goes back to Goltz (1881). Hunter was quick to draw a parallel between the effects on maze learning of progressive peripheral sense privation and those found by Lashley to follow cerebral lesions of increasing size. In attempting to rebut this criticism, Lashley (1931c, 1943) placed great weight on the apparently more severe effects of damage to a cortical projection area than to the corresponding peripheral receptor organ. He also pointed out that whereas the different senses contribute unequally to maze performance, lesions to the different projection areas produce an equivalent loss of habit. As far as the present writer can ascertain, Lashley at no time saw good reason to modify his early conclusion that '. . . the probability of our findings in maze learning and retention being the product of cerebral sensory defects is very slight' (1929, p. 115).

Although Hunter may well have overstated his case, it is possible that Lashley underestimated the role of cortical sensory loss. His arguments, too, are not fully conclusive. Thus we have already seen that the alleged mass action of the sensory areas in maze performance is by no means finally established. Further, it cannot be said that these areas have been proved to be precisely equipotential in maze performance. Lashley and Wiley's own findings raised a suspicion that lesions within the auditory area produce less effect on such performance than do lesions of comparable size in the visual, somaesthetic and motor areas. Further, the results of Pickett (1952) are more consistent with Hunter's theory than with that of Lashley. On general grounds, it is certainly likely that cortical lesions give rise to an integrative defect going beyond mere restriction of sensory information. None the less, the possible role of summated cortical sensory defects in producing impairment in complex behaviour is certainly worthy of further study.

Critical areas: Another possible explanation of Lashley's correlations is the progressive invasion of some area in the cortex critical to the performance at issue. As we have seen, Lashley came to accept this explanation in the case of visual habits but invariably held that it was inapplicable to the results of the maze experiments. As he rightly pointed out, deterioration in maze performance is produced by lesions in widely different parts of the cortex which involve no common area. All the same, one might still argue that maze performance depends upon more than one critical area, as on the theory of 'dispersed specific factors' cited by Morgan and Stellar (1950, pp. 489–50). Yet it is only fair to say that no such critical areas have as yet been isolated, and even if they had, it is not easy to see what functions should be ascribed to them.

The association areas: This is of course a variant of the 'critical area' hypothesis. Important functions had been traditionally accorded to the association areas and it is not surprising that Lashley should have considered that they might bear some special relevance to learning and performance. Yet Lashley claimed that small lesions in the association cortex of the rat result in no significant loss of the maze habit and that the latter could not in any case depend upon a single association area. In general, he was very sceptical as regards the supposed functions of these areas and saw no good reason to believe that higher capacities, such as association or memory, should be ascribed to them (Lashley, 1950).

Interference with trans-cortical connections: Partial isolation of different cortical regions from one another has not been found to affect the learning or retention of maze habits in any systematic way (Lashley, 1942, 1944). Lashley therefore excluded an interruption of this kind as being responsible for the correlations between extent of lesion and relative defect of performance (Lashley, 1943).

Lashley was satisfied that none of the above theories provided a convincing explanation of his results and was therefore obliged to fall back (not, one may surmise, without secret satisfaction) upon the 'residual hypothesis' of mass action. Yet others may well be less certain. As we shall see, there is now strong evidence of a critical area in the rat's cortex concerned with the control of maternal behaviour and also some evidence to suggest that the different regions of the cortex are by no means strictly equipotential in relation to learned behaviour. In primates, the evidence for specific defects in performance following lesions in certain parts of the association cortex is, to say the least, extremely plausible. These defects do not, it is true, involve specific amnesias and we have still got no further than Lashley in his search for the engram. None the less, it might appear that the co-existence of sensory defect with the invasion of certain critical areas (as yet undefined) might provide the most plausible explanation of Lashley's results.

9. 'Critical areas' in the rat

An interesting extension of Lashley's approach is the work of Frank Beach (1937, 1938, 1940) on reproductive behaviour in the rat. Beach was able to show first, that maternal behaviour undergoes progressive dilapidation with increasing size of lesion; and secondly, that a statistical relation may be demonstrated between extent of cortical lesion and percentage of a total group of male rats still capable of copulation. These findings are clearly consistent with the mass action hypothesis in something like its original form (Lashley, 1929, 1931b). Indeed, Beach was led to postulate a facilitatory action by the cortex upon the subcortex of a kind closely parallel to that adduced by Lashley in the case of the striate areas. In short, sexual excitability in the male rat is a function of total cortical mass.[1]

On the other hand, more recent work suggests that, in the case of maternal behaviour at least, not all parts of the cerebrum are equipotential. Stamm (1954, 1955) has been able to show

[1] It must of course be borne in mind that there are important differences between reproductive behaviour and maze habits. For instance, sexual activity in male rats with extensive cortical lesions may be re-aroused by the administration of appropriate hormones.

that small lesions of the cingulate and retrosplenial areas in the rat have a profound effect upon maternal behaviour, as also upon hoarding (Stamm, 1953). The deficit is closely comparable to that found by Beach to follow extensive lesions of the lateral portions of the hemispheres. If there is mass action, then, it must be of a somewhat indirect kind, e.g. a depression of specific activity in the cingulate region provoked by extensive lesions in other cortical areas. At all events, Stamm has provided convincing evidence of a critical area in the control of the maternal behaviour pattern.

A case can also be made out for the existence of critical areas in the control of certain acquired behaviour patterns. It will be remembered that Krechevsky (1935) reported a good many years ago that the effect of a lesion of a given size in a given location may vary with the method adopted by the animal in executing the task. In the case of rats which had adopted 'visual hypotheses' in their attack upon a maze problem, performance was found to be much impaired by lesions placed in the striate cortex. On the other hand, in rats which had apparently adopted 'spatial hypotheses'—i.e. modes of attack based predominantly upon directional clues—performance was found to be selectively impaired by small lesions placed in the post-central areas. Although it might perhaps be possible to explain these results in terms of differential sensory loss, they at least bring out the point that individual differences are of very real importance in understanding the effects of brain lesions on behaviour.

There are a good many scattered observations in the recent literature which suggest that locus of the lesion is an important consideration, even in the behaviour of rats. Pickett (1952) finds that posterior cerebral lesions are without effect on the performance of rats that have learned an elevated maze without the use of vision. Lansdell (1953) has reported that impairment on the Hebb-Williams 'Open Field' test (a kind of rodent intelligence test) occurs principally with extensive unilateral or bilateral posterior brain lesions. Anterior lesions, even if bilateral, are apparently without significance for this performance. These results, taken together with those of Stamm, indicate that care is necessary in interpreting the effects of cerebral lesions and might even suggest that the gross correlations found by Lashley between extent of lesion and degree of impairment have significance only in the statistical sense of the term: they do not necessarily enable one to understand the mechanisms involved in complex behaviour or to conclude that intelligence depends upon an underlying 'mass function' of the brain.

10. *Behaviour mechanisms in primates*

The second stage of the programme which Lashley set himself in 1926 was to test the validity of his findings in the rat for the monkey and other higher forms. Although this was less fully achieved, Lashley did a fair amount of work on primates himself and stimulated many others to concern themselves with this important field. The work of Jacobson, Chow, Pribram and others owes much to his interest and example.

It can hardly be said that Lashley made a direct attempt to verify that his principles of equipotentiality and mass action hold for the primate cortex. Rather was he concerned to labour the inadequacy of more traditional theories of brain function in explaining the experimental findings. For example, he has consistently argued that there is no real evidence that the association areas subserve special functions in learning or other higher processes. In a systematic study of ablation of the visual association cortex in the monkey (Lashley, 1948), no significant post-operative amnesia for habits of differential reaction based upon colour or form could be established. Lashley has also interpreted the well-known experiments of Jacobson (1936) on prefrontal ablation to imply that motor habits are not lost following operation, although they may no longer be executed in proper sequence. More generally, '. . . the so-called associative areas are not essential to the preservation of memory traces' (1950, p. 464).

None the less, it has become clear during the past ten years that lesions in the association areas in primates, even if they do not provoke specific amnesias, are liable to give rise to certain defects of performance of a surprisingly specific character. Thus it has been known since the early work of Jacobson (1936) that 'delayed reaction' in monkeys and apes is much impaired following bilateral prefrontal ablation. Although the severity of the defect depends in part on the conditions of testing, comparable loss of 'delayed reaction' is not found with lesions of comparable size in other parts of the cortex (Chow and Hutt, 1953). Further, the deficit appears to be maximal if the lesion involves the middle part of the lateral surface of the frontal lobes (Pribram and Bagshaw, 1953). It is plain, then, that a defect of this kind bears an intimate relation to the locus of the lesion, even though the function to which it corresponds eludes precise psychological definition.

As regards the posterior association areas, there is good evidence that bilateral lesions involving the parieto-temporal-preoccipital cortex have a marked and selective effect upon differential visual capacity (Blum, Chow and Pribram, 1950; Harlow, Davis, Settlage and Meyer, 1952). This evidence has been well reviewed by Chow and Hutt (1953) and by Milner (1954). Further, it now appears clear that a very similar visual defect may be produced by lesions limited to the temporal lobe alone. This defect is not a function of the mass of temporal lobe tissue destroyed and its critical locus is almost certainly the ventromedial area (Chow, 1952b). It is not known whether the lateral temporal cortex together with the prestriate areas exercise some facilitation upon the ventromedial area or whether, on the other hand, the latter has some special significance as a cortical 'focus' for visual discrimination. In this connection, Chow and Hutt (1953, p. 671) have suggested that there may be separate 'foci' within the association cortex subserving individual behaviour functions, each of which is surrounded by a peripheral 'field'. These 'fields', which may be fairly extensive, may be supposed to overlap to form a common

79

neuronal pool subserving a number of different behaviour functions. This may prove a valuable compromise between the theory of fixed localisation and the principle of mass action.

Lashley has argued that discrimination loss following temporal lobe ablation does not depend upon amnesia for specific habits of reaction (Lashley, 1950, 1952). In this, he is probably right. Indeed, the visual defect appears to be due to lowered 'visual attention' (i.e. responsiveness to visual stimuli) rather than to any true defect of memory (Chow, 1952b). None the less, Lashley's claim that, in view of the absence of both memory loss and primary visual defect, we must be dealing with some quite generalised form of impairment, seems ill-founded. As Milner (1954) has pointed out, there is clear evidence that the defect is specific to vision and certainly does not involve all discrimination learning equally. None the less, it is not true that these findings necessarily entail the rejection of Lashley's position, as Harlow (1953) appears to suppose. In reviewing recent work on primate behaviour, Harlow writes that 'the data ... are strongly opposed to Lashley's theory of equipotentiality ... in that certain tasks (and presumably certain kinds of intellectual abilities) are selectively affected to a statistically significant degree by differential localisation of cortical ablations' (1953, p. 509). But this does not necessarily follow. Equipotentiality and mass action were by no means limited by Lashley to the activity of the whole cortex. The concepts apply no less to any area of the cortex which, when excised, results in impairment or loss of a given complex habit. In the case of the defects of discrimination which have been found to follow limited cortical ablations in primates, it would be necessary to show that the areas concerned are neither equipotential nor subject to mass action before such a conclusion could be drawn. Unless or until this can be done, Lashley's position stands.

At present, work is actively proceeding upon the effects of cortical lesions upon a variety of types of discrimination and some further evidence of differential localisation has already been obtained. This more recent work has been reviewed by

Chow and Hutt (1953) and by Jasper, Gloor and Milner (1956). Beyond stating that it has given evidence of a more pronounced localisation of behaviour defects than might have been anticipated, it is difficult to draw firm conclusions at this stage. It is clear, however, that it is localisation rather than its lack that is again providing the principal challenge.

11. *Brain injury in man*

Although Lashley often touched on the problems of human brain injury (cf. 1929, 1938, 1948), and was extremely well informed in the neurological literature, the present writer has been unable to trace the systematic analysis promised in 1926. This would appear to be the only part of the 'mass action programme' to remain unfulfilled.

As an experimentalist, Lashley was understandably wary of the hazards of clinical study and warned against the acceptance of clinical evidence at its face value. Not without reason, he stressed the inevitable unreliability of much clinical observation and the prevalent tendency on the part of neurologists to 'pigeon-hole' symptoms in logical categories rather than to describe clearly what their patients could and could not do (1929, p. 154). None the less, Lashley was not averse to calling on some of this evidence as providing a measure of support for his own views. In *Brain Mechanisms and Intelligence*, he argued that more recent studies of cerebral function in man are not in the main opposed to his results with the rat; further, the clinical literature provided at least hints of a relation between the extent of brain damage and the degree of deterioration (1929, pp. 154–5). Indeed, apart from more specific representation within the sensory and motor projection fields, Lashley claimed that there is '. . . little evidence of a finer cortical differentiation in man than in the rat' (1929, p. 156).

Although few would nowadays subscribe to the doctrine of cortical localisation of function in its more extreme forms, it is difficult to accept the view that the cerebral cortex in man is in large part equipotential. Although this may perhaps be true of the cortex in early infancy, it certainly does not seem to be so after the development of speech.[1] As has long been known, the hemisphere contra-lateral to the preferred hand bears a special relation to speech and as a general rule injuries to this hemisphere alone give rise to aphasia. It is likely that this lateral specialisation occurs during the earlier years of life and brings about a certain asymmetry of the two hemispheres in relation to intellectual development. Even if one excepts disorders of speech, it is noteworthy that injury to the 'dominant' hemisphere is liable to provoke a pattern of intellectual loss appreciably different from that found to follow comparable injuries of the 'non-dominant' hemisphere (Weisenburg and Mac-Bride, 1935; McFie and Piercy, 1952; Reitan, 1955; Heilbrun, 1956). Even within the 'dominant' hemisphere, moreover, the pattern of defects depends in some degree upon the locus of the lesion. Thus Weinstein and Teuber (1957) have recently shown that loss on a standard intelligence test is most severe and persistent in cases in which the left parieto-temporal region is predominantly involved. One may conclude, then, that there is statistically significant evidence that intellectual defect following brain injury in man is dependent to an important degree upon the laterality and locus of the lesion.[2]

It is curious how little attention has been paid to the issue of laterality in animal experiments. Indeed Lashley himself wrote in his preface to *Brain Mechanisms and Intelligence* that all the lesions to the brain included in this study involved both hemispheres and that unilateral lesions might well '... give results

[1] It has long been known that extensive damage to either cerebral hemisphere at birth or soon after does not necessarily interfere with normal mental growth. Indeed, removal of the damaged hemisphere at a later stage—even if it be the left—fails to produce aphasia or significant intellectual defect (Krynauw, 1950). This certainly testifies to a remarkable degree of equipotentiality as between the two hemispheres in early infancy. But it obviously does not support a conception of brain function in which major emphasis is placed upon total cerebral mass.

[2] None the less, it seems possible that the extent and depth of the lesion are also important considerations. In a survey which appeared too late for discussion here, Chapman *et al.* point out 'the form and degree of the highest intellectual functions was independent of the site of tissue loss within the neopallium and was related closely to the mass of improperly functioning tissue' (1958, pp. 528–90).

of an entirely different character from those reported here'
(1929, p. xi). Although it is well known that unilateral lesions
in animals—unless very extensive—produce little in the way of
behaviour change, it is only very recently that systematic comparisons between the effects of unilateral and bilateral cortical
resections have begun to be made (Denny-Brown and Chambers, 1958). Even so, few—if any—quantitative data bearing
on the problem are yet available. When it is borne in mind
that, in man, bilateral injuries may not only give rise to
syndromes which never occur with injuries to a single hemisphere (Holmes, 1956), but also to disabilities more severe and
long-lasting than those found to follow strictly unilateral injuries (Humphrey and Zangwill, 1952; Faust, 1955), the need
for closer study becomes immediately apparent. *Prima facie*, the
factor of unilaterality or bilaterality of the lesion would seem
to warrant at least as much attention as that of the total extent
of the ablation.

In view of the fact that Lashley's experiments were concerned only with the learning and retention of habits, it might
be argued that comparison of his findings with the clinical
material should be limited to the field of memory. Now it is
certainly true that defects of learning in man are liable to
follow injury to any part of the cerebrum and there is even a
suggestion that the severity of the defect is proportional to the
extent of the damage (Isserlin, 1923; McFie and Piercy, 1952).
At the same time, there is good evidence that severe amnesic
syndromes, which may involve almost total loss of the capacity
to learn, also occur in cases in which the lesion is localised, as
in the case of tumours involving the region of the third ventricle
(Williams and Pennybacker, 1954) or following surgical re-
section of the hippocampus (Scoville and Milner, 1957).
Although the interpretation of these findings is difficult, it is
plain that human learning cannot necessarily be envisaged as a
'mass function' of the brain.

As we have seen, Lashley saw no good reason to believe that
cortical differentiation of function in man is significantly finer

or more precise than in the rat. From a purely neuro-histo-logical point of view, his claim could perhaps be justified (Sholl, 1956). Yet it is doubtful whether the evidence from clinical study could be said to support such a view. Although the dementia associated with diffuse brain disease does appear to be correlated in a rough way with the extent of brain damage, and although extensive brain wounds (especially when deep) are liable to produce more severe mental handicap than small wounds, it is uncertain whether the mass factor as such is the most important consideration. It is well known that extensive removal of cerebral substance in the course of neurosurgical operations, particularly in those involving the frontal lobes, is not necessarily followed by appreciable intellectual loss (Hebb and Penfield, 1940). Further, the range and variety of special syndromes which are associated with lesions of the parietal lobes (Critchley, 1953) inevitably suggest that these areas of the brain contribute in some special way to a variety of special functions, in particular those related to spatial orientation. Although it is always treacherous to argue from 'symptoms' to 'functions', it is almost impossible to rid oneself of the concept of functional localisation in attempting to make sense of the psychological *sequelae* of brain injury. Neither equipotentiality nor mass action suggest themselves as helpful conceptual tools in this daunting task.

12. *Conclusion*

We have tried to give some account of Lashley's views on the nature of cerebral activity, with special reference to the concept of mass action. This concept was derived from a number of experiments which appeared to show that defect in behaviour after operations on the brain is a function of the extent rather than of the locus of the lesion. In the light of this finding, Lashley was led to propose that large areas of the cerebral cortex are equipotential in regard to particular habits and that, within these areas, the efficiency of performance depends upon the total mass of brain tissue remaining intact. Although aware that this relationship between mass and efficiency might be explained in a variety of ways, Lashley consistently advocated a non-specific, dynamic action of the cortex as the most plausible hypothesis. This action was envisaged in terms of facilitation, which might be either trans-cortical or involve an influence of the cortex upon the subcortex. Finally, Lashley took the view that cerebral mass is the most important factor limiting the efficiency of behaviour and that it must be held in some way to underlie the intelligent activities of the organism.

The review which we have given suggests that there has been a good deal of confusion between fact and interpretation, both on Lashley's part and on that of others. Further, even the facts are a good deal less decisive than is commonly supposed. Lashley was himself led to abandon the mass action concept in relation to visual discrimination habits and its application to the results of maze experiments is, to say the least, equivocal. In particular, the claim that the sensory areas of the cortex possess non-sensory functions which contribute to the efficiency of maze performance to an extent proportional to their relative mass cannot be regarded as established. We have also shown that the whole trend of recent work on primate behaviour may be said to favour the differential localisation of specific behaviour patterns. By and large, the mass action hypothesis looks a good deal less plausible today than it did twenty years ago.

None the less, it cannot be said that the existing evidence warrants the abandonment of Lashley's position. As we have said, there is no real evidence that the activity of such limited parts of the cortex as have been shown to be critical to the performance of discrimination habits in primates is not governed by a mass action principle. Localisation and mass action are by no means incompatible, and it is entirely conceivable that behaviour is determined both by 'local' and 'general' cerebral activities. Further, even if everything that could be localised had been localised, we should still be faced with the problem of how

the specialised parts of the cortex—in Lashley's words—
'produce the integration evident in thought and behaviour'.
The mass action hypothesis at least remains as a first attempt to
answer this most challenging problem of all.

PART II

EXPERIMENTAL STUDIES OF
ANIMAL BEHAVIOUR

INTRODUCTION

The examples of animal behaviour which are both most striking and challenging to those naturalistic, objective students of animal behaviour often known as 'ethologists' (particularly those who study birds and reptiles, fish and insects) are those which are found to be characteristic of the species, relatively stereotyped and to that extent, by definition, 'instinctive'. This inflexible instinctive behaviour is so overwhelmingly apparent as to constitute one of the ethologist's most attractive subjects for analysis. So he finds almost incomprehensible the reluctance of the psychologist, both 'human' and 'comparative', to take seriously this stereotyped behaviour and to pay due regard to its major theoretical implications. Practically all naturalists of the nineteenth century, and of course a great many earlier, were concerned with the naming, description and recognition of species; and in as far as they were field naturalists they were aware of the fact that it was possible to recognise their species when alive and under natural conditions by a great many features which had to be omitted when they came to write their specific descriptions—which of course, if they are to be of any use, must be verifiable upon preserved specimens in museum or laboratory. These additional or extra-specific characteristics were very often behavioural:

a method of flight, a characteristic gait or feeding action, some particular trick of display or of taking cover, of fighting or fleeing. To take a concrete example, not only can one find excellent specific characteristics when one studies the vocalisations of birds, one finds that it is often actually easier to identify species of grasshoppers and crickets by the sound they produce than by the shape, colour or proportions of the parts of the body. Many such specific actions can of course be explained solely as the necessary result of certain body structures and proportions, but there are also innumerable examples where such a conclusion can hardly apply and one must assume that these characteristic movements are in some way organised within the nervous system and, since they are specific, are presumably coded in, and the ability to perform them transmitted by, the hereditary mechanism. Such actions as well as being instinctive in the sense used above of specific and stereotyped, are often found also to be instinctive in the sense of internally motivated and internally co-ordinated. At the same time the earlier naturalists also realised that not all examples of this 'unlearned', internally co-ordinated and motivated behaviour were in fact stereotyped and inflexible. Many of those who studied in detail such subjects as the web-building of spiders, the prey-catching and nest-building of the solitary wasps and other solitary and social insects, and the sexual and parental behaviour of the birds and of the mammals, found numerous examples where the behaviour appeared marvellously flexible and adaptable to small day-to-day variations in the circumstances to which the animals are exposed. Many species whose behaviour at first sight appeared to be purely instinctive were found on occasion to adapt their behaviour to cope with such unusual circumstances as the collapse of a nest, the provision of unusual materials for nest making, the acquisition of young of an unfamiliar species, etc., etc. For a long period, however, neither the recognition of fixity nor the realisation of plasticity in instinctive behaviour induced what we should now regard as the obvious next step, namely controlled experiment. At least through a large part of the nineteenth century, and on occasion much later, the attitude of many naturalists was to accept instinct as something 'given', some ultimate unanalysable fact of nature at which one could merely wonder and which it seemed almost impious to expose to critical investigation. Students were indeed for a long time encouraged in this attitude because the physiology and psychology of the day were not such as to give them any clue as to how these examples of behaviour could be further studied and explained; and so there came about an ever-widening gap between the 'naturalists' and the physiologists and comparative psychologists. Much of contemporary psychology and physiology, particularly as the former became more and more physiological, seemed to be quite beside the point to the naturalist since he was, subconsciously at least, all the time obsessed by this tremendous mystery of the origin and perpetuation of these wonderful examples of innate behaviour which obviously had not been acquired by the individual as a result of 'practice' or learning from experience.

The first effective light to be thrown upon the problem of these mysterious actions, here apparently fixed, there apparently plastic, but always adapted as if directed in some mysterious teleological way towards ends that it was desirable that the race or species should achieve, came from the pioneers in what has now come to be known as Ethology, namely Heinroth, Wallace Craig, Lorenz, etc. Their contribution was essentially a very simple one: the view that there are some elements of what we call instinctive behaviour which are indeed extremely rigid and as constant as anatomical structures, but that these rigid and stereotyped actions are often the end-point or climax of a series of instinctive actions the earlier elements of which might be highly flexible. These fixed action patterns, as they came to be called, seemed, when they occurred near the end of a sequence of actions, to be in some essential respects consummatory—in the sense of bringing a complex chain of action to a close. Examples of such are found in the elaborate prey-catching and egg-laying behaviour of the solitary wasps such as *Ammophila* and *Philanthus*, the stereotyped sexual displays of birds and the elaborate, stereotyped and apparently inborn movements of innumerable predatory animals, whether fish, birds, invertebrates or mammals, by which a particular kind of prey is successfully caught or a particular kind of food economically gathered. This simple division into appetitive behaviour and consummatory act opened the way for effective analysis, and very soon what had been regarded as virtually unanalysable was in fact being broken down into smaller elements by simple experiments on the whole living animal, these smaller elements providing the links in the chain which go to make up the extremely elaborate performances loosely called 'instinctive'. Although instinctive actions appeared to be characterised by some sort of instinctive impulse and not simply to be responses to external stimulation, yet obviously instincts, and particularly fixed action patterns, would be useless, indeed highly dangerous, without mechanisms ensuring adjustments to the external situation. Such adjustments of course can often be by means of reflex conditioning, but the original concept of the reflex implied a relatively simple and immediate response to a relatively simple external stimulus. To naturalists accustomed to the complexities of the behaviour of insects, fish and birds in their natural surroundings, it seemed quite obvious that the kind of reflex which the physiologists were then talking about was quite useless as an explanatory concept providing anything approaching a full understanding of the elaborate behaviour which they were observing. It was seen that the responses of the whole animal in the wild were often set off by elaborate constellations of stimuli, involving such features as form, movement, colour, rhythm, etc., etc., together comprising an extremely elaborate stimulus situation—such as the display of a bird of paradise or bower bird, or the song of a lark. For this situation the ethologists produced the concept of the 'releaser', releasers being in fact just such highly elaborate 'external stimuli'. And if one assumed, as seemed unavoidable, that the ability to respond to many such releasers was inborn, then one had also to assume (as McDougall, 1923, had been the first to do) the existence of some corresponding internal organisation which fitted the releaser as a lock fits a key, and which thus, when the key turned, allowed the appropriate behaviour to come forth. Since quite obviously the innate releasing mechanism, or I.R.M. as it came to be called, could not be thought of as residing primarily in the sense organs themselves, it was assumed again that this mechanism must be in the central nervous system. This formulation of releaser and I.R.M. could not of course of itself have made the experimental study of the external initiation of behaviour any easier; but just as with the breakdown of consummatory acts into fixed action patterns, so it was pointed out that the releasers and the innate releasing mechanism might also be analysed by the same kind of experimental procedure and broken down into smaller units and released by specific external situations known as sign stimuli.

This then was the situation about twelve years ago. But since then, experimental research has produced so much new material that the picture just outlined is already appearing crude and out of date. The idea of a relatively rigid consummatory act is indeed theoretically sound and the separation of behaviour into these two categories continues to be very useful; but recent researches, particularly those of the last few years, have made it apparent that the distinction is not hard and fast. Much appetitive behaviour is now seen to contain rigid consummatory acts of what might be termed a minor kind, and many examples are known in which appetitive behaviour and consummatory act vary greatly in degree of rigidity and flexibility, and what appears appetitive in one circumstance may seem to be consummatory in another. Again, both may show some evidence of internal activation, both may show some rigidity and some flexibility. This conclusion has to some extent blurred the original picture drawn by Lorenz which was so attractive in its neatness and simplicity; and it is now harder than it was to envisage the animal's behaviour as governed principally by the need to find the appropriate situation for discharge of its consummatory acts and so relieve what had been thought of as some sort of instinctive tension accumulating around the mechanism for the consummatory act during the period in which it remained unreleased. We are now beginning to realise that animals are not merely motivated by the need to execute a particular piece of fixed behaviour; they are often rather 'searching' for a key environmental situation. Thus there are external situations which, while they may act as releasers in the original sense of the term, serve also to bring previous phases of behaviour to a close. They must therefore be effective in stopping a piece of behaviour even when the original eliciting factors may still be present. They are, in other words, 'consummatory stimuli'. And so we see that the goal of an animal is not always the performance of an action: it may sometimes be the perception of the environmental situation, and the achievement of such a situation may in fact be as effective in bringing a particular piece of behaviour to a close as is the satisfaction of what is more crudely thought of as a physiological need.

Although recent research has thus tended to concentrate on the mechanisms for physiological control and adjustment of the components of instinctive behaviour, and does for the time being tend to play down the evidence for fixed behaviour patterns under the rigid control of an internal drive, yet the essence of the problem still remains, viz. that of explaining the origin and perpetuation, from generation to generation, of the specific action patterns, some of which must have been handed down apparently in perfect precision through countless millions of years. One of the strongest pieces of evidence for the reality and fixity of instinctive behaviour comes from the comparative study of the fixed action patterns of some of the lower animals, and this subject is the concern of the first chapter of this section in which A. D. Blest discusses the present position of the concept of ritualisation. His work is thus in the centre of the modern study of the subject and should be a valuable corrective of those who still find it hard to believe in the fixity of inheritance of many of the actions of the lower animals. Blest himself quotes many of the papers of Jocelyn Crane in this connection. We might here mention her recent study (1957) of the basic patterns of display in the Fiddler Crabs, a genus (*Uca*) found on shores almost throughout the tropics and getting their name from the display, which consists in waving with a particular tempo or motion a single large and brightly-coloured claw. Studying these crabs on tropical shores in widely separated parts of the world, she found that the waving movements of this fantastic claw are characteristic of the species in their speed of move-ment, in the extent to which they are vertical or horizontal, smooth or jerky, etc. She found no gross intraspecific differences in the form or tempo of waving, even from populations in a wide-ranging Pacific species separated from one another by distances as great as 5,000 to 8,000 miles. She says 'many species with highly developed displays did, however, resemble primitive forms when waving at low intensity'. This implies that each species must have a 'typical intensity' of reaction which ensures that when the animal is in its mature reproduc-tive condition the intensity of the response remains constant irrespective of variation in the strength of the eliciting stimulus. Thus there must be some, presumably endocrine, mechanism which has maintained for countless generations an extremely tight control of the rate of movement of the claw. This idea of the typical intensity of a specific action has been developed by Morris (1957*a*) and is undoubtedly one of importance which will require much more investigation in the future.

The term ritualisation as used in the study of the evolution of animal behaviour implies 'the acquisition of a signal or releaser function'. There is abundant evidence that this occurs by the adaptation and development, in evolution, of displace-ment activities and intention movements (such as the appe-titive bobbing movement of a bird about to fly, which origin-ally had no necessary influence on other members of the species) into a social releaser having a profound importance in the social organisation of the species. Innumerable cases are known

93

where this ritualisation (Tinbergen, 1952) has involved the development in evolution of complex morphological structures —*Morphological support* as it is called. Thus, to quote Tinbergen, 'ritualisation very commonly leads to the development of structures which are demonstrated or emphasised by the particular displacement activity of the animal. The best known examples are found among ducks (Lorenz, 1941). Thus the male mandarin duck has, among its dark green secondaries, one in which the inner vane is extremely broadened and orange colored. This huge orange flag stands out permanently, and the displacement preening of the courting drake is nothing but an emphatic touching of this vane with the bill'.

The development of such a conspicuous structure has thus run parallel with an adaptive change, which is both an elaboration and simplification, of the signal movement itself. The garganey provides another example. Here the primary coverts, not the secondaries, have a conspicuous patch, in this case light blue in colour. Accordingly we find that in displacement preening the bill, instead of touching the secondaries or the inner side of the wing, points at or actually touches the blue patch. Here again the change in the movement appears to be adapted to the demonstration or emphasis of a conspicuous structure.

Movements themselves can also undergo changes serving to make them more conspicuous. They can, for example, be exaggerated both by being slowed down and by having their amplitude increased. Instances of this may again be found in ducks, and also in pigeons. Tinbergen points out that, in the process of ritualisation, certain elements are exaggerated whilst other elements tend to disappear entirely. Ritualisation therefore involves addition as well as simplification, often resulting in a form of schematising. He adds 'when a movement changes through ritualization, this means that the underlying nervous mechanism changes and thus becomes different from that of its example. In this way ritualization brings about a kind of emancipation. The further ritualization proceeds, the more the movement loses its displacement character and becomes an element of the drive which uses it as an outlet. It is of course impossible to draw a line between what is "still" a displacement activity and what is "already" an independent element of the drive which uses it, unless one uses arbitrary criteria. The only thing one can say in such cases is that a movement was originally a displacement activity.'

The study of ritualisation thus shows us how 'new' behaviour elements may appear during evolution. Before an activity is ritualised, its nature may be easily recognised. Through gradual adaptive change it may, however, become so different from any other behaviour element that eventually it ceases any longer to be a derived movement and becomes a 'new' type of behaviour. There can be few more convincing pieces of evidence for the reality of inborn movement forms than this.

Blest shows that this problem of ritualisation on the motor side involves changes not very different from those implicit in the evolution of what are usually regarded as the simpler and more understandable patterns of animal locomotion. He thus suggests that there may be examples of animal behaviour easier to analyse and experimentally more tractable than are ritualised movements, and that the study of these examples may yield information essential for their physiological understanding. In addition he gives simple working definitions of ritualisation and several associated concepts which may turn out to be more easily related to direct observation and which are less dependent upon particular behavioural theories than the ethologists' studies of ritualisation have so far seemed to be.

In 1950 Baerends provided many new examples showing how the development of the floral types characteristic of certain groups of orchids can only have taken place by attracting the males of certain ichneumonid wasps which visit the flowers and fertilise them by carrying out the act of copulation with the labellum. The flowers, and especially the labellum, have in consequence come to acquire a striking resemblance to an insect, and it has been possible to show experimentally that, at any rate in some species, the males are attracted to the flowers because of the resemblance of the petals to female individuals of the insect's own species. These orchids are ready to be fertilised long before the female hymenoptera have left the pupae. Since the males usually hatch a week to a fortnight earlier than the females, they are likely to suffer, or so Baerends argues, from a threshold lowering with regard to sexual reactions so that they will, during this period, more easily respond to external stimulation. The orchids need the insects in order to get fertilised, but they do not in return produce nectar or other foodstuffs: they simply make use of a releasing mechanism in the hymenopteron—a mechanism which, since it is essential to reproduction, has such biological importance to the insect that there is no risk of it being lost. The orchids have in fact become parasitic upon the sexual behaviour and releasing mechanisms of certain insect species.

This very striking evidence for the permanence and constancy of both actions and releasers leads us on to the more intensive study of releasers and of some of those difficulties which indicate that, on the perceptual side, the original lock and key concept is no longer adequate. Crook in his chapter on the weaver birds is concerned to analyse the factors upon which the extraordinarily highly organised social behaviour of many species in this group is based. The weaver birds include a great many species, one of which, *Quelea quelea*, is so intensely gregarious as to constitute—now that it has taken to feeding upon man's crops—one of the most serious pests in Africa, if not in the whole world. We find that within the Ploceidae is found every gradation, from species which are essentially solitary to those which show this extreme degree of social organisation. It is already possible to go a long way towards analysing the elementary behaviour factors which serve to control the degree of sociability of these birds into their fundamental components and to show that differences in such things as flock organisation, flock fragmentation and territory and pair formation may in fact be expressions of a few relatively simple specific differences in sexual and social behaviour. Here again we have evidence that these elementary behaviour factors, which go to make up the flock organisation characteristic of given species, must have been precisely stereotyped over very long periods of time. They involve differences in activity cycles, in the tendency to respond to the sight of other individuals of the species by associating with them, in the opposing tendency to repel too close an approach—a tendency which gives rise to another specific character, namely 'individual distance'. Thus we get birds which are gregarious, often intensely so, in many of their activities but which when nesting may still show evidence of a minute 'area of territory' around each nest, invasion of which is strongly resented and which results in a spacing out of the population in the breeding colony which as a whole may be densely crowded but the individuals of which are separated from one another by a constant distance. Crook discusses the social tendency in its relation to problems of flock structure, analysing particularly the activity cycles of the birds, the releasers by which the birds influence one another and the way in which correlation is brought about between the various activities without engendering the complete disorganisation that an uncontrolled social tendency would produce. All the problems and examples he is investigating are essentially examples of the fixed action patterns of the species and the problem of specific releasers.

The original concept of a releaser, implying some central structural organisation keyed to respond to the perception of the releasing stimulus itself, owed its attraction to the way in which it seemed at first to bring order out of a whole chaos of observations relating to the complex situations to which animals were seen to be reacting in the normal course of their lives in the wild. But experimental study soon threw up difficulties; first it was shown that releasers can in fact be analysed into constituent stimulus situations, and secondly what came to be known as supernormal releasers were found. As an excellent recent example of this we may take the work of Baerends (1959) on the response of the brooding Herring Gull to its large

97

spotted eggs. It was found by experiment that the Herring Gull actually preferred eggs that were larger than normal to sit upon and would even accept and try to incubate eggs so large that brooding was impossible. It was also found that eggs that were darker than normal were preferred to normal ones, and eggs that had more spots and darker spots than normal were also preferred. Earlier work by Seitz (1940, 1942) on the releasers which served to co-ordinate the behaviour of many species of fish had shown that the numerous sign stimuli into which an average releaser could be analysed exerted their effect independently. The effect of each could be separately measured, and it was found that the increase in stimulative efficiency of one component sign stimulus could counterbalance a decreased efficiency of another sign stimulus which went to make up a given releaser. That is to say, these single relational key stimuli act merely as a sum of independent elements, and the 'law' which Seitz enunciated as a result of his study came to be known in English by Tinbergen's name of the 'law of heterogeneous summation'. It follows from this law that we should expect that supernormal releasers might be found; and so these more recent studies of the releaser have merely served to confirm what might have been foreseen from the earlier ones, namely that an actual releaser may be in effect that pattern or combination of patterns which, within the limitations set by the general environment, can provide the most stimulative series of sign stimuli, i.e. can come nearest to the supernormal for the largest number of constituent sign stimuli. As an example, it was found by Tinbergen and Perdeck (1950) that the Herring Gulls' yellow bill, with its red spot, is a highly effective releaser for the pecking of the chick, which in its turn elicits the bestowal of food by the parent. Similar studies by R. and U. Weidmann on the Black-headed Gull have shown that it is possible by presenting models of the head and bill of the parent in various shapes, proportions and dimensions to find patterns which are 'better than' nature. Thus a longer bill is much more attractive to the chick than one of normal size, but perhaps the mechanical and structural reasons against the Black-headed Gull's having a very long bill are such as to render the evolution of increased length of bill in this species impracticable. These and a great many similar observations provide us with reasons for doubting the original clear-cut picture of the releasing mechanism as a lock exactly fitted to receive the key of the stimulus situation. Obviously if most or all the usual constituents of the releaser can be varied and 'improved upon' in experimental situations, there is little left of the original releasing concept; and so further experimental analysis has become particularly attractive at the present time. This is particularly true because of the increased understanding, brought about by studies in recent years, of the mechanism of action of the sense organs of animals, and a realisation of the extent to which an understanding of sensory physiology is bound to influence our ideas of releasers. For it is now clear that, as von Uexkull emphasised many years ago, each species is living in a world of its own—a world determined in the first place by the nature of its sense organs. When we thought of a releaser as something very precisely and definitely specific, peculiar to the species, it was difficult to imagine that such precise perceptions, involving form, colour, temporal cycles and so forth, could possibly be coded in the sense organ itself, and therefore it was natural to suppose that the coding must be in the central nervous system. For instance, what kind of visual stimulus gives the strongest response will very likely depend in the first place on the characteristics of the eye of the animal when it is considered as an implement for distinguishing between different groups of wavelengths. Again, certain types of movement can only be effective releasers if the eye has a flicker fusion frequency of such a kind as to allow the animal to see them, and obviously a sound pattern consisting of notes of high frequency very rapidly succeeding one another cannot influence the animal unless the ear has an adequate range of frequency responses and a short enough 'time perception smear'. And all these characteristics of the sense organ will play their part in determining the degree to which different kinds of sign stimulus will appear strong or weak to the animal.

Thus now that we realise that all the elements of a releaser may be acting separately, that their effects can summate and that they themselves are often very simple relational key stimuli, it at once becomes probable that much more than we formerly imagined of the selectivity of the animal for its environment is due to its receptors. That is to say, the complex stimulus relationships which make up the external releaser as we conceive it are, before they cause the animal to respond, subject to a series of physiological 'filtering mechanisms'; and this is the theme of Marler's contribution. He discusses three types: that imposed (a) primarily by the receptors themselves; (b) by the receptors' afferent pathways and the central nervous system as they function together in normal perception; and (c) by a central filtering mechanism (which is perhaps the equivalent of the original concept of the I.R.M.). This subject as a whole, then, is one of particular promise and interest at present and, as Marler points out, is one which is not confined to the study of instinctive behaviour but also arises in connection with learning. In the first we think of a filtering mechanism which is built in or is self-differentiating; in the second case we are considering filtering which arises from the individual experience of the animal. It will not be until a large number of cases have been fully analysed that we shall have any clear idea as to the extent to which learning in this sense is entering into the 'innate' recognition abilities of the different groups of animals.

In conclusion, however, we must emphasise the warning which again we receive from Marler's chapter, that while learning may be playing a hitherto unsuspected part in the perception of what may often be regarded as innately coded releasers, there are innumerable cases in which built-in sensory processes may in fact be playing a far more important role in learning than has previously been thought—even, as he shows, in creatures like mammals, where instinctive processes have been thought to make only a minor contribution to behaviour. Thus it becomes obvious that whether we call ourselves learning psychologists, learning theorists or ethologists, or whatever our label, we have much to learn from investigations of this kind.

IV. THE CONCEPT OF 'RITUALISATION'

1. *Introduction*

Signal movements may be directed towards members either of the same species or of other species present in the same environment as the signaller. Examples of intra-specifically directed signals are to be found in the courtship displays of many vertebrates, while inter-specific displays include the warning postures by which many invertebrates advertise distasteful properties, or frighten away their vertebrate predators. Such displays, whenever investigated, are found to consist of inborn, rigidly co-ordinated, and more or less species-specific motor patterns, which are closely correlated with the morphological structures which they serve to exhibit (Crane, 1941, 1949, 1952, 1957; Lorenz, 1941, 1950; Tinbergen, 1951, 1952; Daanje, 1950; Andrew, 1956c; Marler, 1956a, 1957a; Morris, 1954, 1956a and b, 1957a and b, 1958; Moynihan, 1955b and c; Blest, 1956, 1957b). Careful comparative investigations have yielded the hypothesis that, in the course of evolution, both locomotory movements and acts (concerned with comfort, with heat-regulation, and with the capture of prey) have been selected and modified to produce signals. Such movements have been termed 'derived' and may exist alongside their ancestral activities. The process by which they have arisen has been named 'ritualisation' (Tinbergen, 1952).

Although research on 'ritualisation' under that name has been a relatively recent development, most of the ideas developed by recent workers were implicit in the earlier studies of Heinroth (1910) on the sexual displays of ducks, and those of Hinsche (1928) on the protective displays of Anura, although they were stated by these authors in different words. Not all the subsequent ethological writings have clarified the concept of the evolution of signal movements, which Hinsche saw in terms of the modification and re-integration of simple reflexes.

The broader evolutionary concepts underlying the theory of ritualisation have a still more ancient history which, as Marler (1956a) has noted, can be traced back to the writings of Charles Darwin (1872).

Attempts have been made to give precision to the concept of ritualisation, both by defining the conditions necessary for the process to occur (e.g. Tinbergen, 1952; Haldane and Spurway, 1954), and by analysing the changes in motor patterns believed to have taken place (Daanje, 1950; Morris, 1957b). This paper will attempt to re-examine 'ritualisation' and one of the chief concepts derived from it, that of 'emancipation', in terms of their possible causal background, and in relation to other types of phylogenetic behaviour change. It will also endeavour to show that ritualisation, considered as a distinct evolutionary process, has been over-dramatised by being placed within a particular theoretical framework.

Two facts need to be borne in mind. Firstly, there is no fossil record of micro-evolutionary changes in behaviour, except in such special cases as the termite nests described by Schmidt (1955), so that any statement about specific evolutionary events in a given group needs cautious handling, for it is never likely to be much more than an inspired guess. Secondly, the evidence to be described will be drawn fairly equally from vertebrate and arthropod sources; while the contribution of Marler in the present volume serves to remind us that the outward end-products of the evolution of animal signals are often similar in their functional adaptations, and have in general been mediated by like selection pressures, Vowles suggests that the vertebrate and arthropod nervous systems are probably constructed on rather different principles. Much of the discussion which will follow, therefore, almost certainly contains an unwholesome amount of generalisation.

In evolution, two complementary series of changes must have been responsible for the development of intra-specific displays: those concerning the motor patterns and releasing mechanisms of the *actors*, and those involving the patterns of response and reactivity of the *reactors*. Primarily, this account will discuss the first problem, for relatively little is known about the evolution of changes in responsiveness, and the nature of the selection pressures mediating such complementary changes has not yet received an adequate logical analysis.

2. *The changes accompanying ritualisation*

Daanje (1950) proposed three main categories of change in the process of ritualisation: (*a*) changes in threshold; (*b*) loss of co-ordination; (*c*) exaggeration (changes in extent of movement). These categories are purely descriptive, and they are not wholly separate. For instance, changes in the thresholds of release of individual components could in some cases, perhaps, be regarded as the product of differentially altered thresholds.

Morris (1957*b*) has summarised the overall changes in ritualised bird displays as leading to the formation of acts of 'typical intensity'. Whereas stimuli of varying strength for the release of the unritualised precursors of display movements elicit responses of varying intensity and form, following ritualisation the derived responses acquire an almost constant form and intensity to a wide range of stimulus strengths; this constancy is of obvious importance in any signal movement which serves to elicit a simple response without transmitting any very complex and hence potentially variable body of information. A special case of the development of such constancy will be noted in Section 2*b*, that of the 'stabilisation' of the rhythmic anti-predator displays of moths (Blest, 1957*b*). These increasingly exact channellings of responsiveness may be compared with the analogous genetic processes postulated by Waddington (1952, 1953) to account for the stabilisation of the 'epigenetic landscape' in development.

Such, briefly described, are the gross events of ritualisation. The subsections below will examine the units of change available for selection, adhering, in the first place, to Daanje's scheme of classification.

(a) Changes in the releasing mechanism

(i) *Factors affecting the thresholds of related responses.* Closely related species may differ in the threshold of release of homologous ritualised responses (Blest, 1957b, Saturniid moths; Spieth, 1952, *Drosophila* spp.; Goethe, 1954, *Larus* spp.). In the last-mentioned case such differences are largely responsible for many of the ethological distinctions between species, and it is clear that thresholds may undergo evolutionary change independently of either motor co-ordination or the nature of the 'motivation'.

These changes could be the products of alterations in central or peripheral factors. Reduction in the number of sensory units, or increase in their individual thresholds could both increase the strength of stimulation necessary to allow any response mediated by them. No analysis of homologous vertebrate responses has yet been made; in a different context, the comparative oviposition behaviour of *Lucilia* spp. (Diptera) may be a relevant case, for it involves differential responsiveness to a range of olfactory stimuli (Cragg, 1956; Cragg and Cole, 1956).

It is possible that many features of inborn releasing mechanisms, including even their specificity, may be explained in terms of peripheral factors (Marler, in preparation; Weidmann and Weidmann, 1958). Cases of centrally mediated threshold changes are mostly uninvestigated, but the circumstantial evidence is stronger. Certain species of moth in the genus *Automeris* possess displays in which the fore-wings are protracted from the resting position to reveal eyespot patterns on the hind-wings (Blest, 1957b). These movements are released by tactile stimuli. At slightly higher response intensities rhythmic components appear as well. In some species the thresholds of elicitation of both components are low, in others they are higher; those of the two parts of the displays may vary independently between species, and the rhythmic component may be lost altogether. It is probable that these differences are not primarily due to receptor thresholds. Removal of supra- and sub-oesophageal

A. D. BLEST

ganglia eliminates fully integrated responses, including sustained locomotion, and disorganises the postural reflexes. The operated insects, however, become hyperexcitable for the first components of both parts of the display pattern, which now appear as transient responses to very light tactile stimuli (unpublished observations). It is known that in some insects the excitability of lower reflex responses relates in part to a balance between the facilitating influence which the sub-oesophageal ganglion exerts upon them and inhibition from higher centres (Roeder, 1953).

In vertebrates, central mechanisms have been shown to exist capable of maintaining threshold differences, affecting both sexual and agonistic behaviour. There has been much recent research on the functions of the mammalian rhinencephalon (cf. Rosvold and Delgado, 1956; Weiskrantz, 1956) and of fore-brain structures in fish (Hale, 1956a and b; Noble, 1941; Schönherr, 1955; Segaar, 1956). Although the findings of the earlier authors were anomalous, recent results are more consistent; different experimental methods, notably extirpation procedures and the stimulation of localised brain regions of unanaesthetised animals by means of permanently implanted electrodes, have indicated that fore-brain structures may facilitate a variety of responses. Thus green sunfish with fore-brain lesions will not show the normal aggressive behaviour of the species without intensive supernormal stimulation; yet the separate aggressive acts are normally co-ordinated when they appear after the action of extreme stimuli. The sequence of nest-building behaviour in similarly treated sticklebacks is abnormal, but the individual components are performed normally; since they appear in an inappropriate sequence they are not effective (Schönherr, 1955). Rhinencephalic injury to the savage Norway rat produces animals which are tame and readily handled (Wood, 1956). Karli (1956) has shown that some individuals will consistently kill white mice, while others as consistently ignore them. Killers may be converted to non-killers (albeit at the cost of overall lethargy, i.e. 'tameness') by bilateral amygdalectomy, while non-killers may be converted to killers by bilateral frontal lobotomy. Although these very crude and histologically poorly analysed procedures are difficult to interpret, they again suggest the existence of higher central systems maintaining a balance between the excitation and inhibition of certain responses. Such systems could, in principle, provide a basis for the 'fixation' of central thresholds at different and specifically characteristic levels.

Inter-specific differences in the tendency to perform aggressive behaviour are known to exist between closely related birds and fish. A comparative study of fore-brain lesions in the ethologically well-known sticklebacks, for example, along the lines already provided by Schönherr would probably prove rewarding.

(ii) *Changes in the dominant sensory modality.* Within groups at the lower taxonomic levels, the relative contribution of different sensory modalities to the general control of behaviour may undergo modification in the course of evolution through primarily peripheral changes. Thus the sub-families of Salticid spiders may be arranged in a series showing increasing reliance upon visual stimuli for general adjustment to the environment. This change has involved far-reaching modifications of structure, locomotory and prey-catching behaviour, and in the relative importance of the different sensory items releasing the male courtship displays (Crane, 1949). Much of the latter parochial change, though not all of it, may most plausibly be interpreted as secondary, for it is clear that the trend towards increased visual dependence is associated with fundamental and probably earlier changes in predatory behaviour.

(iii) *Facilitation and inhibition by environmental variables.* In addition to threshold changes *per se*, there may also occur differences between species in the facilitating or inhibiting effects of non-specific environmental stimuli. Two sets of comparative data may be cited.

In the Drosophilae of the *D. melanogaster* species-group, one species, *D. auraria*, cannot mate in total darkness, its near relative *D. rufa* is somewhat inhibited by the absence of light, while *D. montium* and *D. simulans* are greatly inhibited by dark-ness; *D. ananassae*, *D. takahashii* and *D. melanogaster* are not affected (Spieth and Hsu, 1950). Similar differences within another species-group have been found by Wallace and Dob-zhansky (1946). Neither study reveals which components differ between species; the block in the case of flies unable to mate in the dark might be wholly perceptual and due to the absence of a visual releaser under such conditions. Nevertheless, in another group, the Salticid spiders, with a predominantly visual courtship, the environmental conditions necessary for the release of the males' display differ between species and show in certain North Temperate cases a clear relation to the ecological conditions under which the individual species live. Thus, faced with a mature female, males of *Evarcha falcata*, *E. arcuata*, *Neon reticulatus* and *N. valentulus* will display at quite low light in-tensities, the first two species in twilight. By contrast, *Euophrys petrensis* and *E. aequipes* will display only in strong direct sun-light. Species of *Evarcha* live in dense undergrowth, *Neon* spp. are litter forms, while the two species of *Euophrys* inhabit bare pebbly areas on sandy heath and mountainside, and have their brief period of maturity during the spring, when the necessary light conditions are available (unpublished observations).

Although it was originally believed that intra-specific displays were largely protected from environmental selection pressures, on the grounds that they could only be influenced by the responding individuals of the same species (Lorenz, 1941; Delacour and Mayr, 1945), examples such as these, and those given by Stein (1956) and Cullen (1957) demonstrate that environmental selection pressures must, in fact, play a con-siderable part in the evolution of ritualised displays. Stride (1956, 1957) has analysed a particularly complex case, in which it would seem that the releasing mechanisms for the male courtship of the West African butterfly *Hypolimnas misippus* possess a number of features apparently adapted to the need for avoiding the release of inappropriate responses to other species of butterfly present in the same habitat.

(b) *Changes in co-ordination*

Co-ordination changes are difficult to analyse on the basis of the rather crude descriptions of behaviour which are all that is usually available. Ideally, each ritualised act to be considered would be broken down into a number of components, and these would then be related to homologous components in the ancestral activity. This has not been done at a precise analytical level so far in any group studied. The most complete quantitative account of ritualised acts involving a number of species is probably that of Andrew (1956*c*) who describes the pattern of tail flicking in nearly 200 species of passerine bird, this study itself being a continuation of that of Daanje (1950).

The two types of co-ordination change distinguished by Daanje will here be joined and re-classified. The attempt to classify them is rather artificial. Morris (1957*b*) lists the following types of change which may broadly be considered to come under the heading of co-ordination change:

(1)   Intensity change.
(2)   Increase or decrease in speed of performance.
(3)   Omission of components.
(4)   Changes in component co-ordination.
(5)   Changes in sequence of components, i.e. in the order of their performance.
(6)   Differential exaggeration of components.
(7)   Development of rhythmic repetition.

To these may be added an eighth:

(8)   The transfer of signal function from one set of effectors to another.

Obviously these categories are arbitrary, and are not wholly independent. Space does not allow a detailed discussion, nor is the type of information needed to evaluate their status readily available. A few comments about certain of them are, however, worth making, if only to emphasise those points about which least is known.

(i) *Intensity changes.* The term 'intensity' tends to be used in a variety of ways by different authors. For example, it can denote features as varied as rate of performance, persistence, or the extent of the movement involved in performing a given component of display. In the last case, the extent to which differences in intensity may be due to differences in receptor thresholds, in central nervous patterns of activity, or to differences between the structure of peripheral effectors, such as those affecting muscle attachments in relation to the lever systems of which they form a part, has not yet been determined in any simple case.

(ii) *Increase or decrease in speed of performance.* Overall speed changes have been described in both vertebrates and invertebrates as an accompaniment of ritualisation (Morris, 1954, 1958; Blest, 1957*b*). The identification of a speed change must usually imply that the form of the complete pattern remains more or less unchanged.

(iii) *The development of rhythmic repetition.* A movement which in its primitive form was a single discrete act is repeated a number of times, so that the ritualised derivative consists of an elaborate rhythmic performance. This process is believed to have occurred in the evolution of certain courtship displays of birds, such as the rhythmic dances of Ploceids (Morris, 1954, 1957*a*, 1958), and has clearly played a very important role in the evolution of bird song (Thorpe, 1956 & 1958*a* and *b*). It may also have been responsible for certain components of the protective displays of Saturniid moths, though here it is likely that the rhythmic displays are derived from slowed-down flight movements that have been subjected to component selection. Some of these displays, in addition, incorporate movements which seem to be derived from the brief run which normally precedes take-off in a moth about to fly (Blest, 1957*b*). Here the movements are repeated in phase with those of the wings; such a process can be regarded as a phylogenetic analogue of the 'magnet effect' of von Holst (1936). It is formally intermediate between this category and

(iv) *The transfer of signal function from one set of effectors to another.* The most radical transformation of this type has been described by Faber (in press) as having occurred in the course of the evolution of the songs of Orthoptera. A whole motor pattern is transferred with time sequence intact from one set of effectors (tegmina and femora) to another (mandibles). Both motor patterns result in stridulations whose acoustic properties are very different, although the rhythmicity of the movements is similar in the two cases.

This last category has only been demonstrated to have occurred in the evolution of rhythmic signals. Other changes in such responses may be shown to have taken place, although it is perhaps not worthwhile to attempt to categorise them. One such set of changes has been termed 'stabilisation'. In the rhythmic displays of certain Saturniid moths (see above), the mean periods of the individual cycles of the rhythm, and their mean amplitude, both relate directly to response strength, as measured by the number of cycles performed in response to a single stimulus. This is due to the fact that within each such response the movements show a progressive decline in amplitude coupled with marked deceleration. Within phyletic lines there is a tendency to achieve displays in which the deceleration is reduced, and the amplitudes held more nearly constant throughout the course of the response. These changes appear to be linked, for they have not yet been found to occur independently (Blest, 1957b, 1958b). Oscillograms of Orthopteran songs suggest that they too may possess different degrees of 'stability' in this sense (Haskell, 1957).

It may be noted here that comparative neurological studies of co-ordination have not yet been made on ritualised responses per se, but two studies are relevant, both concerning signal movements, and may be briefly mentioned.

Firstly, Pringle (1954) and others have studied sound production by Cicadas neurophysiologically and have obtained a remarkably detailed picture of the relative roles of peripheral structure and patterns of central nervous activity in determining the forms of the final acoustic signals.

Secondly, Boycott (1953, and in preparation) has described

an interesting situation in the chromatophore responses of four species of cephalopods, *Sepia officinalis*, *Loligo vulgaris*, *Argonauta argo* and *Octopus vulgaris*. Since the chromatophore muscles are innervated without the intervention of any synapses directly from the chromatophore lobes of the brain, the overt colour changes of the animals' surfaces are direct transcriptions of central nervous events occurring in those lobes. The colour responses differ in complexity between the four species. The patterns produced are simplest in *Loligo* and *Argonauta* and extremely complex in *Octopus* and *Sepia*. In apparent correspondence to these differences, a distinction can be made between regularity of fibre pattern in the neuropil organisation of the posterior chromatophore lobes of *Octopus* and *Sepia*, and more random arrangements in those of *Loligo* and *Argonauta*. It has not, unfortunately, proved possible to quantify these structural differences.

(c) *The concept of 'emancipation'*

These foregoing types of evolutionary change have been considered in terms of a fourth: the process of *emancipation* (Tinbergen, 1952) whereby, it is postulated, ritualised responses become 'freed' from the causal factors which mediated their phylogenetic precursors. Tinbergen, adopting the term from Huxley (1923), originally used it with reference to the ritualisation of displacement activities: '*Ritualisation of a displacement activity, resulting in an increase of the difference between it and its original example, has, of course, neurophysiological implications. When the movement changes through ritualisation, this means that the underlying nervous mechanism changes and thus becomes different from that of its example. In this way ritualisation brings about a kind of emancipation. The further ritualisation proceeds, the more the movement loses its displacement character and becomes an element of the drive which uses it as an outlet.*'

Two features must be noted about this passage. Firstly, it is so phrased as to imply that any change in the neurophysiological mechanism is by definition to be counted as emancipation;

secondly, the concept is placed within the framework of an hierarchial causal model of instinctive behaviour which carries over the concept of 'nervous energy' from the earlier hydraulic model of Lorenz (1950). It is axiomatic that, with the special exceptions noted above differences in overt behaviour must reflect differences in pattern of neural activity. Clearly, if this definition were to be accepted literally, 'emancipation' would become, for practical purposes, synonymous with 'ritualisation'. In fact it is never so used. Writers do not, however, always make clear upon what principles they are basing their assessment of the degree of difference between a ritualised act and its ancestral movement.

Many authors distinguish levels of 'emancipation', and the criteria which they use may differ. Some of the difficulties may be illustrated by these simple examples:

(1) The sexual displays of male Fringillid, Ploceid and Estrildine finches are derived in part from locomotory intention movements (Hinde, 1955–56; Morris, 1957a; Andrew, 1957). Yet the courtship displays appear in a sexual context and are mediated in part by the endocrine systems of reproduction. Originally associated with other causal systems, they are now part of the complex of sexual behaviour. In the Tinbergen scheme, it could be said that the display movements have become encompassed by a system of motivation new to them.

Similarly, the mobbing displays of small passerines overtly retain signs of the approach and escape tendencies from whose conflict the displays are believed to have been evolved, although the releasing stimuli and various properties of the response have changed (Hinde, 1954). Some authors, probably unwisely, have described such acts as partially emancipated (e.g. Morris, 1956b).

(2) The protective displays of Saturniid moths have been derived from flight movements, yet the present relationship between the overt acts of flight and of display is such as to suggest that inhibitory relationships exist between the causal mechanisms which mediate them (Blest, 1957b, 1958b; Bastock

and Blest, 1958). Such displays are sometimes considered to be fully emancipated.

But, on the whole, few examples are as simple as these. Many displays are derived from intention movements, displacement activities, or compromise behaviour (Andrew, 1956a) given in conflict situations, when two incompatible tendencies are competing for expression. It is claimed that the ritualised displays derived from conflict postures may be influenced by three sources of motivation: the novel, emancipating 'drive', and the two 'drives' mediating the ancestral conflicts (Morris, 1954, 1956b).

This additional complexity allows other criteria of emancipation. Morris (1956b), for example, writes: '*The courtship dance of the three-spined stickleback has clearly become ritualised and serves as a sexual signal to the female. Also it is now sexually motivated, although it originated as an ambivalent movement of alternating aggressive and sexual intention movements. . . . If a courting male is unduly aggressive, Tinbergen's analysis has shown that the component of the dance leading the male* towards *the female is more pronounced than usual. If the male is unduly non-aggressive, then the component leading the male* away *from the female is more pronounced.*' He cites another, similar example, the male courtship dance of the zebra finch (Morris, 1954), and concludes that both displays must be regarded as semi-ritualised (i.e. partly emancipated) because they are still subject to the influence of their ancestral drives. In this case, therefore, the criterion is not the acquisition of novel endocrinological competences or encompassment by a new motivational system, but the degree of persistence of phylogenetically older motivating influences. In view of these conflicting attitudes to emancipation and its definition, it is desirable to examine the concept more closely in simpler instances in which the possibility of these postulated multiple persistences does not occur.

It is worthwhile to look for a set of concepts which are not dependent upon either the hydraulic or the hierarchial models of behaviour for their definition, for the following reasons. The model of Tinbergen (1951) was an attempt at causal explanation. Corresponding to the supposed hierarchy of overt behaviour, Tinbergen postulated a parallel hierarchy of nervous mechanisms. The relationships between the nervous centres of this system, arranged so as to distribute cascades of 'nervous energy' to their appropriate acts, were predominantly excitatory; inhibitory processes, though physiologically respectable, were not allowed a very prominent place. Further, the model allowed an inadequate role to exteroceptive 'feedback' stimuli, to proprioceptive control, and to the more complex properties of reflexes (Hinde, 1956). Such a model was bound, in its nature, to dramatise the role of emancipation; the evolutionary transfer of motivation for an act could only take place at a higher level in the system, since the hierarchy at its lowest level consisted of individual reflex acts (designated as centres) linked ascendingly to decreasing numbers of intermediate and ultimately higher 'centres'. Implicit in this scheme, therefore, is the concept of emancipation as a *major* change. Further, the model induces an additional complexity: displays may be ritualised from derived activities, e.g. displacement behaviour, which must, on the Tinbergen model, have its genesis at the higher hierarchical levels. The elaboration of the incidental rechannellings of nervous impulses which explained displacement activities in the earlier models into separate centres mediating ritualised responses was difficult to conceive either formally or neurophysiologically. It followed that there was a tendency to believe that the emancipation of patterns derived from 'allochthonous' and 'autochthonous' acts might be dissimilar processes (Tinbergen, 1952).

Finally, to conceive of emancipation as a process necessarily operating at higher hierarchical levels as a major change tends to imply that it is a late event in the evolutionary sequence, likely to occur after the changes in threshold, co-ordination etc. Such a view is implicit in the discussion of Baerends, Brouwer and Waterbolk (1955) and Baerends (1956). Nevertheless in Saturniid moths, at least, there would seem to be no relation between the evolutionary status of the motor co-ordination of protective displays and the relationship between them and their ancestral activity, flight; on the basis of any of the criteria that have been used, all these displays are equally emancipated (Bastock and Blest, 1958; Blest, 1958*b*).

The position may perhaps be made more clear if the types of objective evidence upon which assessments of emancipation are based are considered. In practice, the statement that a ritualised act is emancipated generally depends upon some combination of the following observations:

(1) That the ritualised act does not necessarily occur in the same context as the act which is presumed to be ancestral to it. Even in the absence of further analysis such an observation at least implies that the releasing factors for the two activities are probably different, or that there are differences in the necessary internal conditions for their elicitation.

(2) That there exist inverse quantitative relationships between the ancestral and derived acts. Thus it may be found that repeated performance of the ancestral act lowers the threshold for, and increases the strength of its derivative. This is the case for the flight responses of Saturniid moths, and the protective displays which have been derived from them (Bastock and Blest, 1958; Blest, 1957*b*, 1958*b*).

(3) That the two acts show total quantitative independence. In this, admittedly hypothetical, case the performance of one act would affect neither the strength nor the threshold of the other.

Conversely, the most conclusive evidence for stating that an act has not undergone emancipation is the observation that repeated performance of the ancestral activity raises the threshold and/or decreases the strength of the derived act.

These criteria are not dependent upon any particular system or theory for their validity, for they could equally well be used in the framework of a hydraulic model, of an hierarchical scheme such as that of Tinbergen (1951), or within the terms of reference used by Sherrington in interpreting the relations existing between different reflexes mediated by the spinal cord (Sherrington, 1906; Creed *et al.*, 1932). What is established is not a particular hypothetical causal structure, underlying the ritualised act and that from which it is supposed to have been derived, but the degree of causal independence which they exhibit.

In the majority of cases the evidence given is of the first kind —that is, evidence of context (e.g. Moynihan, 1955*a*, 1958); indeed, the last-named paper attempts to use evidence of context to provide an argument of great complexity concerning the relative contribution of two 'drives' to the motivation of a large repertoire of postures in two species of gull (Laridae). The alternative type of demonstration which involves, in effect, experiments on the patterns of extinction shown by groups of innate acts, has been performed with the protective displays of Saturniid moths (Blest, 1957*b*, 1958*b*; Bastock and Blest, 1958), with the mobbing display of the chaffinch (Hinde, 1954), and, if the work of Crane (1949) and Precht (1952) is considered jointly, with Salticid spiders.

There are, further, other criteria which may be based upon changes in the endocrinological relations of the acts. Male sexual displays in the higher vertebrates, whatever their derivation, are primed by male sex hormone. The exact sites of action of hormones affecting behaviour patterns are still uncertain; there is both direct (Fisher, 1956) and indirect evidence (for example, Hinde, 1958*b*) pointing to the action of sex hormones at the higher integrational levels as well as at specifically competent peripheral structures, sensory feed-back from which may allow the performance of new items of behaviour (Lehrman, 1955, 1956). Some of the effects of sexual hormones may not prove to be qualitatively very different from those in which single acts are primed at the reflex level through the induction of neuromorphogenesis as, for example, in tadpoles subjected to strictly localised thyroxine treatment, which induces selectively the growth of some neural elements and the involution of others (Kollross, 1942, 1943; Pesetsky and Kollross, 1956). The degree of emancipation of a ritualised sexual display might, in principle, be assessed by the specificity of its responsiveness to hormonal priming as compared with the ancestral act.

To summarise, then, the impression gained from a consideration of the different ways in which the term 'emancipation' may be used is that it may cover several processes, and that while it has certain uses, it would be better if authors were to state precisely what they knew about the relations between a ritualised activity and its ancestral act, rather than to label it with a general term having no very exact meaning.

3. *Some limits to the process of ritualisation*

There is some evidence that the direction taken in the ritualisation of a signal may be influenced and perhaps limited by the nature of the material upon which selection has to act. Evidence exists that components which may be actually disadvantageous are retained in certain ritualised displays.

The inter-specific protective displays of Saturniid and Sphingid moths often incorporate wing-quivering movements; such components are most frequently found in the groups having the higher frequencies of wing-beat. There is ample evidence (Blest, 1957b) that these protective displays have been evolved from flight movements; yet flight itself, and wing-shivering especially, release attack from small passerine birds, and certainly do not evoke fear (Blest, 1957a).

Andrew (1957) describes the intention movements of flight of certain passerines. They show inter-specific differences both in form, in their quantitative characteristics, and in the contexts in which they occur; these differences correspond rather closely to the accepted taxonomic divisions. In a few cases the behavioural groupings cut across the taxonomic arrangement. For example, ground-living insectivores, irrespective of family, give an especially elaborate and highly stereotyped tail-flick. This otherwise surprising correlation may be explained with reference to the intermittent cursorial locomotion, frequently interrupted by flight, which is one of the terrestrial insectivore's primary adaptations to its feeding habits.

Hinde (1955–56) notes that in passerines wing-raising is commonly associated with aggressive behaviour, and wing-quivering and drooping with sexual displays, and comments on the lack of any very obvious reason for these widespread linkages. He further argues that structural and mechanical factors may affect the forms taken by displays, and that the widespread similarity of certain display components in many groups of birds, within which they have probably been independently evolved on more than one occasion, can be related to mechanical necessities implicit in the original movements from which they have been ritualised.

The apparently independent evolution of inter-male fighting and other behaviour in *Drosophila* may in part be attributed to their possession of homologous genetic complements, though this, of course, has not been proved directly (Spieth, 1952). The selection pressures from the environment which may influence the evolution of intra-specific displays have been noted above. It may be added here that for inter-specific displays the environment must supply the greater part of the selection pressures; the behaviour patterns produced may bear little relation to systematic status at the lower taxonomic levels (Schmidt, 1955; Blest, 1957*b*; Marler, 1957*a*) and convergence is often extreme.

4. *Some complex cases*

The phylogenetic analysis of signal movements involves the heuristic assumption that all signal movements must have a derivation from more ancient patterns with a less specialised function. The more complex the movements under consideration and their contexts, the greater the part likely to be played by axiom in the analysis, and the less by the observation of direct similarities between ethologically 'primitive' and 'advanced' species. Nevertheless, it is perhaps striking that in no case has it been proved totally impossible to suggest the basis of a complex ritualisation on the basis of reasonable comparison between related species. Two examples may be quoted, and are particularly appropriate since both have, in the past, been the subject of much anthropocentric speculation.

The process of pair-formation and the maintenance of the pair-bond in bower birds involves the performance of an extremely elaborate behaviour pattern by the male, in which he collects a variety of inanimate objects and assembles them into a construction, oriented, in some species at least, with respect to the sun; in or near the construction he displays to his mate. Both the major fabric of the construction, and the small objects with which it is secondarily decorated, are characteristic of the species (Marshall, 1954). These bowers have led naturalists variously to credit the birds with intelligence, insight into their actions, aesthetic appreciation and fetishism. Marshall has shown that bower construction may plausibly be explained in terms of the modification of patterns associated with social, territorial and sexual behaviour, and that no new ethological categories need be proposed for them.

A second case is that of the communication dance of the honey-bee. Worker bees on returning home from a new source of food are able to communicate to their fellows information about the direction of the source, its richness, and its distance from the hive (von Frisch, 1950). These different items are presented in a dance which contains components corresponding to each of them. The evolution of such a complex ritualised pattern would not be easy to conceive unless analogous, albeit less highly developed behaviour patterns could be shown to be present in other, non-communicating insects. Such a prediction has proved to be justified. There is reason to believe that the dance as a whole may have been evolved from searching behaviour performed primitively at the food site itself (Dethier, 1957), for such behaviour is found in many insects. The transformation of a sun (polarised light) orientation into an equivalent gravity orientation within the hive is not, in principle, unique to the honey-bee; Vowles (1954) and Birukow (1953) have shown that similar transformations may be performed by ants and by beetles under appropriate artificial conditions. Crude 'dances' may be performed by blowflies (*Phormia regina*) after feeding, and their intensity and persistence may be shown to relate to the sugar concentration of the food, and to the distance flown between feeding and performance. Dethier (1957) has shown that those features of the bee's dance which are simultaneously dependent upon flight distance and sugar concentration (and hence allow confusion between a rich source at a distance and a dilute source near the hive) may be plausibly accounted for in terms of the mechanisms known to regulate central taste-thresholds in insects. One feature of the dance alone is solely dependent on flight distance. This is the number of waggle movements of the abdomen during each straight run of the dance, the frequency of the movements being such that they cannot be counted accurately without the use of special equipment. The relationship between the distance between hive and food source and the number of oscillations as most recently and accurately determined (Steche, 1957) is not far removed from linearity over much of its course. But even this sophisticated relationship is not unique to the bee. A similar condition has been found in a Hemileucid moth (*Automeris aurantiaca*) which performs a special rhythmic movement on settling into the rest position from any other activity (Bastock and Blest, 1958). The number of oscillations performed by adult moths once emergence is completed is influenced by two main factors only: the age of the individual from eclosion, and the preceding flight duration, to which it bears a linear relationship (Blest, 1958a, 1959). Yet it is certain that the adult moths have no social behaviour and cannot be using this at present rather enigmatic act for communication. The number of oscillations performed on disturbing the moths and allowing them to re-settle remains fairly constant for at least 1 hour at 29° C. after initially settling from flight, and this stability of the response is another feature which the waggle phase of the bee dance must clearly possess. Thus all the important features of the actor's side of the communication dance are found to be present in other insects, in which their potentialities for social regulation have remained unrealised. As with much simpler ritualisations, the mechanism mediating the behaviour of the reacting bees and the factors allowing the evolution of their responsiveness are still quite obscure.

5. *Conclusions*

Some general conclusions about the concept of ritualisation may now be stated. Firstly, the components into which the process of ritualisation has been split in the past are not very satisfactory. There is considerable overlap between the categories, which are not mutually exclusive, and this type of conceptual difficulty is seen especially to affect those components such as co-ordination changes, whose description is in the first place difficult.

Since there is virtually no palaeontology of behaviour, the evidence for any particular detailed sequence of events having taken place is invariably tenuous. This limitation makes statements about the course of ritualisation in any one group somewhat hazardous. If it is felt necessary to define 'ritualisation', probably the best working definition which may be achieved is also the vaguest—that it is the evolutionary process responsible for the existence of inter- and intra-specific signalling movements. It is clear from the foregoing review that (1) all the available types of change are not likely to contribute to any single example of ritualisation; (2) at the causal level, outwardly similar results may be produced, in different groups, by mechanisms which may well prove to have little in common. For instance, a number of criteria of emancipation may be used which can be shown to have equal application to ritualisation processes in insects and in birds, in so far as they concern overt behaviour; but it may be doubted whether the neural events involved are the same.

Further, many of the changes in co-ordination, etc. which take place in ritualisation are probably similar to those involved in the evolution of the diversified patterns of locomotion found in the major terrestrial phyla. In so far as the analysis of the differences between patterns of co-ordination are concerned, the work of Manton (1950, 1952*a*, 1952*b*, 1954) is yielding a very detailed picture not only of arthropod movements themselves, but of the systems of musculature responsible for them. A less detailed summary of mammalian locomotory patterns has been given by Maynard-Smith and Savage (1956). The conclusion that, at this macro-evolutionary level, the patterns of neural activity must be of extreme plasticity is not likely to be contradicted by any study of the lesser and more specialised changes of ritualisation. It is even possible that in the course of the evolution of flight behaviour, changes rather similar to those qualifying as emancipation may have taken place in the relationships between the 'ancestral' (locomotion) and 'derived' (flight) activities. In this case, however, their evolutionary importance has been vastly overshadowed by that of the complex servomechanisms necessary for the regulation of flapping flight (Maynard-Smith, 1952; Weis-Fogh, 1956; Pringle, 1957). Very probably, the basic problems in causal analysis involved will be clarified through work on the experimentally more accessible patterns of locomotion, rather than on ritualised signal movements themselves.

6. *Summary*

1. This chapter has re-examined the main categories of change accompanying the evolution of signal movements, with particular reference to what is known of the causal mechanisms underlying them. Present knowledge is largely confined to the motor patterns of the signals themselves; little is known of the mode of evolution of the special patterns of responsiveness shown by the reacting animals to which intra-specific displays are directed.

2. Several different categories of change have been proposed in the past. These are here broken down into the following major groups:

(a) Changes in the releasing mechanisms responsible for the release of the signal movements.

(b) Changes in co-ordination.

(c) Change in the physiological relation between the ancestral act and the motor pattern of the signal which has been derived from it. This is the process which has been termed 'emancipation'. It is shown that the concept of emancipation can be defined in terms which do not necessitate the adoption of a particular theoretical scheme.

Each of these groups is examined in some detail, and a few complex cases are discussed.

3. It is also suggested that the changes involved in the process of ritualisation are, on the motor side, not very different from those implicit in the evolution of the patterns of animal locomotion which may, indeed, prove experimentally more tractable.

4. Simple working definitions of 'ritualisation' and 'emancipation' are provided, which may be related to empirical observation, and which are not dependent upon the special features of any of the current behaviour models.

V. THE BASIS OF FLOCK ORGANISATION
IN BIRDS

1. *The social tendency and problems of flock structure*

There can be few naturalists who have not at some time in the course of their studies paused to admire and reflect upon the elaborate movements of bird flocks and remarked with what precision their gyrations are carried out. In this chapter we will attempt to analyse the various factors responsible for the organisation of bird flocks with a view to producing both concepts and analysis that may be readily subjected to empirical testing.

The first characteristic of gregarious birds is the fact that individuals seek out others by the performance of distinctive behaviour sequences that bring them into the flock. It was at one time generally assumed that some kind of social 'instinct' was responsible for this behaviour. Tinbergen (1951), however, denied this by concluding that there were no special activities that could in their own right be called social which were not part of some other 'instinct'. 'An animal is called social when it strives to be in the neighbourhood of fellow members of its species when performing some or all of its instinctive activities ... in other words when these instincts are active the fellow member of the species is part of the adequate stimulus situation which the animal tries to find through its appetitive behaviour.' Tinbergen did admit, however, at least implicitly, that the individual may 'strive' or 'try' to be in the neighbourhood of its fellows.

More recently, however, there have been several accounts indicating that Tinbergen was mistaken in his view. For instance, Moynihan and Hall (1953), in discussing the question in relation to the Spice Finch (*Lonchura punctulata*), remark that there are indications that the birds' gregariousness is really more than an aspect of other 'drives'. The birds appear to do more than sleep, feed and preen together. 'They certainly look as if they just want to be together.' In this species, and also in *Quelea q. quelea*, the behaviour of separated birds was marked by vigorous attempts to rejoin their companions. An escaped *Quelea* remains near the aviary and moves up and down along the wire in accord with the movements of the birds inside, without respect to their actual activity, whether feeding, preening or washing. Keenleyside (1955), studying the schooling behaviour of fish, considered the stereotyped nature of the activity, varying in the form it takes in different species but typical within a single one, as indirect evidence for controlling mechanisms in the C.N.S., and suggested the existence of a separate schooling instinct. It seems, then, that in both birds and fish there is indeed a very strong social motivation (i.e. 'that which induces an animal to act,' Marler, 1956*a*, p. 5). (See also Morris, 1956*b*, p. 84; Thorpe, 1956, p. 266.) There is, however, evidence, in fish at least, for the development of the schooling response by a learning process and there are difficulties about distinguishing this from the possibility that an inborn recognition of species-specific social releasers may be maturing (Thorpe, loc. cit.). For this reason, although we may conclude that evidence is good for an independent social motivation in fish and birds, we should be unwise to call it necessarily unlearned or instinctive in the sense of Tinbergen (loc. cit.). In addition, recent criticisms of the use and validity of such terms as drive and I.R.M. (Hinde, 1955) give rise to the belief that a consideration of the phenomena in this terminology would be fruitless. Hinde (loc. cit.) has proposed the term tendency, defined as 'the readiness to show a particular behaviour as observed under natural conditions', and we may apply this to the observable striving to be with the other members of the species. We may therefore call it the Social Tendency and employ the term with the same strictness as has been followed by recent studies in which the older drive concepts have been abandoned (Hinde, 1955, 1956; Marler, 1956*a*; Andrew, 1957; Tinbergen, 1957).[1]

Once a bird flock has come into being it is maintained by certain social mechanisms in a more or less constant degree of dispersion relative to the environment. Hediger (1950) pointed out that among animals loosely termed social there are two types: the 'contact' animals that maintain touch contact with each other, particularly during resting activities, and 'distance animals' which, even when resting, rarely come into contact with each other, but maintain 'individual distances' between them. Flocks of contact birds are found, for instance, in the Artamidae, Coliidae, Hemiprocnidae and Estrildidae (Emlen, 1952) and have been studied in any detail only by Moynihan and Hall (loc. cit.) and Morris (1954, 1957*b*) for some Estrildidae. Distance species are, however, found in many families, including the Laridae, Scolopacidae, Fringillidae and Ploceidae, and it is this type of flocking organisation with which we shall concern ourselves particularly here. (Flocks made up of integrated family groups in which elaborate personal relations are established, as in geese, for example, naturally show a more complicated type of organisation, requiring a rather different type of analysis from that developed here. Nevertheless, much of the discussion will be of relevance to such species.)

The problems of flock organisation were studied in Cambridge and in the Senegal, and birds both in aviaries and in the wild were observed and some experiments performed (Crook, 1958). The problems were first sub-divided in terms of the apparent response mechanisms that served particular types of social relationship between individuals.

(a) *The response mechanism subserving Group Synchronisation*

Stationary flocks normally show some degree of synchronisation of maintenance activities such as feeding, drinking, sleeping,

[1] The drive concepts referred to here are those used in ethological theory. In American learning theory the term also appears but in a very different analytical context, in which its use is unaffected by Hinde's criticisms based on ethological grounds.

hopping about, washing and preening. There is an inter-relation between cycles of these activities in individuals (see below) that produces a group cycle. Two factors contributing to this have been previously recognised (Lorenz, 1935, 1937; Nice, 1943; Morris, 1956*b*):

(i) The 'following reaction' whereby one bird follows the movements of others and thus moves more or less synchronously with them.

(ii) 'Social facilitation'—the immediate copying of the behaviour of one individual by another.

It might be said that the following reaction is in fact a special case of social facilitation in that the following bird is copying the flying away behaviour of another individual. However, it seems that, apart from any tendency to copy, the tendency to keep close to other individuals is operating (i.e. the social tendency). In addition, the following reaction is directionally oriented, and is usually a direct reaction to a precise pattern of stimuli such as the sudden exposure of wing flashes or the sound of flight cries (Hinde, 1953). In a stationary flock it is social facilitation alone that functions in group synchronisation, whereas in group movements it is the following reaction that plays the predominant role.

(b) *The response mechanisms subserving Group Integration*

Flock cohesion is maintained during movement in spite of the frequently different activities of component members. Thus during the progression of a foraging flock of tits, for example through a beech wood, different individuals may at any one time be eating, drinking or hopping about. The flock is maintained within certain bounds, however, by the behaviour of birds left behind which soon fly after it and rejoin it. Previously Hinde (1952) had used the word integrated to refer either to the density of the birds (i.e. their spatial distribution) or to the 'strength of the social bonds holding them together'. There are, however, occasions when analysis requires a clear distinction between them. Here, then, integration is used to refer to flock cohesion during movements and may be measured in terms of the speed of the following reactions, the units being thus temporal ones. In measuring following reactions we obtain measurements that may be used as an indication of the 'strength of the social bonds', and hence of the social tendency, in the absence of aggressive responses which are present as a factor in measurements of spatial distribution (see below).

(c) *The response mechanisms subserving the 'over-dispersion' within flocks*

A species population may be distributed in its environment in three main ways: at random, in an 'over-dispersed' manner or in 'aggregations' (or 'congregations'[1]). Over-dispersion, a term used by Hinde (1952, 1955–56) following Salt and Hollick (1946), means that individuals are more evenly distributed in space than they would be if distributed at random. Aggregation, however, means that they are dispersed less than they would be if at random. The term 'dispersion' (Lack, 1954; Tinbergen, 1957) denoting the state of being dispersed as opposed to crowded or distributed at random is synonymous with over-dispersion.

Individuals of gregarious species are normally congregated into flocks or herds occupying, at any one time, only a portion of the habitat or range. However, within each flock the individuals composing it may be distributed more or less at random or over-dispersed within the area covered by it. Thus resting flocks of contact species are over-dispersed internally even though the individuals may be so close as to touch one another. When active, however, the internal distribution may be relatively random. Flocks of distance animals, however, show over-dispersion internally both when resting and when active, though certain randomising effects may also be observed.

---

[1] Congregation is a term for an aggregation of animals not necessarily brought about by ecological factors but rather by the action of positive social responses (Allee, 1931).

When flocks of different distance species are compared, different average distances separating the individuals composing them may be found, and these may be sufficiently constant as to be quite species-specific (see below). The nature of the tendencies active in maintaining flock over-dispersion has not previously been analysed, though Emlen (1952) has discussed the problem in terms closely following those of Craig (1918). As will be seen below, the internal over-dispersion of flocks is due primarily to the interaction of two opposing response tendencies, the social tendency and the tendency to respond aggressively when approached closely.

In considering these problems, synchronisation and integration will be taken together due to certain marked correlations between the responses concerned, and the internal over-dispersion of flocks discussed separately. Examples will be taken particularly from related Ploceine species (Weaver Birds) from West Africa. These birds are specially suited to the enquiry since they show marked differences in their social behaviour in the wild. *Euplectes afra* breeds in large territories and gives courtship flight displays above them, and in the dry season has a loosely cohering flock structure. *Sitagra melanocephala* breeds in loose colonies and gives courtship displays only in the colony area, especially in the neighbourhood of the nest. *Quelea q. quelea* breeds in extremely dense colonies in which all display is restricted to the small area around the nest site, and in the dry season it forms flocks remarkable for their density and integration. (See also Crook, 1959, and in press.)

2. *Flock Synchronisation and Integration: the role of the following reaction and of social facilitation*

(a) *The group activity cycle*

When the maintenance activities of isolated individual caged birds such as *Quelea*, *Euplectes afra*, many Waxbills (Estrildinae), Spice Finches (Moynihan and Hall, 1953) and Chaffinches (Rowell, personal communication) are observed for periods of a few hours, the recorded activity sequences are found to take the form of a cyclic alternation between activity and rest. According to one classification (Crook, 1958), in the active period hopping about, flighting and feeding occur particularly, while in the rest period we find sitting alert, sitting drowsily, preening and sleeping. Four groups of tendencies are responsible for such a cycle: they are the tendencies to feed and drink, to wash and preen, to sleep and, lastly, to show restless movement. All occur in the absence of companions, though the last is more noticeable when other birds are present.

Feeding, drinking and sleeping are primarily responses to deprivation and their appearance a function of the time elapsed since their last performance. Preening behaviour is observed most usually on awakening from sleep and immediately prior to sitting drowsily. Andrew (1956b) considered it to arise primarily as a response to peripheral stimuli—foreign matter on the plumage, feather disarray, etc.—which increase with time during which the activity is not performed. Absence of other motivation is an apparent pre-requisite for its appearance since it is easily suppressed by other tendencies and rarely interrupts other activities. Sleep, too, except presumably in cases of extreme fatigue, tends to appear when other motivations are low and the associations between rest and preening seem largely due to this.

Individual members of a caged group of *Quelea* tend to do the same things at the same time: thus, in any one arbitrarily chosen observation period, the majority if not all of the birds will be eating, sleeping or washing. Figure 6 is a sample recording during such a two-hour observation period noting the birds' activities every five minutes. By comparing this with a similar graph for *Euplectes afra* (Figure 7) it will be noted at once that in the former the majority are performing a given activity at the same time, whereas in the latter the activity is spread among the birds over several observation periods. This is particularly marked where feeding is concerned. There is then, in both these species, a synchronisation of activity cycles bringing them closely into phase with one another, but the rapidity of the diffusion of change of phase within the flocks may vary and appears to be characteristic of the species. Activity cycles of *Euplectes* are frequently out of phase and may come

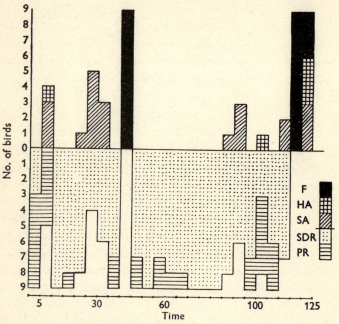

Fig. 6. Sample recording of Group Activity Cycle of *Quelea quelea*. Nine birds observed for 125 minutes.

Key: F = Feeding
HA = Hopping about
SA = Sitting alert
SDR = Sitting drowsily
PR = Preening
W = Washing

back into phase only through the presentation of an external stimulus producing a common reaction in each individual, after which they usually continue to act in unison for some time. Among *Quelea*, however, individual cycles are kept more constantly in phase by extremely marked following reactions.

132

In the wild, the appearance of small-scale cycles in exceedingly large flocks of gregarious birds (e.g. the *Quelea*) is largely masked by the over-riding mass activity of the group. The cycles are then slower and the rest phases bunched into the hot early afternoon, with long restless feeding periods before and after. In species which live in smaller flocks, however, quite rapid cycles may be observed in the wild (e.g. Cordon Bleu *Uraeginthus bengalus* in the Senegal valley). In general, activity cycles of captive birds are inevitably to some extent artificial, but nevertheless very useful for revealing certain features of flocking behaviour.

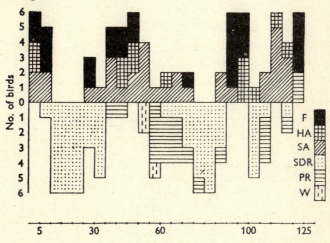

Fig. 7. Sample recording of Group Activity Cycle of *Euplectes afra*. Six birds observed for 125 minutes.
Interpretation of Key as in Fig. 6.

(b) *Following reactions*

Among related Weavers, following reactions may show considerable variation in strength. Some measures were made, using groups of caged *Quelea* and *Euplectes afra*, by observing the birds' movements during visits to food bowls. This produces a quantitative estimate of the degree of flock integration in this particular context for the two species. Food was constantly available and the normal feeding period was observed. The birds followed one another singly or in groups to the bowl and fed together, maintaining an individual distance while doing so. Very little difference was found in the actual time intervals between the descent of the initiating birds and following birds in the two species. The number of individuals following together in groups after the initiation of the movement was, however, significantly different (Table 1) since among *Quelea* several birds might move at the same time whereas *Euplectes* individuals always flew down singly. The percentages of the total group integrating at the bowl were also significantly different. In the *Quelea* group almost the total number integrated, whereas in the *Euplectes* groups only a proportion did so (Table 2).

Table 1. *Number of individuals moving together in each following reaction after movement by the initiating bird*

Species	Number of birds	Number of observations
Quelea q. quelea	1	20
,,	2	5
,,	3	2
,,	4	2
Euplectes afra	1	17
,,	other figures	nil

Table 2. *Percentages of total numbers of birds in an aviary integrating at food bowls during feeding periods*

Species	Percentage of group integrating at food bowl	Mean %	Number of observations
Quelea q. quelea	100		4
,,	90	98	1
Euplectes afra	60		2
,,	50		1
,,	30		3
,,	17	35	3

(c) *Social facilitation*

There has, unfortunately, been a great deal of fruitless discussion about the term social facilitation. Crawford (1939) proposed it to cover 'increments in the frequency and intensity of responses already learned by the individual, shown in the presence of other individuals usually engaged in the same behaviour'. Nice (1943) used this term, as defined above, to cover the phenomenon Lorenz (1935, 1937) had called the 'inducing of reaction by contagion' and pointed out that it was a 'matter of suggestion and not of conscious imitation'. Armstrong (1951), evidently dissatisfied, proposed and defined the term 'Mimesis' as the 'reproduction by one animal of the instinctive behaviour patterns of another'. Hinde (1953) strongly criticised the definition and application of this term throughout Armstrong's paper and in later papers in which the phenomenon was discussed (Moynihan and Hall, 1953; Morris, 1956b) it was no longer named at all. It is difficult to see quite what the terminological difficulty has been. Crawford's (1939) term as used by Nice (loc. cit.) seems quite satisfactory so long as the question as to whether the behaviour under discussion is innate or learned is left an open one. It will therefore be used here meaning that phenomenon observed when the performance of an activity by an individual stimulates the immediate performance of the same activity by its neighbours.

In *Quelea* social facilitation is found particularly in the synchronous performance of preening and washing activities and it is thus more characteristic of the rest phase of a group activity cycle than of an active one. Moynihan and Hall (1953) remarked that there was often a queue of their Spice Finches waiting for a bathe. After the first bird had washed it would begin the marked wing vibration typical of preening after washing. Other birds would follow suit at once, even though they had not themselves washed and their plumage was quite dry (see also Andrew, 1956b). This particular social facilitation was not, however, noted among those Ploceine species which were slow at integrating as a group at the wash bowl—*Euplectes afra*, for example.

In considering social facilitation it is most important to consider the manner in which the individual cycles come into phase. If they are already in phase, the social facilitation of an activity cannot always be assumed. Thus should all the birds be seen hopping about at the same time, we might assume simply that they were all in the same phase of the cycle rather than that some social interaction had occurred. It requires careful observation to decide whether a given behaviour state of an individual is in fact functioning as a stimulus to another. In the case above, an obvious stimulus-response has occurred. It is not, however, always so obvious. Social facilitation can occur only when birds are in approximately the same 'mood' and there must be definite observational evidence that the 'facilitated' bird was in fact looking at or had heard the 'stimulating' bird. The criterion for its occurrence is speed of reaction taken together with sufficient data indicating perception of the provided stimuli.

In a more detailed example, some *Quelea* were kept in three 3 m. cube aviaries spaced at intervals of a couple of metres from one another. It often occurred that washing behaviour was begun by the birds in one aviary. The birds in the neighbouring aviaries would very quickly move to their respective wash bowls and begin bathing too, so that—within the space of a moment or two—all three aviaries would be full of birds washing and preening after washing. Morris (1956*b*) having observed the same phenomena among his caged Estrildidae, concluded that there must be a tendency on the part of individuals of communal groups to copy, regardless of whether this involved 'generalised locomotion or specialised activities'. In cases where birds fly to *different* wash bowls in their *separate* aviaries and carry out *synchronised* washing, we have very clear evidence for social facilitation and a chance of quantifying it.

The releasing stimuli for social facilitation are apparently more complex than those of the following reaction and apparently consist in the complete behaviour of another individual; it seems unlikely that such a response will prove to be innate. Possibly some learning occurs in which the activities of other birds are associated with the individual's past performance of the same sequence. The closeness in phase of the individual's activity cycle to that of the performing bird might then allow the same behaviour to occur, thus synchronising the activity with that of other birds, even when the activity (such as washing dry feathers) appears to be functionless. In such cases we may hazard that the individual was not sufficiently motivated for actual washing, but that, in the absence of opposing tendencies, facilitated movements could be made.

(d) *Individual versus group activity*

Both the following reaction and social facilitation are to a large extent dependent upon the mood of the individual and its correlation with the phase of the activity of the group. There were many occasions during which preening by one bird elicited no response in others. They simply remained sitting drowsily. In such cases it was clear that the motivation of the other individuals was not such as to allow them to follow the set example, for the activity cycle of the preening bird was slightly in advance of those of its companions.

Similarly with the following reaction, Moynihan and Hall (1953) remark that during their observations on the Spice Finch a 'conflict between individual desires and gregariousness was often painfully evident'. A bird would sometimes leave the group in order to feed or drink. If other members of the group were slow in giving the following reactions, or if they failed to give them, the separated bird would show an obvious uneasiness. It would complete the activity it had begun rapidly and in a somewhat disorganised manner, or even abandon it entirely and return to the group. Similar observations were made on *Quelea*, but they were not noticed in the less social *Euplectes*. Thus the following reaction requires similar motivation among a group of birds at the same time if it is to occur. Very often a conflict between the social tendency and other tendencies was clearly apparent.

(e) *The correlation between flock synchronisation and integration*

Flock synchronisation is produced by the co-action of the social tendency maintaining flock cohesion, through the following reaction, and the tendency to copy in social facilitation. Both tendencies result in bringing the differing activities of individuals into phase. Flock integration is, however, brought about by the social tendency alone. In many species the following reaction is greatly reinforced by the presence of social calls or visual signals which modify the releasing stimulus considerably. The principal effect is an increase in the speed of diffusion of a reaction through the flock. Integration and synchronisation are likely to perpetuate each other. Thus a synchronised stationary flock is likely to show rapid integration, while rapid integration allows a quick re-establishment of synchronisation. We may summarise in a diagram (Figure 8).

It is clear that the approach here followed produces a type of analysis which it would not be at all difficult to quantify. The speed of integration of groups, the time taken between the stimulus of a departing bird and the following reaction of a companion, and observations on the speed and number of birds integrating at wash bowls in separate but neighbouring aviaries all provide useful measures which may also provide a quantitative basis for the comparison of specific sociality in a family.

3. *Group over-dispersion. Spatial distribution within bird flocks of distance species*

(a) *Individual distance*

In flocks of distance birds, over-dispersion is maintained by the existence of what Hediger (1950) called an 'individual distance': that area around an individual within which the approach of a neighbouring bird is reacted to either with avoidance or with attack. Recent studies have greatly clarified the component factors that maintain this distance and analysis at the present level is fairly complete.

Burckhardt (1944), Conder (1949) and Crook (1953) all recorded variations in the individual distance of Black-headed Gulls (*Larus ridibundus*), depending upon the type of activity

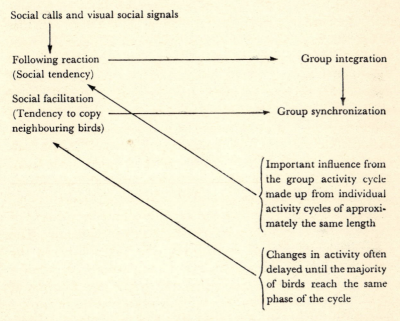

Fig. 8.   Scheme showing the relations between factors in flock organization.

being performed by the group. It tends to be larger when the birds are feeding over dispersed foods and much reduced when the food is closely aggregated. Reduction appears to occur only when the bird's attention is so closely rivetted to some activity that the display of individual distance aggression is thereby inhibited.

A number of experiments have been carried out in order to quantify differences in individual distance with the type of activity being performed and to compare individual distance between related species of differing degrees of sociability.

Marler (1956b) quantified the individual distance behaviour of chaffinches (*Fringilla coelebs*) brought close together during feeding. Two movable hoppers were brought increasingly near to one another until on 100% of encounters (i.e. when two birds were present simultaneously at the hoppers) a fight occurred. The results revealed a zone rather than an abrupt threshold distance around the feeding individual, across which the probability of aggression increased. A convenient measure was the 50% distance at which there was an even chance of two birds fighting or tolerating one another. (Large differences were found between groups of females (7 cm.) and males (21 and 26 cm.). Females with breasts dyed red were treated as males.)

Individual distance behaviour is also shown away from food or other potential sources of aggressive competition. It is, for instance, found upon open perches in the aviary. Birds tend to group upon such perches but approach within a certain distance is not tolerated. It was therefore decided to attempt to measure the phenomenon without the use of food bowls. Aviaries were prepared in which only a single long perch, divided into 4-cm. units by black bands, was supplied. It was thus possible to record distances between perched birds. Tests were then carried out using *Quelea q. quelea* and *Euplectes afra* and another related Ploceine, *Sitagra melanocephala*. In making comparative tests between these species great care was taken to ensure exactly similar conditions in the aviaries, in which the same number of birds of the same sex were kept.

(b) *Measurable distances using the single bar perch*

As soon as the behaviour of a small flock of birds was observed in the prepared aviary it was at once clear that a number of different measures might be taken.

(i) *The arrival distance.* When a bird alights on a perch it may land in the centre of a line of birds, or at the end. If it lands in the centre it fills an established gap between two individuals. When this happens the birds on either side often spread out. This moving outwards from the centre means that the moving birds approach others, and a shunting process along the perch occurs so that the size of the gaps between them is adjusted to its former size. If the newcomer alights at the end of the line, it then frequently moves closer to its neighbour. We can thus measure the distance from another individual at which a new arrival lands (the Arrival Distance) and also the extent to which either it or its neighbours adjust through subsequent closure or spacing.

(ii) *The settled distance.* After arrival, and when adjustments have occurred, the resultant distance is the Settled Distance, at which no further approach or avoidance is observed. The settled distance is therefore not necessarily the individual distance (i.e. the minimal approach distance tolerated) but rather that distance established after interaction between the individuals has occurred.

(iii) *Distance after departure.* Often when individuals go hopping about other birds may be left sitting alone in relative isolation. If such an individual is not asleep or sitting too drowsily, it then frequently shows closure upon other individuals. Care was taken not to measure distances shown after the departure of neighbours, for these clearly reveal nothing of the tendencies bringing about spatial distribution.

The observed behaviour:

Two tendencies (cf. Emlen, 1952) are immediately recognisable in the aviary:

(i) The tendency to approach other birds (the social tendency).

(ii) The tendency to show aggressiveness once an approach tolerance threshold—expressed in terms of distance along the bar (the individual distance)—has been passed.

Birds are constantly coming and going on the measured perch. Closure (social tendency) often occurs so that an individual comes within the individual distance of another. At once intention movements of attack can be observed, which may develop into threat or into an actual lunge. Sometimes, however, the approached bird merely hops a little away, though usually without any show of flight intention movements, the relaxed posture being maintained throughout. Often it is the approaching bird which may separate, thus showing spacing after arrival. There is an interaction between the two birds, and which one moves appears to be determined by their relative dominance. In the Senegal the birds had not had time to have learnt a peck order of any kind and the interaction was thus immediate.

Some differences between arrival distances (here distance on alighting after flight) and settled distances were measured. A group of *Sitagra melanocephala* was used (Table 3). In the particular cases measured, only spacing was in fact seen and the measures were made when all the birds were 'sitting alert' and showing some 'hopping about'. Closure occurred when the prevailing activity of the birds was different (see below). The

Table 3. *Spacing in* Sitagra melanocephala *following arrival distances of* 1, 2 *and* 3 *units*
(*Ten birds in aviary, Senegal*)

Arrival distances	No. of observ- ations	Settled distances	No. of observ- ations	Spacing	No. of observ- ations	Mean spacing (following arrival distances of 1, 2 and 3 units)
3	1	4	1	1	1	1 unit
2	18	4	8	2	8	
		3	8	1	8	1·5 units
		reactor fled	2			
1	10	3	6	2	6	
		2	3	1	3	1·6 units
		reactor fled	1			
Mean 2·02 units =8·08 cm.		*Mean* 3·5 units =14·0 cm.		*Mean* 1·5 units =6·0 cm.		*NB.* 1 unit =4 cm.

table shows that the mean arrival distance was 8·08 cm. and the mean settled distance 14·0 cm., the mean spacing being 6 cm. It shows further that the nearer a bird lands to another, the greater will be the resultant spacing.

The relations between individual distance, approach tolerance, arrival and settled distances, closure and spacing, may be expressed in a diagram (Figure 9).

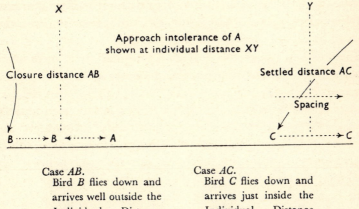

Case *AB*.	Case *AC*.
Bird *B* flies down and arrives well outside the Individual Distance (here *X–Y*) of *A*. Closure follows and the Settled Distance *AB* is established	Bird *C* flies down and arrives just inside the Individual Distance (here *X–Y*) of *A*. Spacing follows and Settled Distance *AC* is established

The line of flight on arrival is shown by unbroken black curve

Fig. 9. Diagram of the relations between measures taken upon a marked perch. Three birds *A*, *B* and *C* are present. The arrival of *B* and *C* outside and inside the ID of *A* is shown.

(c) *The measurement of settled distances: variations with phases of the group activity cycle*

Preliminary observations made it clear that the settled distance was the easiest to measure. The birds were closely observed and the settled distances read off in the 4-cm. units from the marked perch. Preliminary observations had also shown that the distances varied with differences in the phase of the group-activity cycle of the birds. It was therefore decided that the phases would have to be judged by observation and settled distance measured only during established phases of the group activity cycle. In this way accurate estimation of the change of settled distance with the changes in the phasing of the activity cycle could be made. An activity phase was considered established when the majority of the birds were doing the same thing. Any measurements taken in what subsequent observation proved to be the onset of a transition period were abandoned. Four phases of the group activity cycle could be judged accurately enough for use. These phases were:

1. Sitting drowsily (SDR) — All or the majority of the birds sleeping or sitting drowsily.
2. Sitting alert (SAPr) — With preening and occasional hopping about.
3. Sitting alert (SAHA) — With hopping about and a little preening.
4. Flying about and hopping about (FAHA)

In addition, a further group activity, Fright, from a known alarm source, was used. During activities FAHA and Fright the birds were so energetic that no settled distance could be observed. In these cases arrival distances were measured and compared with the settled distances of other phases. The results following statistical analysis showed that the settled distances recorded during each activity phase were significantly different except when SDR and SAPr (in any case the most similar activities) were compared. The distances were smallest during SAHA, larger during SDR/SAPr, and largest during FAHA. In addition, the variation in the distances recorded was greatest during SDR/SAPr and FAHA but much less during SAHA. During Fright the birds always crowded together into dense little bunches in which no aggressive behaviour was ever observed.

We may interpret these results in the following way. During SAHA both individual distance aggression and the social tendency are equally present and a balance of forces revealed in the normal distribution of the measures without great variation. Thus an almost perfect over-dispersion within the flock is observed.

During SDR and FAHA, however, two other factors have to be considered:

(1) During SDR many individuals go to sleep. Other birds may then approach within the waking individual distance without being pecked simply because the approached individual cannot notice. Hence during SDR groups of birds, often sitting extremely close together, shuffle apart with squeaks and pushings of the beak as soon as they awake. In addition, sleeping birds do not notice the departure of neighbours to a position further along the perch. Being asleep, they show no social tendency and the great variation is due to the absence of group integration during this phase of the cycle.

(2) During FAHA the birds are constantly coming across objects to investigate, flitting to and fro in the aviary, picking at a wire here and bill wiping there. The intense activity makes measurement difficult. The birds tend to land well apart, closure then follows, or else they are off again round the aviary. The variation here is due mainly to the fact that the social tendency, though present, is partly masked by the individual rather than group activity. FAHA usually precedes a feeding period. Activity inevitably spreads the birds about when the interest concerns dispersed objects and the spatial distribution that results includes this randomising factor in addition to the social tendency and individual distance aggression (cf. Contact species groups below).

Closure is most usually seen after intense activity when individuals once more move closer to one another. Spacing is commonest when the birds are concentrated together (e.g. as

in SAPr or SAHA) and alert, and it is under the latter conditions that individual distance encounters are most frequently
observed.

During Fright a special situation develops. Individual distance aggression is always absent when the bird's attention is
fully occupied with danger (as during 'scares' in the wild) or
with potential danger (mobbing), and under these conditions
the social tendency, being completely unopposed by individual
distance aggression, produces extreme concentration.

Spatial distribution in bird flocks is thus not only a function
of the balance between the social tendency and individual distance aggression but also of the presence or absence of attention
to the movements of other members of the group.

In comparing measures taken with different species it was
found that in general the same changes may be observed with
the activities of the groups, but that the overall size of both
arrival and settled distance vary apparently specifically. In
Euplectes afra, for example, although close concentration could
be induced during Fright, the distances remained much
greater than those for *Quelea*, and the same applied for other
activities. *Sitagra melanocephala* also showed measures in general
larger than those of *Quelea*. Such results would be expected
from field observations on the social life of the species. We may
suppose that not only are the degrees of territorial congregation
during breeding and in the extent of spatial distribution within
flocks correlated and of survival value, but that the measures
we may make upon them, under controlled and identical
circumstances in aviaries, will also show species-specific
characteristics. In comparing results from different species the
measure:

$$\frac{\text{Number of birds in a group}}{\text{Total of settled distances between them}}$$

denoting the Index of Line Dispersion, is particularly convenient.
In practice, the mean of a number of calculated indices for
groups of the same number of individuals showing the same
activity phase should be compared. The nearer the index

approximates to unity, the closer the birds are congregated upon the perch. In comparing the *Sitagra* and the *Quelea* the mean index of figures during SAHA worked out at 0·38 and 0·68 respectively. The use of such indices may contribute much to a comparative understanding of degrees of sociality among species in families such as the Ploceidae, Icteridae and Fringillidae where wide variations in flock and breeding densities may be found and be helpful in the study of their respective phylogenies.

4. *The development of individual distance responses in young birds*

In many birds, e.g. Chats (*Saxicola*), many Turdidae, the Wheatears (*Oenanthe* spp.), Nightingales (*Luscinia* spp.) and the Thrushes (*Turdus* spp.) as well as the Snow Bunting (*Plectrophenax n. nivalis*), Song Sparrow (*Melospiza melodia*) and Chaffinch (*Fringilla coelebs*) (Heinroth, 1924–33; Tinbergen, 1939; Nice, 1943; Marler, 1956a) the young scatter among the surrounding cover as soon as they leave the nest. Such rapid dispersion and concealment is likely to be of survival value in protection against predators. In the *Quelea*, however, in their dense colonies sheltered by the spiny armour of the thorn scrub, a 'huddling' response is retained after leaving the nest so that the young fledglings remain squatting together shoulder to shoulder in families some days after departure. (Fledgling wrens also remain in family groups after leaving the nest (Armstrong, 1955).) The survival value for *Quelea* is clear in that such behaviour keeps the young within the protection of the thorns until they have become adept at flying.

Young *Quelea* at this time show a simple activity cycle in which close concentration, during either feeding by a parent or when resting, is followed by a separation of a metre or two in hopping about the twigs during phases of activity. On the arrival of a parent they clump together, touching each other, while begging for food, and while resting they press their sides together and make every effort to maintain tactile contact and at the same time may preen mutually upon each other. They thus show an alternation between clumping and spacing during activity, the latter not being brought about by any individual distance aggression but by a simple process of drifting apart. After eight days the young begin feeding on fallen seeds below the bushes. They feed rubbing against one another and hopping over one another's backs without any aggressive encounters occurring, but soon after the onset of this new feeding method the first individual distance peck may be seen. Unfortunately it was not possible to remain in the thorn bush long enough to complete these observations, but it seems likely that there may be some temporal correlation between the onset of using the beak in picking up food and in using it against a neighbour. Its use in one context for the first time may facilitate to some extent its use in another. Andrew (personal communication) notes that pecking in curiosity begins among young Yellow-hammers (*Emberiza c. citrinella*) on the ninth day and at their fellows—still in curiosity—a little later. Really aggressive pecking was in this case, however, seen only about the 29th–30th day.

The difference between the social behaviour of young and adult *Quelea* is thus most striking, and a study of the developmental transition from the one to the other is badly needed. Pending further data, we may suggest that individual distance aggression begins either as a result of a maturation process around the same time as, or just after the first employment of the beak in food pecking, as work by Marler (1957b) and Guhl (1958) has suggested, or it may originate largely or partly as a result of learning in a competitive situation. Group feeding without individual distance is a highly competitive business. The birds crowd densely together with the beaks all pecking at seeds in the same general area, and accidental pecks at one another may well bring about separation and hence facilitate individual feeding; a rapid learning of such an effect would be of considerable value to the individual and to the species.

Once this behaviour was learnt, an individual would then peck at an approaching bird as soon as it came within a certain distance, probably determined by the reach of the bird.[1] The result of course would be over-dispersion within the group. Moreover, it would seem probable that such a reaction, once learnt, would appear in other situations in which the birds are active and at the same time closely congregated.

5. *Flock organisation in contact species*

Activity cycles of contact species such as *Estrilda melpoda* are closely similar to those of juvenile *Quelea* with respect to mutual preening, clumping and spacing. It seems then that for contact species natural selection has favoured the retention of clumping behaviour, a reaction apparently present in all nidicolous young as a kind of thigmotropism, into the adult life and that mutual preening is linked with this. We have then to make a distinction between intra-flock spatial distributions brought about by spacing without aggression (as in adult Estrildidae and juvenile *Quelea*) and those maintained principally by individual distance phenomena (as in the chaffinch and adult *Quelea*). Thus while the spatial distribution during active phases in contact species is produced by spacing without aggression, in distance species spacing due both to drifting apart and to aggression is found. During rest phases in contact species the individuals huddle, but in distance species they simply sit close to one another, never approaching except when a neighbour is asleep, when the bird next door may creep in close beside it as if bemoaning its lost youth. The exact survival values of these differing types of social structure cannot yet be determined, but with increasing comparative study on species of different types and degrees of sociality their significance will undoubtedly soon become clearer to us.

[1] Body size is thus a probable factor influencing individual distance but it is clear that in adult flocks other factors are also important in maintaining the different kinds of flock dispersion.

VI. THE FILTERING OF EXTERNAL STIMULI DURING INSTINCTIVE BEHAVIOUR

In an historic paper on the experimental analysis of instinctive behaviour, published in 1938, Lashley drew attention to the restriction of instinctive responsiveness to certain specific external stimuli. Some kind of filtering mechanism seems to be implied, through which effective and ineffective stimuli are differentiated. This filtering process has been appreciated earlier, and led Lorenz in 1935 to postulate a series of 'release mechanisms' (originally 'innate release mechanisms') which would only permit certain specific stimuli to evoke a response. These two papers established a new point of view which has stimulated work in many directions in the twenty years intervening since their publication. Lashley emphasised the need for serious effort to define the adequate stimulus, not only in studies of instinct, but equally in studies of reflexes and of learning. While he tended to concentrate on learning, Lorenz and his associates have mainly studied instinctive behaviour. Many of the results were reviewed in 1951 by Tinbergen, who has played a leading part in these investigations.

Emphasis has been placed by certain workers on the role of the actual receptors as stimulus filters and in another part of this book Barlow (Chapter XIII) has discussed some of the physiological processes which may be involved. A re-examination of the ethological data suggests that much can be explained by peripheral filtering, more in fact than has yet been suggested. However, there remain cases which seem to imply a central filtering process, and it is the aim of this chapter, after reviewing the specificity of stimuli which evoke instinctive responses, to try to assess what kinds of stimulus filtering are taking place.

1. *Specificity of stimuli evoking instinctive responses*

The actual specificity of stimuli evoking instinctive responses has been called into question. Lashley tended to the view that a relative lack of specificity is the rule, citing the early work of Stone (1922) and Beach (1937) on rats. Subsequent work has fully substantiated their conclusions that both sexual responses of the male rat and maternal responses of the female rat to her young are evoked through stimulation by several sensory modes, no one of which is indispensable (Beach and Jaynes, 1956). Among many other examples, Tavolga (1956) has shown that male courtship in a fish, *Bathygobius*, is evoked by a combination of stimuli—visual, chemical and auditory.

This situation contrasts with the demonstration by Tinbergen and others of highly specific responses in, for example, the fighting behaviour of male sticklebacks (ter Pelkwijk and Tinbergen, 1937) or the courtship of male spiders (Drees, 1952). Unlike the previous examples, these responses can readily be evoked by models and so are more accessible to detailed experimental analysis of the stimuli concerned.

There seem to be two different kinds of investigation, whose results do not necessarily conflict. The studies of Beach and others establish that the same response can be evoked by stimulation of different sensory modes. However, the specificity within any one of these modes has not been investigated in detail and it is possible that the range of olfactory or auditory stimuli which evoke, say, the strongest retrieving response of the lactating female rat is as narrow as the visual stimuli evoking maximal attack in sticklebacks (though the specificity may vary in different sensory modes). The difficulty of eliciting the response with models suggests that a complex pattern of stimuli is concerned, and the specificity might prove to be greater than in some of Tinbergen's examples.

Similarly, in his *Bathygobius* studies, Tavolga (1956) found that while relatively unspecific visual stimuli would elicit male courtship, the only adequate olfactory stimulus came from the female's ovarian fluid; other body fluids were ineffective.

Whether substances from other species evoke the response is not recorded, but nevertheless the restriction of responsiveness to certain chemical stimuli is clearly demonstrated.

Accessibility of the same response to stimulation by several sensory modes is widespread. The begging of the herring gull chick can be evoked by auditory as well as visual stimuli (Tinbergen and Perdeck, 1950). Chemical stimuli co-operate with visual and tactile cues in evoking spider courtship, and may play a role in the stickleback. However, the approach of the male silkworm moth to a distant female depends on olfaction alone, and here this is the only sense which could be used, the others being too insensitive to operate at this range. Within the physical limits of the situation imposed on each animal, it seems likely that all available senses are used in detection of the appropriate stimulus situation. We may conclude from the evidence that there is also variation, in different animals and responses, in the specificity of evoking stimuli within one sensory mode, which can be correlated with the functional requirements of the situation, the specificity being sometimes high.

2. *Circumstances favouring specific responsiveness*

In some of the circumstances in which instinctive responses are evoked only by highly specific stimuli, we can see the advantage to the animal. Some insect parasites thrive only in certain hosts, and the response to a specific stimulus from the host or from the host's normal environment has obvious selective advantage (Thorpe and Jones, 1937). Animals whose diet is restricted to specialised foods or prey respond to stimuli of a range and specificity appropriate to their biological needs, as is true of some phytophagous insects which feed only on plants providing the same chemical stimulus (Dethier, 1947*b*).

Escape behaviour is sometimes evoked by specific stimuli. Some night-flying moths will either flee or go into catalepsy in response to the high-frequency calls of bats which hunt them. Although they have some response to sounds ranging from

10–200 kc./sec. the strongest response occurs between 40–80 kc./sec., corresponding with the frequencies used by many species of bats (Schaller and Timm, 1949; Griffin and Novick, 1955). Similarly with chemical stimuli, some inter-tidal gastropods flee from substances from tube feet of predatory starfish, and ignore those from herbivorous species (Bullock, 1953). In birds the visual stimuli from a flying hawk cause fleeing behaviour in ducks and geese (Lorenz, 1939). In each case, the response to specific stimuli can be related to a particular and dominant source of danger to the species concerned.

Above all, the specific responsiveness is favoured in reproductive behaviour because, if it is to result in viable offspring, it must be directed at members of the same species. It is here that we find some of the most extreme examples, like the silkworm moth which will respond by approach only to the substance produced by the scent glands of the female and certain closely related compounds (Butenandt, 1955). There are other ways of restricting responsiveness to members of the species, e.g. by moving in the same habitat, as seems to occur in the potato beetle (Hellwig and Ludwig, 1952) or by moving at a particular time of day, as suggested for two species of moth (Schwinck, 1953). But often this is achieved directly by confining responses to specific stimuli from other members of the species.

3. *Circumstances favouring unspecific responsiveness*

In contrast with the previous group, there are situations where a restriction of responses to certain specific stimuli might place the animal at a disadvantage. Species capable of surviving on a range of foods may be best served by relatively unspecific responsiveness. Young grain-eating birds, for example, peck at any small object which contrasts with its background, especially if it is shiny. Similarly, a wide variety of predators including ants (Vowles, 1955), dragonfly larvae (Baldus, 1926), spiders (Bristowe, 1941), fighting fish (Lissmann, 1932), toads (Eibl-Eibesfeldt, 1952), owls (Räber, 1950) and domestic cats will seize as prey any small moving object (with size limits varying with the species concerned), again making use of the most widely shared characters of possible prey objects. A highly specific stimulus would also serve if it satisfied this criterion and such a case has been described in *Hydra* (Loomis, 1955), an omnivorous predator, whose feeding response is evoked by a particular substance, glutathione, which is a common constituent in the tissues of most animals. The effective stimulus seems to represent a common denominator for all possible food objects.

Unlike the specific parasites mentioned earlier, some are able to live in a variety of hosts. The wasp *Trichogramma* is known to parasitise the eggs of more than 180 different species. In contrast with the specific stimuli for host selection in many parasites, *Trichogramma* will attempt to lay eggs in anything that protrudes from the ground surface, is firm enough to walk on, falls within certain size limits, with no dimension more than about four times greater than any other. Odour, colour and surface texture are irrelevant, as long as the object is not wet or sticky (Salt, 1935). As a result of these broad limits, the most improbable objects may be selected for attempts at egg-laying, like a drop of mercury or a chip of glass.

In avoidance behaviour as well, response to unspecific stimuli has value, since most animals are exposed to a wide variety of dangers. Here the basic stimulus seems to be little more than a sudden change in the intensity or quality of stimulation of any of the sensory modes (Thorpe, 1944), which could again be regarded as a common denominator of potentially dangerous situations.

The impression we form is that animals respond to stimuli with the greatest specificity which is compatible with their biological needs in the given situation. In so far as one can generalise, neither extreme specificity nor extreme unspecificity of responsiveness is the rule in instinctive behaviour, the common condition being somewhere between, shifting in one direction or the other according to circumstances. However, even in response to the least specific stimuli, there is some degree of stimulus-filtering and this applies to all instinctive responses, if not to all behaviour, ranging from the simple question of which sensory mode is stimulated, to responses to highly specific chemical substances, sounds and visual configurations.

We may note in passing at least two other circumstances in which responses to relatively unspecific stimuli take place. First, the distance receptors may be too insensitive to permit perception of specific stimuli at long range. This seems to be relevant in the reproductive behaviour of some spiders and butterflies. Male butterflies (see below) will make the first courtship approach to a variety of unspecific visual stimuli, and it is only at close range that precise identification occurs. It is unlikely that either the simple ocelli of short-sighted spiders or the compound eyes of insects are capable of form-discrimination at a distance. So although the unspecific responsiveness is a disadvantage, it is the best which can be achieved with the animal's sensory apparatus.

Unspecific responsiveness also occurs in intermediate links of a behaviour chain, where overall specificity is required but is ensured by responses to specific stimuli earlier in the sequence. This applies to the 'following' responses of some young birds, evoked by relatively unspecific stimuli (Hinde, Chapter VII of this book), but ensured of specificity in nature by the parent's attachment to its own nest and eggs. The same probably applies to the egg-retrieving behaviour of gulls, and even to the copulatory behaviour of some species, where the most specific stimuli are concerned with pair formation, while actual mating can be evoked experimentally by less specific stimuli.

4. *Mechanisms for stimulus filtering*

It was originally thought that the 'release mechanism' postulated to perform the filtering process must reside somewhere in the central nervous system (Lorenz, 1935). However, Lehrman (1953), Schneirla (1956) and others have pointed out that the actual receptors may play an important role, especially in arthropods. This requires a revision of the original concept, which visualised a unique filter mechanism for each example of specific responsiveness. In some cases at least it is clear that the processes of normal perception are responsible. The specificity of the responsiveness of honey-bees to 'flower patterns' seems explicable in this way and Schneirla has a similar explanation for the stimuli which evoke the pecking response of the young herring gull (Tinbergen and Perdeck, 1950).

The examples of this kind of filtering, taking place in the receptors and afferent pathways as they function in normal perception, now form a substantial list. We can recognise two rather different types of filtering within the group. In one, the receptors respond much more strongly and in a discontinuous fashion to certain specific stimuli, or will respond to them alone. Here we can regard the filtering as peripheral. In the other type, the receptors respond quite well to a wider range of stimuli but still respond more strongly to some than to others. To appreciate the significance of this second type of filtering, we must consider the mode of experimentation used in studies of responsiveness.

Tinbergen and others have used the method of presenting different models for a standard number of tests, either singly or with a choice of two, and scoring the effectiveness of the stimulus by the proportion of presentations which evoke a response. A low score may occur either because the ineffective or non-preferred stimulus lacks certain pregnant characteristics, or because it is less conspicuous or fails 'to catch attention'. There is danger of a circular argument here, but the point being made is that the relatively greater 'attention-catching' property of some stimuli is evident whenever they impinge on the receptors, whatever the context. Thus the effectiveness of these stimuli in evoking responses is a result of the normal functioning of the perceptual pathways. Unlike the previous type, the filtering mainly takes place centrally or in the afferent pathways (Barlow, Chapter XIII of this book) but need not imply additional mechanisms to those used in normal perception.

5. *Peripheral stimulus filtering*

(a) *Examples from insect hearing*

Certain adult insects will give courtship responses to artificial sound stimuli as they normally would do to the specific song. Manipulation of such artificial songs reveals that a wide range of sound frequencies will evoke responses, extending far beyond the spectrum of the species song, as long as the type of amplitude modulation conforms to the species pattern.

One aspect of this specificity of responsiveness, the 'irrelevance' of frequency and the relevance of amplitude modulation, is directly related to the structure of the sound receptors. Over most of the range of insect tympanal organs, the pattern of afferent nervous activity has no relationship to the frequency of the incident sound, though amplitude variations are clearly reflected (Pumphrey, 1940; Haskell, 1956). The filtering out of information about frequency evidently takes place in the actual receptors: that is to say, the receptors must be relatively insensitive to it. In the same way the human eye might be said to 'filter out' information based on ultra violet and on the polarisation plane of light.

The sensory hairs of some insects do respond synchronously with low frequency sounds. It has been shown that the females of different species of mosquito have a distinctive wing tone. In one species at least the male has a rather specific response to the frequency of the female wing tone of his own species (Roth, 1948). On the basis of observation of the movement of sensory hairs on the male antenna when exposed to different sound frequencies, it is suggested that the receptors are constructed to resonate mechanically to the frequency of the female wing tone. If this interpretation is correct, the filtering process is again peripheral.

(b) *Examples from insect chemoreception*

The female silkworm moth produces a volatile substance which, if a wind is blowing, attracts the male from considerable distances (Schwinck, 1955). The approximate identity of the substance is known to biochemists, and experiments with a wide range of compounds established that the male responds only to the female secretion, and certain closely related primary unsaturated alcohols (Butenandt, 1955). In electrophysiological studies on the responses of silkworm moths to chemical stimulation, action potentials were measured from the isolated male antenna. There was only slight responsiveness, except to the female secretion and the other effective substances discovered by Butenandt, which all evoked strong responses (Schneider, 1957). We have, in fact, an extreme case of stimulus filtering in the receptors, here specialised to respond only to a very narrow range of chemical stimuli from the opposite sex.

A blowfly which tastes a chemical solution on the sensory hairs of its mouth parts may either drink, if sugars are present, or withdraw its proboscis if solutes other than sugars are present (salts, acids, alcohols) (Dethier, 1955). Electrophysiological studies show that the tip of each labellar hair receives two neurons, one of which responds to sugars and mediates the feeding response, while the other responds to non-sugars and inhibits feeding (Hodgson and Roeder, 1956). Comparative study of the effects of a range of carbohydrates also yielded a positive correlation between the electrical responses of the sugar receptors and the occurrence of feeding (Hodgson, 1957a).

Hodgson (1957b) also has evidence from other arthropods of specific responsiveness to different groups of chemicals, for example, a crayfish with sensory hairs responding to amino acids—and it may be that specific peripheral responsiveness is widespread in insect chemoreception, as seems to be true of mammalian taste receptors.

6. *Central stimulus filtering by the mechanisms of normal perception*

(a) *Examples from insect vision*

In addition to the studies already mentioned on flower patterns which attract honey-bees, there are two other demonstrations of stimulus filtering in insects of a similar type. The first is an analysis by D. Magnus (1954) of stimuli evoking the first courtship approach of the males of a European butterfly *Argynnis paphia*. By exhaustive studies with mechanical models, he found three stimulus parameters which result in increased responsiveness: rapid movement, larger rather than smaller size (within certain limits) and a particular orange-yellow colour. Here we are concerned only with the first two (see also p. 162).

The attractiveness increased with the rate of movement to a maximum of about 125 presentations of a revolving model per second and then declined. The change was presumed to coincide with the flicker-fusion frequency of the butterfly's compound eye. Subsequent electrophysiological measurements gave a value of 150 (Magnus, 1958). If we accept the view that the compound eye is largely concerned with perceiving different degrees of flicker (Wigglesworth, 1953), then a more rapid flicker will be more 'conspicuous' and the stimulus filtering is an accompaniment of normal vision in this species. The same may be said of the increased responsiveness to larger forms, which will stimulate more ommatidia.

In a more recent study of stimuli evoking the courtship approach of male houseflies, G. Vogel (1957) reached very similar conclusions. The attractiveness of the models again increased with the rate of movement up to 270 flickers per second and then declined. This value corresponds well with Autrum's (1950) figure for the flicker-fusion frequency of a blowfly of 265 per second. The greater attractiveness of a figure with a spiked outline is consistent with this flicker effect. The value of contrast with the background as an effective stimulus is seen both here and in normal vision. The effect of size of the model varied with the body-length of the subject, the optimum being roughly half as long again as the fly itself. If this is a function of how many ommatidia are stimulated, once again we need postulate no special filtering mechanism.

(b) *Examples from vertebrate vision*

The elicitation of prey-catching behaviour in many lower vertebrates by a simple moving form of a certain size implies a stimulus filtering process not unlike that involved in normal perception (Schneirla, 1956). This author suggests that the effective stimuli might be summarised as those from small conspicuous objects, the movement contributing to increase the degree of stimulation of the retina. Barlow (1953) has demonstrated a mechanism in the frog retina which may account both for the strong effect of movement and for the effectiveness of images of a certain size. The diameter of the receptive field of each ganglion cell is such as to be nicely filled by the image of a fly at 2 in. distance. He suggests that these retinal units may in fact be adapted to act as 'fly detectors', the stimulus filtering again taking place peripherally.

Few exhaustive studies have been made of stimuli evoking species-specific responses in vertebrates, but we have two careful studies of young birds (Tinbergen and Kuenen, 1939; Tinbergen and Perdeck, 1950). Thrush nestlings beg most readily to a form which moves, falls within certain limits of size, and is placed in the upper half of the visual field. The first two items again seem to imply no unique filtering process, though it must be admitted that we know no details of normal vision in birds of this age.

More elaborate are the stimuli for the pecking response of young herring gulls, normally evoked by the parent with food in its bill. Tinbergen and Perdeck list the following parameters of the optimal visual stimulus situation. The shape should be 'low', 'near', and 'moving', with a 'certain shape' 'pointing downwards', 'with a red patch at the tip' as on the parent's bill and with 'something protruding from the outline'. Schneirla (1956) suggests as the simplest explanation that the red spot is effective according to its degree of prominence as a moving object. Red is evidently conspicuous to many vertebrates which have colour vision, and the abundance of red, orange and yellow oil droplets in the retina of most diurnal birds shifts the zone of maximum sensitivity towards the red end of the spectrum (Walls, 1940). Movement will add to the conspicuousness. The elongate shape of the most effective stimulus object may draw attention to the spot at the tip and the same can be said of nearness and of discontinuity in the otherwise smooth outline. We are left with 'lowness' and 'pointing downwards' to explain. The latter is discussed below. Lowness may have a quite different explanation, resulting from the greater ease with which a bird can peck downwards.

R. and U. Weidmann (1958) have recently made a parallel study on another species, the black-headed gull, whose bill is entirely red in colour, to test the possibility that normal perceptual filtering may be involved in determining the specificity of responsiveness. The optimal stimulus situation is similar to that found in the herring gull. The role of contrast of the coloured spot against its background is again established, and the greater effectiveness of a vertical orientation of the model, as compared with a horizontal model, may be partly related to the phenomenon of barrel distortion which makes radial movement of a retinal image more conspicuous than tangential movement. The Weidmanns' conclusion is that at least some of the stimulus filtering in this context can be explained by the process of normal perception.

Conclusion. It seems probable that many aspects of the stimulus filtering observed in species-specific behaviour are effected by the receptors, afferent pathways and central nervous system, functioning as they do in normal perception. The best substantiated examples come from visual responses of

insects and young birds, and from olfactory responses of insects.
The last in particular may be expected to provide many more
examples of highly specific responsiveness which is determined
in the actual receptors. There are, however, cases of specific
responsiveness which seem to imply a special filtering mechanism, and these are discussed in the next section.

7. *Specific responsiveness implying a special filtering mechanism*

As originally conceived by Lorenz, an animal possesses a set of
release mechanisms, one for each case of specific responsiveness.
Although we have seen that the normal functioning of the
perceptual pathways can explain some of this stimulus filtering,
there are cases which seem to imply special mechanisms for
ensuring highly specific responsiveness in a certain context.
There are two kinds of evidence to consider. One comes from
response to stimulus configurations which are so elaborate and
specific that it is difficult to see how the filtering could take
place in any other way than through a special mechanism. The
other, more satisfactory, kind of evidence comes from demonstrated changes of responsiveness in the same animal, from one
specific stimulus quality to another, accompanying a change in
motivation.

Some of the stimuli for courtship responses of arthropods
discussed earlier include elements which cannot at present be
explained in terms of normal perception. In studies of two
different genera of butterflies the strongest response was given
to models coloured like members of their own species (Peterson
et al., 1952; Magnus, 1954), which suggests that a specific filter
may be involved. However, before one can be certain it is
necessary to know how sensitive the eyes of each species are to
colours. In addition, these studies were made on animals which
lived in the natural state and so might have acquired some
aspects of their specific responsiveness through conditioning.

In the examples discussed previously it seems sufficiently

unlikely that learning is involved in the development of the details of responsiveness that no special point has been made of the history of the experimental individuals. In fact some of the studies were made on experienced animals (e.g. the *Argynnis* experiments), others on naïve animals (e.g. the studies on silk-worm moths, houseflies, young herring gulls and black-headed gulls).

However, in seeking evidence for special stimulus filtering mechanisms which are built-in, we approach so close to the kind of specific responsiveness acquired by learning that it is necessary to confine consideration to studies of animals raised in a controlled environment. As a result, many important studies conducted on animals in the natural state must be passed over, such as the demonstrations of specific responsiveness of grasshoppers to the amplitude modulation pattern of the species song (Weih, 1951; Haskell, 1956; Loher, 1957; Perdeck, 1957), studies on fiddler crabs (Crane, 1957), butter-flies (Stride, 1956), dragonflies (Bucholtz, 1951, 1955), fireflies (McDermott, 1917; Buck, 1937) and gulls (Baerends, 1955). From studies on animals raised in restricted environments, with reduced chances of conditioning, there are some examples of responses to more or less elaborate configurations of form and tone evoked at the animal's first experience of them.

O. Drees (1952) studied with the aid of models the stimuli evoking male courtship in a species of jumping spider raised in isolation. The strongest responses were given to visual patterns with a certain size and orientation, divided into cephalothorax and abdomen, with legs having a particular size and orientation with respect to the body and with the species pattern of black and white stripes on the abdomen. Outside the breeding season the response disappeared and all models evoked either prey catching or avoidance, according to size. It was clearly shown that the courtship response is not simply evoked by a broken pattern but has a truly specific responsiveness, which is difficult to account for except in terms of a special mechanism for stimulus filtering.

Two species of birds, song sparrows and chaffinches, when raised in restricted environments, give a strong 'mobbing' response to owls when seen for the first time (Nice and ter Pelkwijk, 1941; Hinde, 1954). Experiments with models reveal a specific responsiveness to certain owl characteristics, notably the general outline, the colour pattern and the presence of a beak and a pair of eyes. Again a special stimulus filtering mechanism seems to be implied.

The specificity of the responses of some courting butterflies to colours has been mentioned earlier. Two studies have been made on animals raised in a controlled environment, and in each case experimentally naïve males responded most strongly to models with the species colour (Ilse, 1941; Crane, 1955). Both species also had different colour preferences in other contexts. The tropical *Heliconius* gave the strongest courtship response to orange-red and the strongest food-search response to yellow. Cabbage-white butterflies responded to any colour except green while feeding, to yellowish-white while courting, and in the case of the female ready to lay eggs and both sexes while looking for a resting place, to green and blue-green. It is possible that some change in the peripheral mechanism of colour vision takes place during different activities, and this should be investigated. Alternatively, a special central stimulus filtering mechanism may be involved.

Discussion. It is evident that regrettably few studies have been made of the details of specific responsiveness in animals raised in controlled environments. However, on the basis of what we already know, there seem to be at least three kinds of stimulus filtering. One takes place primarily in the receptors, the best example being the olfactory response of the male silkworm moth. The second type involves the whole perceptual pathway, up to and including the central nervous system, filtering as it does in normal perception. Examples here are the courtship responses of the male houseflies and the pecking responses of young gulls. The third type seems to imply a special central filtering mechanism which functions only in certain restricted situations, as in the owl-mobbing responses of birds or the court-ship of jumping spiders. While the neurological mechanisms underlying the first are relatively simple, and for the second type are beginning to be understood, the basis for the third type of filtering is unknown.

The general functions required of the sense organs will have an important influence on the type of stimulus filtering which takes place. If the function is simple and restricted, as in the male silkworm moth, which seeks the other sex after it emerges, without feeding, and then dies, highly specific responsiveness can be imposed at the level of the receptors. Where more versa-tile sense organs are required, the problems of stimulus filtering become more complex, and it is here that we must seek for special filtering mechanisms such as Lorenz originally visualised.

Although this discussion has been concerned with instinctive behaviour, Barlow and Broadbent (Chapters XIII and X of this book) have indicated the similar problems which arise in learned behaviour. In one case we visualise a stimulus filtering mechanism which is 'built-in' or 'self-differentiating' (Ewer, 1957). In the other case it arises through individual experience of the animal. Thus the contrast between them lies in the differences in ontogeny of the mechanisms, while the actual process of filtering appears, at the behavioural level, to be quite similar, though the number and specificity of the filters arising through learning is infinitely greater than in instinctive be-haviour. It may be that the same basic mechanisms are in-volved, as Barlow suggests, and discoveries in one field may have important implications in the other.

One such implication for the approach to learning problems is implied by the previous discussions. It is clear that some degree of stimulus filtering is widespread, and perhaps universal in instinctive behaviour. Much of this filtering takes place as a result of normal functioning of the perceptual pathways. If this is true, the effects must also be imposed on any sensory learning in which these same pathways are involved. Thus, built-in sensory processes may play a more important role in learning than has been previously thought, even in creatures like mammals where instinctive processes are sometimes thought to make only a minor contribution to behaviour.

8. *Summary*

1. After surveying the evidence it is concluded that some specificity of responsiveness is universal in instinctive behaviour.

2. The degree of specificity varies with the context and in a manner which can be regarded as biologically appropriate. Often the effective stimuli represent a common denominator of all functionally appropriate stimulus situations.

3. Exceptions in which relatively unspecific responsiveness occurs without advantage to the individual are seen in (*a*) intermediate links of a behaviour chain where ultimate specificity has been ensured by an earlier link; (*b*) animals whose receptors are not sufficiently sensitive to permit specific responsiveness in the circumstances.

4. The physiological mechanisms for stimulus filtering are considered. Three types are discussed: (*a*) imposed primarily by the receptors; (*b*) imposed by the receptors, afferent pathways and central nervous system as they function in normal perception; (*c*) apparently imposed by special, central filtering mechanisms, active only in certain contexts and corresponding with the original Lorenzian concept of a 'release mechanism'.

PART III

STUDIES OF PROBLEMS COMMON TO THE PSYCHOLOGY OF ANIMALS AND MEN

INTRODUCTION

We are now coming to realise that there is hardly any aspect of the behaviour of animals which may not have some reference to problems in human behaviour. It is now seen that the responses of animals are more complex and need more profound analysis than had been thought and that the concepts of the reflex and of reflex conditioning in their original form are often inadequate to the task. The chapters in this part range over a very big field of fundamental psychological problems; but it is hardly stretching definition too much to say that the whole Part can in fact be regarded as in one way or another dealing with problems raised by the phenomenon known as imprinting. Although the idea which it enshrines can be traced back to Spalding (1873) and James (1892) the name imprinting was originally given by Heinroth to a type of learning characteristic of the development of the following response of the young of nidifugous birds, such as geese, ducks and rails. Heinroth in 1910 found that young geese reared from the egg in isolation reacted to their human keepers, or to the first relatively large moving objects that they saw, by following them as they would their parents. Studies shortly after those of Heinroth led to the suggestion that imprinting

(1) is a learning process confined to a very definite and brief period of the individual's early life; (2) once established is often very stable and in some rare cases perhaps irreversible; (3) is a process which, though completed early, comes later in the life cycle to affect various specific reactions as yet undeveloped, such as those concerned with sexual and social behaviour; and (4) is learning which is generalised in the sense that it leads first to an ability to respond to the broad characteristics of a situation, though later it may enable finer discriminations to be achieved. On the whole, it may be said that subsequent research has tended to emphasise the importance of conclusions (1) and (4) and to suggest that perhaps (2) and (3) apply only, if at all, in very exceptional cases.

There are a number of reasons why imprinting is now regarded as a particularly significant phenomenon and why it in fact creeps into so many studies of behaviour, even though the name itself may not be used. We may review them briefly here.

1. Because, on the face of it, imprinting is 'unrewarded' learning (i.e. without any of the conventional rewards), it raises in an acute form the whole problem of reward or reinforcement. For instance, the young bird following its parent or parent-substitute continues to do so often with a dramatically intent enthusiasm even though it may receive no nourishment, warmth or other comfortable attention from the object it follows. All it appears to be doing is attempting to maintain a constant spatial relationship with a moving object, and such an 'objective', considered as something which brings about learning, obviously raises difficulties for many conventional theories of learning, e.g. that of Hull.

2. Imprinting is a special means of developing and changing motivation—doing what can in fact be described as 'priming the drive'. Both Hinde and Watson in the chapters which follow refer to this aspect of the matter, and the former quotes Weidmann's (1956) emphasis on the important difference between the behaviour of an imprinted and a non-imprinted

duckling, for whereas the former searches for the parent object if it loses contact, the latter, even if tame, shows no such appetitive behaviour. It has lost, or never gained, this urge to be near a parent. Watson is dealing in essentially the same manner with the problem when he accepts Deutsch's conclusion that in both learning as well as in innate behaviour the animal is regarded as searching for a cue or for stimuli. In this connection, however, we should do well to heed the warning contained in Vince's essay (Chapter IX) when she shows us how difficult it is to be certain whether learning ability really changes in a specific way with age or whether the changes we observe may not in fact be due to other variables. Beach and Jaynes (1954) conclude that 'there may be critical periods in development but much of the evidence for them is of doubtful reliability' and Vince emphasises the need for a more profound analysis of learned behaviour into its component elements before the real conclusion can be arrived at. She points out that behaviour in young animals can obviously be affected firstly by the general level of activity on which the degree of responsiveness may directly depend, and secondly by the ability to control actions by precisely timed inhibition. Tasks requiring ability of the kind which is determined by a high degree of activity and therefore of responsiveness are likely to be easier for younger juveniles, while those requiring the second type of achievement, namely the type of control manifested by internal inhibition, are likely to be easier for the older ones.

3. Imprinting is a special and sometimes very rapid process for restricting the stimulus situation setting off a particular response. By its means a response initially set off by a 'broad' or 'generalised' stimulus situation may come to be evoked by a much 'narrower' or more specialised stimulus. Thus Hinde points out that in responding to potential predators it is better for a bird to be safe than sorry, and we do in fact find that responses to harmless objects are common—that is to say, the response is set off by a broad or generalised stimulus. On the other hand, the pair formation of passerine birds must only

occur with a conspecific female and the rarity of hybrids in nature shows us quite dramatically how restricted must be the stimuli which are governing the establishment of the pair bond. Thus the degree of specificity in responsiveness will in the long run depend upon the consequences of inappropriate responses on the one hand and on the elimination of the appropriate ones on the other. 'If the former are disastrous specificity will be high; if the latter (as in responses to predators) it will be low.'

The Chapter by Hinde (VII) and some of the others in this section will be found to give a number of special and illuminating examples of the adaptive adjustment of instinctive behaviour by experience. A recent study which bears very closely on this general problem is that of Wells (1958) working with newly-hatched cuttlefish of the genus *Sepia*. His study discloses behaviour having both resemblances to and differences from the standard examples of imprinting. Young *Sepia*, when they hatch, will always attack and eat 'prawns' of the genus *Mysis* when these are offered. The movements for attacking *Mysis* are highly stereotyped from the beginning, but the delay between the presentation of the *Mysis* and the reactions of the *Sepia* decreases with experience, and this decrease is shown to depend upon the number of attacks already made, regardless of the age of the animal or whether it had been allowed to feed after having attacked. It appears that there is a kind of facilitation occurring somewhere between the retina and the motor centres, a process which leads both to a reduced delay in attack and a higher probability of attack in a given visual situation. Concomitantly, there is an increase in the range of patterns of stimulation that can evoke attack. Thus in the first period of life, attacking is self-facilitating and results in a widening, not a narrowing, of the adequate stimulus situation. At a later stage the animal, probably as a result of the increased development of the vertical lobe of the brain, appears to learn once again to restrict attacks, but this time not to *Mysis* only but to certain other types of prey as well. An example such as this shows how the study of imprinting in the widest sense leads not merely to questions such as that of the 'priming of the drive' but also straight into the problems of perception.

The close relationship between the study of human and animal learning is brought out very effectively by Vince (Chapter IX) in regard to the control of movements and by Broadbent (Chapter X in relation to perception. Both these authors refer to forms of the concept of inhibition and perhaps a few words of explanation of these will be helpful here. The former uses Pavlov's general term 'internal inhibition'. Pavlov (1947), in developing his theory of this subject, always assumed that inhibition is an active state which counteracts the excitatory effect of conditioning. As Spence (1951) shows, he first thought of this as dependent upon the lack of the unconditioned stimulus, although later he modified this concept and assumed the existence of an inhibitory state developing *pari passu* with the series of presentations of the conditioned stimulus, and that this 'state' or 'centre' is weaker and less stable than is the excitatory condition. Broadbent in his chapter refers to Hull's terminology in which Pavlov's *original* 'inhibitory state' is represented by reactive inhibition. This is regarded as building up, as in Pavlov's later view, by the activity of a response system, so that when it becomes linked to a new stimulus it gives rise to conditioned inhibition. Thus in general Pavlov's 'inhibitory state' is equivalent to reactive inhibition plus conditioned inhibition. Broadbent shows how important it is for some types of study of human behaviour, particularly work on the question of continuity theory, that animal experiments should in future be designed bearing in mind the results on men and that until this is done little further progress may be possible with certain types of investigation.

4. Imprinting is an example of a sensitive period in learning and few human psychologists will have any doubts as to the importance of this problem to their studies, particularly in relation to educational psychology. With humans, as with some animals, the close of a specific learning period for a simple type

of task may result not from a waning of the ability to achieve
the simpler type of performance but may be the expression of
the acquisition of more complex abilities, and of an increased
exploratory and manipulatory tendency. Again, Hinde (Chapter VII) shows that in many respects the development of the
relationship between human mother and child resembles that
occurring in birds and there is no doubt that a properly
cautious comparison between the two may be instructive to
students both of human and animal learning. Chapters VIII
and IX are also concerned with these and similar problems.

5. The initiation of cyclic or rhythmical behaviour (Chapter
VIII) shows behavioural similarities to some examples of
imprinting and this is obviously an immensely important aspect
of both animal and human activities, for recent work has shown
that some of the twenty-four-hour rhythms of renal function
and other physiological variables in man respond to experimental treatment in such a way as to suggest that they are
based on intrinsic and imposed timing mechanisms co-existing
and co-operative in the same individual at the same time in a
very similar way to those which have been investigated in
animals.

6. Finally, imprinting is primarily concerned with parent-
offspring relations, that is to say it is a phenomenon of social
animals and so is particularly relevant not only to the study
of birds but also of mammals, not excluding human beings. It
has been related to some aspects of the child-parent relation in
man which has been the object of much investigation in psychology.

Besides these various aspects of study which can be linked,
sometimes closely, sometimes loosely, with the question of
imprinting, there are one or two other points which the present
section brings out which seem worthy of mention here. Broadbent, as has been indicated, is particularly concerned with the
controversy between continuity and discontinuity theories.
These theories have many implications, but for our purposes
the question is whether, in a learning situation, reinforcement

affects differentially only those aspects of the total stimulus situation which are being responded to at the moment. If this is true, it would mean that the development of discrimination should be a discontinuous process ('discontinuity theory'). If, on the other hand, all perceptible aspects of the total stimulus situation are affected, the acquisition of discrimination should be continuous ('continuity theory'). Broadbent in his chapter puts very cogently the difficulties of a continuity hypothesis and emphasises by contrast the strength of the discontinuity view from observations with both animals and men. He suggests that neither accounts for the fact that 'not everything is learned' (as of course it should be on J. B. Watson's view). Some sort of 'filter' or discontinuity theory is admitted and this brings us back to a problem that recurs again and again throughout this book—the problem of the nature and location of the filtering mechanisms which must exist somewhere between the centre and the periphery. Such problems have already been examined in Part II by Marler and come to the fore again in the Chapter by Barlow (Chapter X). Broadbent goes further and shows very persuasively the advantages of some form of filter theory by referring to Péon's recordings from the cochlear nucleus of a cat when it sees a mouse, and gives cogent reasons for regarding the Law of Effect and Hull's formulation of reinforcement theory as inadequate to cope with the fact that not everything is learned. He shows in effect that a concept involving filtering of information giving selective perception offers a valuable alternative to reinforcement theory, and this chapter combined with others in the book serve to emphasise that such a formulation has much to offer to both those concerned with learning and those concerned with the study of instinct. So once again we are back, and looking from a new angle, at this central problem of imprinting and its relation to reward and reinforcement.

In the final chapter in this part Watson considers the same subject from yet another aspect. He points out that in practice we often cannot answer the question as to what are reinforcement conditions without making initial assumptions about the drive state. The 'drive' account of a motivational mechanism suggests that what occurs during learning is the connection of stimuli with responses. If, on the other hand, we suppose that the goal is in fact certain special stimuli—that is (to put it in 'ethological language'), the animal is seeking the perception of consummatory stimuli rather than the release of consummatory acts, we then tend to regard learning as more in the nature of association between cues. Coming by way of evidence that either not all learning is dependent upon reinforcement or that reinforcement itself does not always consist in drive reduction, he provides many new and convincing arguments for the view that exploration is not solely explicable as an expression of the primary drives. He indicates a valuable new viewpoint by suggesting that when an animal is satiated for its primary incentives, that cue is selected for the goal of the behaviour which is least associated with other cues. He concludes that the consummatory response theory which was implicit in the work of Lorenz and is more precisely expressed in that of Sheffield and of Tinbergen, cannot be a generally applicable explanation and that indeed no simple drive reduction hypothesis will meet the facts. In works such as that on the study of learning of rats fed by means of a fistula, he shows that two entirely different motivational mechanisms must be involved; namely that both taste and the reduction of hunger are separate reinforcement factors. This chapter then, like so many others in the book and particularly the others in this part, serves to emphasise, with point after point and instance after instance, the relevance of studies of animal behaviour for understanding of the processes of human learning.

VII. THE ESTABLISHMENT OF THE PARENT-OFFSPRING RELATION IN BIRDS, WITH SOME MAMMALIAN ANALOGIES

1. *Introduction*

Many of the responses normally shown by young birds to their parents can at first be elicited by objects superficially quite unlike an adult of their species. Thus the newly-hatched young of many species will follow a wide variety of moving objects, and may subsequently direct towards objects which they have followed much of the behaviour normally elicited by their parents. Begging, similarly, may be evoked by quite inappropriate objects. Under natural conditions, however, these responses come to be elicited only by the particular parent of the bird concerned—the range of effective stimuli having become restricted in the interval by learning.

Although studied by earlier writers (e.g. Spalding, 1873; Heinroth, 1910), modern interest in the establishment of the mother-offspring relationship in birds is largely due to the work of Lorenz (1935, 1937), who called the learning process involved 'Imprinting' (German 'Prägung'). This term has, however, been used in a number of different ways. Sometimes it refers only to the learning which occurs when a young bird follows its parent. Learning during following is, however, often difficult to distinguish from that which occurs during feeding or brooding by the parent, and so 'imprinting' is often used simply for the learning of the parental characteristics by young birds: it is in this sense that it is used in this essay. More recently some authors (e.g. Thorpe, 1945, 1956) have extended the use of the term to include learning with similar characteristics which occurs in other contexts.

Lorenz claimed that imprinting is a special form of learning, differing from 'associative learning' in four ways: (i) It is

175

confined to a definite and short period of the life cycle. (ii) It is irreversible.

Further, since birds imprinted on a foster parent sometimes direct their adult social behaviour towards individuals of the foster parent species, Lorenz added that: (iii) It is often completed long before the response concerned has itself become established. (iv) It involves a learning of species rather than individual characteristics.

Later evidence, however, has shown that imprinting is not unique (Fabricius, 1951; Hinde, 1955; Thorpe, 1956): its peculiar characters are, as we shall see, as much a product of the context in which it occurs as of the learning itself. Nevertheless its study raises many points of fundamental interest. Not only is it of special importance in the understanding of the perceptual side of instinct (Thorpe, 1956), but the dramatic and apparently dysgenic manner in which young birds can be made to respond to bizarre objects requires a functional explanation, and this can be obtained only by a study of other aspects of the parent-offspring relationship.

The purpose of this essay then, is to examine some aspects of the establishment of the parent-offspring relationship in order to see how far they can be related to features of behaviour in other contexts, and whether they can be understood in functional terms. The discussion is limited primarily to nidifugous[1] birds but a few comparisons with other birds and with mammals are included in order to illustrate how the peculiar characteristics of the behaviour can be related to the functional context in which it occurs.

2. *The diversity of objects to which imprinting is possible*

The young of many nidifugous birds will follow a very wide variety of objects. Thus Fabricius and Boyd (1954) found that ducklings would follow objects varying from a match-box to a

---

[1] Nidifugous species are those whose young are capable of walking or running soon after hatching, while nidicolous species are those which first have a more or less prolonged period in the nest.

walking human; and Hinde, Thorpe and Vince (1956) that young moorhens would follow a large canvas hide as well as they would a life-size model of a moorhen. In many species the only essential visual characteristics of the object seem to be certain very wide limits of size, shape and speed of movement, and reasonable proximity to the ground.

In addition to visual stimuli, following is usually elicited by auditory stimuli from the parent, such as the clucking of a broody hen. Although the auditory factors may be more effective than the visual ones (Lorenz, 1935; Fabricius, 1951), the auditory responsiveness is little more specific than the visual. Thus Collias and Joos (1953) found that repeated tapping on a table top would attract young chicks, and Ramsay's (1951) experiments suggest that the response of some ducklings to the calls of their parents is even less specific than this (see also Ramsay and Hess, 1954; Fabricius and Boyd, 1954; Nice, 1953; Collias and Collias, 1956). Imprinting to the characteristics of the parent's call occurs in the same way as that to its visual characteristics (Brückner, 1933; Ramsay, 1951; Klopfer, 1959).

Although the range of stimuli which can elicit following is wide, it is not unlimited, and in some species objects which are especially conspicuous, or which resemble the natural mother, present an optimum stimulus situation (see e.g., Ramsay and Hess, 1954; Collias and Collias, 1956; Jaynes, 1956). There is thus a stronger tendency to follow some objects than others, and imprinting therefore occurs more readily to some objects than to others.

Another example comparable with the following response of nidifugous birds is the feeding behaviour of nidicolous ones: once the eyes are open, young passerines will gape to a wide variety of objects—forceps, spatula, fingers and so on—as well as to the parent (see e.g. Holzapfel, 1939). Tinbergen and Kuenen (1939) found that the sign stimuli releasing the gaping reaction of young thrushes of about ten days of age are as follows: 'the object (the parent bird) has to move, it may have any size above about 3 mm. in diameter, and it must be above the horizontal plane passing through the nestling's eyes' (Tinbergen, 1951; see also Tinbergen, 1953). The stimuli evoking anxiety responses from young bullfinches are equally generalised (Kramer and von St. Paul, 1951) and the precocious adult behaviour of young birds (Nice, 1943) and mammals (Beach, 1948) is often directed towards functionally inappropriate objects.

Although these juvenile responses are elicited inappropriately only rarely in nature, the diversity of stimulus situations which can evoke them is surprising. The individual characters of the parent are learnt, but imprinting involves also a learning of specific ones. Why, if the young are to be equipped with responses to visual stimuli at all, are these not more precisely defined? In other contexts the stimuli eliciting bird behaviour often seem to be highly specific—for instance hybrid matings are extremely rare in nature, even in parasitic species where the parental characteristics cannot have been learnt.

In the first place, a broad range of eliciting stimuli is, of course, not confined to juvenile behaviour. A number of studies have shown that when responses are given appropriately more or less independently of previous learning, the eliciting stimuli are usually confined to a few features of the object (review by Tinbergen, 1951): since the adequate stimuli are simple, responsiveness may be lacking in specificity. Indeed, the analogy of a lock opened by a key, which has often been used for the mechanism by which the animal responds to such 'sign stimuli', has led to an exaggeration of its selectivity—it is often at first a rather inefficient lock which can be opened by a variety of keys, and only later, as a result of learning, is the range of adequate stimuli decreased. This was recognised by William James (1892) who emphasised the role of 'habit' in limiting the range of stimuli able to elicit instinctive behaviour, and Freud (1910) who placed the emphasis on maturation. Thorpe (1956) has recently discussed this matter in detail, suggesting that the diversity of appetitive behaviour can be accounted for only by successive modifications of the adequate releasing stimuli 'as a result of particular stimulus situations in which the response did or did not lead to a clear approximation to the goal situation'.

In practice, the degree of selectivity varies with the behaviour in question. In responding to potential predators it is better to be safe than sorry, and responses to harmless objects are in fact common. On the other hand, for a male passerine, pair-formation must occur only with a conspecific female, and hybrid pairings are indeed rare in nature. Again, in nest site selection, it is advantageous to use the best site available, and also to be able to make do with a poor site if a good one cannot be found: in the Great Tit, at least, appropriate mechanisms occur (Hinde, 1952). The degree of specificity in responsiveness thus seems to have been selected for in evolution, the optimum varying with the behaviour in question. In general this optimum will depend on the one hand on the consequences of inappropriate responses, and on the other hand on the consequences of the neglect of appropriate ones. If the former are disastrous, specificity will be high; if the latter (as in responses to predators) it will be low.

Now in adult birds appropriate responses (courting, copulation, threatening, etc.) are given to the display postures of other individuals no matter which way the latter are facing. This does not necessarily imply an instinctive responsiveness to all possible images which, say, a soliciting female could present. More probably an instinctive responsiveness to a female-in-a-particular-posture is coupled with a sense of size and pattern constancy developed, in part at least, through previous visual experience.

In a young bird, no learning of this type has occurred. The parent may appear in many shapes, sizes and even colours according to its posture, distance, background and the light conditions; a responsiveness confined to objects within a limited range of shapes and sizes would be too selective, for appropriate responses might be eliminated. It is of course important that this should not happen, for the young of nidifugous species would soon perish if they failed to follow their parents; and, for nidicolous young, there is always some degree of competition between the young in the nest for food which puts a premium on a quick begging response. Thus in following and begging behaviour the optimum degree of selectivity must initially be low. Presumably the rarity of inappropriate imprinting in nature is due partly to the co-operation of auditory stimuli, and partly to the rarity of objects, other than the parents, which are adequate to release the response in question.

Thus the most surprising feature of the following response—the wide range of stimuli by which it can be elicited—represents a difference only of degree from other instinctive responses, especially those of young animals. Furthermore a possible functional reason why such a condition is especially necessary in following and begging behaviour can be envisaged.

3. *Ambivalence towards the eliciting object*

Fabricius (1951) was the first to stress that objects eliciting following in young nidifugous birds may also elicit fear. His view has been confirmed for ducks by Ramsay and Hess (1954) and for moorhens and coots by Hinde, Thorpe and Vince (1956). Ducklings often also show aggressiveness towards the stimulus object. Observations on the young of nidicolous species show similarly that objects which elicit begging also evoke fear. Similar ambivalence is of course common in adult birds—for instance in agonistic behaviour, throughout pair-formation, courtship and copulation, and in many responses to predators. Its occurrence in the following response is probably accentuated by the broad range of stimuli eliciting both following and fleeing at this age. Most birds have a tendency to flee from strange objects and for young birds, accustomed to only a limited environment, most objects are strange.

The ambivalence can readily be understood on functional grounds. There is strong selection pressure for each parent to care for its own young, and in fact most nidifugous birds soon learn to recognise their own young and attack strange ones.

Reciprocally, therefore, it will be to the advantage of young nidifugous birds to flee not only from potential predators but also from strange adults. With nidicolous species, which often suffer considerable losses from nest predators, fear of strange objects near the nest is clearly likely to be adaptive. In practice, though fear responses to the parents' alarm calls, etc., may be present from before hatching, the tendency to flee from strange objects usually appears rather later than the tendency to follow. Ambivalence is thus at first absent and later any tendency to flee from the parent can become habituated while it is still weak. Of course, if this were not the case, imprinting would be impossible.

4. *The role of imprinting in decreasing the range of objects eliciting following*

As we have seen, there is a gradual restriction in the range of objects eliciting these juvenile responses. Several processes probably play a part in this:

(i) There is initially a weak tendency to respond to a wide range of conspicuous objects and sounds. Social facilitation from other members of the brood may also play a part in inducing following (Collias and Collias, 1956).

(ii) This tendency is primed by practice and conditioned to the characteristics of the particular objects which have elicited the response.

(iii) Responses elicited by auditory stimuli are conditioned to visual characteristics of the object, and vice versa.[1]

(iv) Differentiation occurs in the usual way as the object elicits a response in different environments, and other objects are seen but not responded to. (In experiments on the following response of moorhens and coots conducted under constant conditions little differentiation occurred and generalisation

---

[1] Dr P. Klopfer (1959) has shown that at least one species of hole-nesting ducks is able to learn to respond to an auditory stimulus in the absence of associated visual or motor stimuli. This ability was not shared by any of the six species of surface-nesting ducks tested.

was marked.) Further, as the object is perceived from different directions and angles, and as the senses of size, shape, etc., steadily develop, the individual peculiarities of the eliciting object will become relatively more important than its characteristics as (e.g.) a moving object.

(v) The fleeing response to the object becomes habituated during the period when the tendency to flee is low. Strange objects seen later, however, elicit fleeing rather than following. Differentiation thus becomes more marked.

Brooding, feeding and other activities normally directed by the parent towards its young are not essential for imprinting of the following response, though they are additional factors in nature. The learning which occurs during following therefore does not depend on any of the conventional rewards used in learning experiments. If the learning is to be explained in terms of a reinforcement theory, then reinforcement must lie in some aspect of keeping near the object. Thus moorhens and coots, trained under conditions such that the following response could be generalised to several objects, would continue to follow under massed practice conditions only if the object was familiar. The amount of following waned little in successive tests (1 per min.) with familiar models, but rapidly with unfamiliar ones. With models which had not been seen before but which nevertheless resembled in some respects the familiar ones, the response often waned temporarily and then recovered to the initial level as the object became familiar. The waning, which resembled that seen in experimental extinction, can be regarded as due to the absence of the consummatory stimuli presented by the familiar model which normally act as a reinforcement (see Section 6). There is evidence that the energy expended by the following bird in keeping near the familiar object may also be a critical factor (Ramsay and Hess, 1954). Thus the response of coots to a familiar but silent and stationary model soon wanes (Hinde, Thorpe and Vince, 1956). However, the following response is itself complex (Jaynes, 1956) and there is great need for further analysis of the precise factors necessary for learning to occur.

5. *The rigidity of imprinting*

Lorenz regarded imprinting as totally irreversible, describing it as having an 'absolute rigidity' which is never found in 'associative learning'. There has been some controversy about this, due partly to the varied conditions under which experiments were performed (Thorpe, 1956), but it is now clear the case for irreversibility has been exaggerated. Even in geese it is not absolute. Thus Steven (1955) found that a Lesser White-fronted Goose (*Anser erythropus*) reared until between seven and fourteen days old under natural conditions and then captured, soon became imprinted on man: other cases are cited by Thorpe (1956; see also Jaynes, 1956). In other species imprinting is even more reversible (see, e.g. Fabricius and Boyd, 1954; Hinde, Thorpe and Vince, 1956). However, once imprinting to a parent object has occurred, that object is preferred to others, and it may be extremely difficult to get a young bird reared by its own mother to accept another, even of the same species.

Such a degree of rigidity is not, however, confined to the following response. Once a bird has started to build vigorously in a nest site, it will disregard all others: in monogamous species, once pair formation is accomplished, other individuals of the opposite sex rarely elicit pairing behaviour. It was just this which James described in his 'Law of inhibition of instincts by habits'. 'A habit, once grafted on an instinctive tendency, restricts the range of the tendency itself, and keeps us from reacting on any but the habitual object, although other objects might just as well have been chosen had they been the first comers.'

6. *The sensitive period and the rapidity of imprinting*

Lorenz stressed that imprinting is rapid, and confined to a brief period in the organisms's life. Of course learning to follow or beg from a particular object can occur only during the period when these responses are part of the bird's repertoire, but the sensitive period for imprinting is much shorter than this—in partridges, for instance, only a few hours (Lorenz, 1935; see also Ramsay and Hess, 1954; Fabricius and Boyd, 1954; Jaynes, 1957). Functionally, rapid learning is of course essential for nidifugous species (Thorpe, 1956). Amongst nidicolous young, which spend some time in the nest in any case, the learning of the parental characteristics is apparently much less rapid.

The general problem of sensitive periods in learning is considered by Thorpe in the next chapter. Here we are concerned in evaluating the sensitive period as a characteristic of imprinting and defining the factors which limit it. In doing so we must distinguish between the period during which the responses in question can be elicited by new objects, and that during which a learning of the characteristics of the eliciting object occurs. *Some* learning must surely occur whenever a response is made. Indeed, the following response of coots can be generalised to new objects even when they are sixty days old, and the characteristics of the new object become learnt. It is thus not the learning capacity but the elicitation of the response by new objects which is limited in time, and the factors involved here differ between species.

We have seen that the eliciting object arouses ambivalent tendencies. With nidifugous young, under normal conditions the tendency to follow is at first stronger than that to flee, and fleeing from the parent is alleviated by habituation. Later, however, the tendency to flee becomes stronger and a strange object no longer elicits following (Hinde, 1955). Similarly, if the nidicolous young are taken from the nest when or soon after their eyes open, they beg readily to forceps and similar objects, but if taken a day or two later they are at first afraid of them, and will not beg until either the fear is habituated or they are so hungry that it is overcome. This effect, stressed by William James (1892; see also Verplanck, 1955) on the basis of Spalding's (1873) experiments with chickens, is certainly important in moorhens and coots (Hinde, Thorpe and Vince, 1956). It probably also plays a role in limiting the sensitive period of ducks (see, e.g. the records of Fabricius, 1951; Ramsay and Hess, 1954; Collias and Collias, 1956).

The relation between the appearance of fleeing and the end of the sensitive period is, however, by no means constant (Jaynes, 1957) and in some cases imprinting apparently becomes impossible even in the absence of a tendency to flee. Weidmann (1956) found this with mallard ducklings and it seems also to be true of geese. Such cases are covered by James' 'Law of transitoriness' of instincts: '. . . if, during the time of such instinct's vivacity, objects adequate to arouse it are met with, a habit of acting on them is formed, which remains when the original instinct has passed away; but . . . if no such objects are met with, then no habit will be formed; and, later on in life, when the animal meets the objects, he will altogether fail to react . . .' This is essentially similar to the explanation of the sensitive period advanced later by Fabricius (1951)—namely that the internal motivation becomes low so that the 'innate sign stimuli' alone are not able to elicit following unless 'secondary sign stimuli acquired at the imprinting co-operate.'

Weidmann (1956), however, emphasises an important difference between the behaviour of an imprinted and a non-imprinted duckling—whereas the former searches for the parent-object if it loses contact, the latter (even if tame) shows no such appetitive behaviour; it has lost (or never gained) this urge to be near a parent. A similar difference can be seen in moorhens, coots and chicks. Thus imprinting to an object involves not only an increased specificity in responsiveness and an increased tendency to respond to the object when it is presented, but also an increase in the intensity of the appetitive behaviour shown when the object disappears. It is helpful here to think in terms not only of conditioning to an object, but also of priming, perhaps comparable with that occurring in the fanning of the Three-spined Stickleback (*Gasterosteus aculeatus*) where the eggs both elicit the behaviour and 'prime the drive' (van Iersel, 1953).

7. *Modifications to consummatory stimuli*

Most nidifugous birds show a special type of appetitive behaviour, consisting of searching with repeated 'distress' calls, when they lose contact with their parents. Collias showed that distress calls are given by a newly-hatched chick allowed to emerge on a table, apparently because of loss of contact with the egg-shell and cooling. During the period immediately after hatching the frequency of the calls is diminished by warmth, contact, and also by the clucking of the parent. At this stage (five to ten minutes after hatching) a moving object has no effect on the distress calls, but later (one hour after hatching) a nearby moving object will inhibit them (Collias, 1952). Later still, when the chicks have become imprinted on a particular moving object, this will inhibit their distress calls when a strange one will not.

The imprinting which occurs when the moving object is followed thus involves modifications not only to the stimuli which elicit the following response, but also to the consummatory stimuli which 'switch off' the appetitive behaviour of the lost chick. This can be compared with the facts of secondary reinforcement: the stimulus situations which act as reinforcements in learning situations (whatever their mode of action may be) are also stimuli which produce a decreased tendency to perform the immediately preceding behaviour.

8. *Imprinting and the mother-offspring relation*

Under natural conditions, the behaviour patterns operating between parent and offspring are complex and diverse, and there are innumerable opportunities for associative conditioning to the mother during following, brooding, feeding, etc. As a result of these a personal knowledge of the parent is built up in which the eliciting stimuli for the various discrete responses are united—a 'parent-companion' in Lorenz's (1937), not von Uexkull's, sense.

The role which the following response plays in this varies

between species. If the young are cared for in the nest for several days before following the parent (e.g. moorhens and coots), they may learn to recognise their parent individually before they leave the nest. Indeed, in some nidicolous species little following occurs even in the early fledging period, the young hiding in cover while the parents forage (e.g. *Fringilla coelebs*, Marler, 1956*a*). On the other hand, if following is almost the first thing they do, the associated imprinting will contribute largely to the individual recognition of the parent (e.g. geese). No doubt such specific differences are partially responsible for the conflicting evidence about the irreversibility of imprinting during following.

In captivity, the various juvenile responses may be attached to diverse objects, so that conditioning to a parent-companion does not occur. Thus the young moorhens and coots studied at Cambridge obtained warmth from an infra-red lamp, food from forceps, and followed various models. This suggests that the 'parent-companion' is merely a consequence of the mother's presenting stimuli for the various juvenile responses, and not of any inherent mechanism in the young.

However, as we have seen, the same appetitive behaviour (distress calls, etc.) is given by chicks as a result of various types of deficit—loss of contact, cold, 'loneliness', etc. The moving object is only one of several stimulus situations consummatory for this behaviour, all of which are normally combined in the mother. Thus, although she satisfies several types of deficit, these are expressed by one type (or perhaps later several very similar types) of appetitive behaviour in the young.

Since the behaviour of parent to young and young to parent is potentially ambivalent, individual recognition of the parent-companion is important functionally. This is especially so in social species, where there are possibilities of confusion between broods and parents are often actively hostile to strange young. In territorial passerine species, on the other hand, mixing of broods may occur without aggression (Hinde, 1952). The occurrence of 'greeting ceremonies' in many species, such as geese, is functionally related to inter-family aggressiveness.

Imprinting is often said to be 'supra-individual' learning, i.e. the learning of species characteristics rather than those of an individual. This is based on two facts. First, that sexual attachment to objects similar to the parent-companion sometimes occurs in adult life. This is considered later: it is sufficient here to say that such attachments may be consequences of imprinting but are hardly characteristics. Second, the initial imprinting is to the broad characteristics of the object, so that generalisation to similar objects occurs readily. This, however, merely implies that the learning process is incomplete. In the case of nidifugous species, the important thing for the young bird is to learn the characteristics of his own parents—and especially that he should follow his own parents and not strangers.

9. *Imprinting and later social behaviour*

Birds reared by foster parents may later direct their adult social behaviour towards similar individuals, even in preference to conspecific ones. Examples of this are now known for many species. Heinroth recorded owls and ravens which could not be bred because they responded sexually to their keepers instead of to each other, and Portielje described a similar case in a bittern (Lorenz, 1935). Lorenz himself has described cases in ducks and geese, and Räber (1948) in a domestic turkey. A similar transference occurs in some nidicolous species (e.g. budgerigars and jackdaws, Lorenz, 1935).

Lorenz (e.g. 1937) has given many examples showing that the various social patterns may be directed inappropriately to different extents. Thus Grey Lag geese imprinted on man would follow him swimming or walking, but ceased to respond to him in any way whatsoever the moment they took flight in pursuit of another Grey Lag. Similarly a tame jackdaw, whose normal imprinting was prevented experimentally, used Hooded Crows (*Corvus cornix*) as flock companions, courted human beings, and showed parental behaviour towards a young jackdaw. Apparently imprinting is essential in establishing the stimuli for some social responses, while in others it is of no consequence.

An interesting example of the interweaving of imprinted and instinctively recognised stimuli is given by Räber (1948). A hand-reared turkey cock subsequently courted men, attacked women, but copulated with objects the size of a turkey hen lying motionless on the ground. Apparently the discrimination between men and women was due to the handbags and flapping clothes of the latter: men with brief cases were attacked. Räber suggests that such objects correspond to the wattles and drooping wings of a male turkey and thus turn a member of the species (recognised as a result of imprinting) into a rival. Such cases are, however, rare. More usually all the functionally related patterns are directed to the same object.

Lorenz used such examples as evidence for the peculiarity of imprinting. 'The process of acquiring the object of a reaction (e.g. the pair-forming responses—R.A.H.) is in very many cases completed long before the reaction itself becomes established.' Two points, however, must be considered here. First, there has so far been no study in which the imprinted bird was isolated from its parent-companion and similar objects between the pre-flying stage and adulthood. We thus know as yet nothing about the role of later learning in producing the unnatural adult behaviour (see also Steinbacher, 1939). Second, even if imprinting does result in abnormal sexual attachments, this is a property of the make-up of the organism, or of accompanying effects, rather than of the imprinting process itself (Hinde, 1955).

As yet nothing is known as to why objects similar to those which elicited juvenile behaviour should later come to elicit adult social behaviour.[1] Labelling the phenomenon as a characteristic of imprinting is little help in finding out: a more

[1] Verplanck (1955), having discussed the importance of fear responses to the followed object in the juvenile, goes on to suggest that 'when adult behaviours appear, proactive inhibition should lead to interference (by persisting fear and flight components of behaviour) with response to those objects that had *not* been followed prior to the appearance of the fear and flight behaviour in the animal's early days.' This theory is inadequate on two grounds: (a) Conspecific individuals evoke agonistic responses more than individuals of other species; (b) it is a negative hypothesis, and does not account for the positive response to a potential mate.

profitable line may be to look for similarities between the drive-states of adult and infant which would help to account for the transfer.

10. *Some mammalian analogies*

We have seen that many of the properties of the following-response, and of imprinting, can be understood in functional terms through a consideration of the selective forces acting on parent and offspring. Similar forces will, of course, operate on the parent-offspring relation in groups other than birds. In this section, therefore, are given some brief comparisons between mammalian, and especially human, behaviour and the features of the mother-offspring relation in birds.

In considering them it is essential to remember that: (*a*) the basic structure of the telencephalon of birds is very different from that of mammals; (*b*) the synapsid and diapsid reptiles, from which mammals and birds arose, have been distinct since at least the Permian; and (*c*) parental care has evolved independently in the two groups. Similarities in behaviour, therefore, are not very likely to indicate similarities in underlying mechanisms. Nevertheless, the development of the human mother–child relationship bears many similarities to the analogous process in birds.

Proximity to the mother-figure for much of the time is of great importance to the human child from an early age until three to five years. During this time the individual character-istics of the mother are learnt gradually—as with birds, many of the responses appropriate to the mother can at first be elicited by quite a wide range of objects. This is of course well known in the case of feeding, but even smiling can be evoked by quite crude masks in about six weeks old babies (Spitz and Wolfe, 1946; Ahrens, 1954), though later it becomes for a while more restricted to the mother. Even the following response, present in the human child as well as most other mammals (e.g. Hudson, 1892), can be elicited in young children (say twelve months old) by anyone walking away: it is, however, merely incipient to strangers, more marked to siblings, and persistent only to a parent. This lack of specificity in responsiveness is not limited to social behaviour: for example, the reflexogenous zone for sucking is at first large, shrinking during the first ten days of life (Pratt, 1954); and it is clear that 'play' objects often elicit the behaviour appropriate to 'real' ones in human children (Klein, 1932) and other mammals (Holzapfel, 1956). No doubt this lack of specificity in responsiveness is to be interpreted partially in perceptual terms.

As with birds, the behaviour of the human child towards its parent is to some extent ambivalent. This is clearly evident in behaviour at ten months, and in some circumstances earlier.[1] Actually fear of strangers begins to appear in young children at five to nine months, rather later than the age at which personal recognition of the mother develops (Bowlby, 1955, 1958). The developmental sequence of first social responses to the parent, then fear, and later aggression, is thus similar in birds, anthropoids (e.g. McCulloch and Haslerud, 1939) and humans.

Opinion has been divided as to the extent to which the establishment of the mother–child relationship depends on tangible rewards: while some maintained that the relationship is entirely dependent on the satisfaction of the primary physical needs, others believed social factors to be equally important (Bowlby, 1955). Evidence is now accumulating in favour of the second view: the factors involved are certainly complex but include physical contact, caressing, etc., which is sometimes perhaps facilitated by the reflex behaviour of the child (Stirnimann, 1941; Wolfle (1949); Gesell, 1954); smiling, which, although at first perhaps non-social, is important after the first month (Bühler, 1933; Spitz and Wolfe, 1946; Ahrens, 1954); sucking (as

[1] In recognising the importance of ambivalence in mother-child relations there is no need to subscribe to the complex view of the emotional life of the neonate and infant, held by some psychoanalysts (e.g. Klein, 1932, 1952), which seems so improbable in view of the undifferentiated and undirected nature of emotional behaviour in young children (e.g. Sherman, 1927; Bridges, 1932) and their poor powers of perception (e.g. Piaget, 1929).

witness the use of dummies for comforters); echoing maternal words (Stengel, 1947) and so on. The apparent importance of these social patterns is reminiscent of the greeting ceremonies of geese and other species, and of the way in which young nidifugous birds will learn to follow a particular model without receiving any of the conventional rewards. Learning during following is, however, itself of minor importance in the human species, since the child is unable to follow until long after individual recognition of the mother is possible.

The human infant's crying is functionally equivalent to the distress calls of a nidifugous bird. During the first month there seems to be no constant difference between the cries given for 'hunger', 'attention', etc. (Lynip, 1951), although later they become differentiated. The presence of the mother provides one of the sets of consummatory stimuli for this appetitive behaviour.

Just as the establishment of the mother-offspring relation in nidifugous birds is limited to a certain sensitive period, so in the human case there is some evidence that children who have been unable to establish an adequate relationship in early life are less able to do so later (references in Bowlby, 1952a). The gradual maturation of a fear of strangers is no doubt a hindrance to the formation of a relationship after the normal period, but there is also evidence that the various behaviour patterns which contribute to the relationship each have a sensitive period of their own which may be limited in the manner described in James' 'Law of the transitoriness of instincts' referred to above (see in this context Sears and Wise, 1950; also Scott *et al.*, 1951), and that priming plays an essential part.

Thus in many respects the development of the relation between human mother and child resembles that occurring in birds, while some of the differences appear to be related to other factors in the life history, such as the age of locomotion. It remains to stress again that the similarities are the result of similar selective forces, but not necessarily of similar mechanisms.

11. *Conclusion*

The establishment of the mother-offspring relation in birds, then, does not involve a special form of learning. Since its characteristics are peculiar only in degree—many of them, such as the generalisation of responsiveness, the sensitive period and so on, being shared by other forms of instinctive behaviour—the study of imprinting has general implications, and provides a challenge to any over-simplified view of instinct. The limitations of the conception of a rather rigid 'innate releasing mechanism' become apparent once it is recognised that 'imprinting' as found in the offspring-parent relationship of nidifugous birds is not without parallels elsewhere. Many of the characteristics of imprinting in birds can be understood in functional terms, and similar learning processes are found in functionally similar contexts in other groups.

Note added in proof

Since the above was written a number of papers on the parent-offspring relation in birds have appeared. A particularly important contribution has been made by Hess (1959), who also cites other studies. Harlow (1959) has made a detailed study of affectional responses in young rhesus monkeys, and has shown that conventional rewards are of little importance.

VIII. SENSITIVE PERIODS IN THE LEARNING OF ANIMALS AND MEN: A STUDY OF IMPRINTING WITH SPECIAL REFERENCE TO THE INDUCTION OF CYCLIC BEHAVIOUR

The object of this chapter is to consider what may at first sight appear to be two very different problems, both of which are of major interest to ethology. I am taking them together because I want to suggest that they have some features in common and that perhaps a study of these common features may in due course throw some further light on both. These two problems are Imprinting and Endogenous Periodicities in Behaviour.

It has long been a commonplace belief that most human beings can learn most things better when they are young than when they are old. But beyond the doubtful and very general impression that mature young brains are quicker at learning and more retentive than are old ones—a conclusion which the recent studies of Welford have shown to be at least doubtful—there has not until recently been much evidence that individual human beings pass through particular periods during which their minds are particularly receptive for particular kinds of impression. The primary questions are two: *Firstly*, is it true that learning-ability, whether specific or general, is restricted to a certain period or periods of the life? *Secondly*, if so, what factors, peripheral or central, sensory or motivational, delimit such periods?

Richardson (1933), as the result of a mass study of a school population by means of a special intelligence test technique, concluded that there was a steady increase in some basic factor underlying intelligence between the ages of 6 and 14, and that the absolute variability of intelligence increases with age during that period. But although this increase in variability with age might be explained by assuming a change in specific factors underlying intelligence, the results do not seem to provide any definite evidence that this is so. Nor, at the other end of the life-span, is there any convincing evidence for pre-senile waning of specific learning abilities.

It is, however, obvious that interest, and therefore attention and consequently the motivation for learning, although enormously influenced in detail by social custom and tradition, are governed in a very general way by age and sex. Some such broad assumption is indeed the essence of educational methods such as that of Montessori. A small boy may display astonishing facility in remembering motor index marks or engine numbers, aircraft silhouettes or the characteristics of British birds; a small girl may be correspondingly entranced by dolls' dresses, hairstyles or the points of show ponies; whereas their parents, and themselves when adult, may take note of none of these things. Moreover, at one age the acquisition of food may be the dominant incentive, at another the establishment of constant and satisfying relations with the parent and family, and at yet another the acquisition of information about and control of the inanimate or non-human environment. Later still the heterosexual interest becomes more fully overt and so a new form of social behaviour is manifest. But none of these indicates the existence of what is actually meant by the term sensitive period. In order to obtain evidence for the existence of sensitive periods in the development of learning ability we want to know whether one particular age group is better equipped mentally for learning to recognise, for instance, the complexities of visual pattern, another for learning to co-ordinate muscular movements, yet others for mastering tonal association and analysis, the symbolism of language, or the more abstract symbolism and the deductive reasoning involved in mathematics—and so on. In this chapter we will consider the evidence for the existence, in humans and animals, of restriction of learning abilities of this nature, and the bases for any such restrictions.

That maturation of brain mechanisms, sensory mechanisms and effector mechanisms of the child must be determining the onset of the ability to learn both mental and manual tasks seems well established. The slow and irregular rate of maturation of the inhibitory aspect of behaviour, especially 'Internal Inhibition', which gives rise to regulation and control of newly-developed responses is particularly important and is discussed fully in Chapter IX of the present book. Luria (1932) suggested that the child's cortex, being 'functionally weak', is unable to repress the 'large masses of excitation' provided by external stimuli. Work such as that of Piaget, Valentine and Gesell, to mention only three examples, is also very relevant here. For present purposes evidence that maturation may also result in the *termination* of the ability to learn a particular kind of task, may—in other words—bring a particular sensitive period to a close as it prepares the organism for some new type of achievement in learning, is much more significant and much more difficult to obtain.

In human beings the classical conditioned reflex technique has given some information. A considerable number of straightforward conditioning experiments have been carried out with very young infants, much of the work being summarised by Kantrow (1937). In general, it may be said that although even some pre-natal conditioning undoubtedly occurs, new-born infants are conditioned with difficulty and conditioned responses once established are often unstable and tend to wane unpredictably. As the infant grows older, greater ability is shown and the retention of conditioned responses developed in early infancy has been observed over periods ranging up to 7 months (Munn, 1954). With older children there have been a great many studies, many of them carried out in Russia, usefully summarised by Razran (1933). Marinesco and Kreindler (1933) found that with children between the ages of 15 and 30 months many of the characteristic phenomena established by the classical animal conditioning experiments could be paralleled in the human being.

Goodwin, Long and Welch (1945), again using the classical conditioned reflex technique—in this case pairing the withdrawal of hand or foot to an electric shock, with a light or sound stimulus—showed a growth in the powers of generalisation as between groups of memorised words during the age range 7 to 10 years. However, too many imponderable factors must have played their part in the production of these results[1] for them to be regarded as definite evidence for the onset of a particular sensitive period for the acquisition of a language, although the study of deafness (see below) gives much more definite information.

Mateer (1918), using fifty normal and fourteen sub-normal children between the ages of 12 and 90 months, obtained evidence indicating that within the age-range 12–48 months susceptibility to the conditioning process increased with age. In children of 7–19 years, however, a different result was obtained by Osipova (1926) in a series of experiments in which withdrawal reactions were conditioned. She found that the number of trials required to secure a given degree of conditioning increased rather than diminished with age.

The classical conditioned reflex technique is, of course, extremely artificial and it might be expected that more significant and meaningful results would be obtained by the use of 'instrumental conditioning' or trial-and-error learning.

In the study of Dernowa-Yarmolenko (1933), in which a thousand children between the ages of 8 and 19 years were investigated, the subject was told to lift his hand when the experimenter did so. The experimenter then tapped on the table with a pencil and two seconds later raised his left hand. After ten paired presentations of tapping and hand-raising had thus been made, it was noticed that the children began to raise the hand in an anticipatory fashion after a tap but before the experimenter had raised his hand. Ninety per cent of the 8-year-olds showed this response. With increasing age the percentage responding gradually fell until the late teenagers were responding very little, and with them it was found that many more than ten repetitions of the paired presentation were

[1] See also the general criticisms by Vince in Chapter IX of the present work.

required to produce a similar level in response. Razran (1935) has suggested that laboratory experiments of this type may become more difficult as the individual ages, not because of decrease of susceptibility but rather because the subject becomes more self-conscious and has a greater mastery of verbal associations, and this makes him less naïve and less willing to play the passive part required of the subject in such experiments. It thus seems likely that the greater the development and co-ordination of responses to the outer world, the less likely is a subject to show a high performance in simple conditioned response tests.

Hebb (1942) has established the curiously suggestive fact that early brain injury may lead to specific defects in particular specific learning capacity at some precise subsequent stage in a child's life. Hebb's examples of this concern speech defects. He found that destruction of tissue in the infant's cortex outside the speech areas will prevent the development of verbal abilities, even though the same destruction may not greatly affect these abilities once development has occurred. Bartlett (1932) and Russell Davis (1957) have each described the occurrence of functional word-deafness developing in later life as a result of trauma at about the age of 2 years—as if the shock had occurred at some critical period in the development of the mechanism conferring the ability to organise the kind of auditory impressions necessary for learning human language—although the hearing organs themselves are unimpaired.

It seems from the work of Fry and Whetnall (1954) and others, summarised by Russell Davis (1957), that there is a period of special 'readiness' to learn to listen between 9 and 15 months, and of special readiness to learn to speak between 12 and 14 months. The learning of speech by imitation is at its height from 12–18 months; just as there is (according to many specialists in education) a period of readiness to learn to read at 6–7 years. It also appears that children learn particularly easily to control urination at about 24 months old. 'If the family environment is disturbed during these periods the child appears subsequently to be resistant to learning. Once he has "missed the bus" he does not usually catch up' (Russell Davis, *in litt.*).

More subjective and seemingly less scientific types of observation do seem to offer some evidence for the kind of process we are discussing. Thus Bowlby (1952*b*) considers there is a very strong case for believing that in children the conceptualisation on which adequate social response and, even more interestingly, conceptualisation involving time, is dependent on appropriate reactions with a mother or mother-substitute during the first five years of life. Bowlby believes that a separation of a child from the mother figure for six months or more during the first five years of life stands foremost among the causes of later delinquency. Children thus separated tend to develop into affectionless characters, unable properly to love or feel guilty, apparently having no conscience, and in particular without a proper concept of time. This lack of time concept is a striking feature and seems to result in an inability to benefit properly from past experience and to be motivated by future goals. It is as if there is a particular period in the early life at which these concepts can be acquired, and if they are not then acquired no later experience can fully compensate for the deficiency.

It has been assumed by more than one group of workers that reminiscence,[1] like eidetic imagery, is correlated with age and that both are more prevalent in children than in adults. There has been much work on this, work admirably summarised by Munn (1954), but in no case does the prevalence of reminiscence amongst children appear statistically greater than amongst adults. The accounts of eidetic imagery are at first sight more convincing but even harder to evaluate, and it seems that little can safely be based upon them.

We can, then, sum up the situation in children by saying that the time of onset of specific learning abilities is undoubtedly closely related to maturation of the nervous system, sense

---

[1] The phenomenon of improvement in memory for a partially learned task shown during the period commencing some hours after the last trial, as compared with the period *immediately* after it.

organs and effectors. The close of the specific learning period for a simple type of task may result not from a waning of the ability to acquire the simpler performance but because acquisition of more complex abilities, perceptions and skills, and the tendency actively to experiment with and explore the environment, render the subject less willing to restrict his attention to the simpler situation and to play the more passive roles. But there are these tantalising but as yet rather isolated and scattered observations which suggest that besides such effects there do exist specific brain mechanisms ready to be activated during and only during a particular period of the life span of the child and that if they are not properly activated at the right time subsequent activation is difficult or impossible, resulting in permanent disabilities in later life.

To turn now to the waning of learning ability in later life. It is of course well known that in people entering on senility there is a great loss of short-term memory, but the senile period apart, the evidence is most doubtful. Gilbert (1941) showed there was very little difference between people aged 20 and 60 as far as the immediate memory span for numbers was concerned. It is true that Kirchner (see Welford, 1956) and Kay (1953) found that although older subjects were almost as successful as younger ones when required to press that one of twelve morse keys which was underneath the lighted bulb in a row of twelve as the lights came on one at a time in irregular order at two-second intervals, they were far less successful when the task was changed so as to involve remembering what had been the situation four seconds before (i.e. two moves back). Welford considers, however, that the difference between Kirchner's experiments and the immediate-memory-span tests, such as those of Gilbert, lies in the fact that 'in the latter all information is taken in before any responding action is begun ... and that some process involved in responding exerts a serious interfering effect upon the information stored in short term memory'. The hypothesis that this interference increases with age could account for the relative differences of older people as the amount to be learnt increases. 'The difficulty seems to lie in the subject holding enough information in short-term memory to obtain a conceptual framework in which the necessary processes can be carried out. There are also physiological factors to be borne in mind, particularly the lowering of signal to noise ratio in the sense organs and perhaps in the whole nervous system in old age' (see Crossman and Szafran, 1956, and Gregory, 1956). It may be that because of failing sense organs it is more difficult for older than for younger people to obtain, within a given time, the necessary information upon which conceptualisation can be based. And so we seem forced to the conclusion that there is extremely little satisfactory evidence for pre-senile deterioration of the brain mechanisms mediating any of the specific learning abilities in adult human beings.

Animals, like men, have to achieve three kinds of primary adjustment. Firstly, they have to recognise periodicities in time; this conferring the possibility of acquiring new rhythms of action. This also facilitates the co-ordination of simple actions into more complex sequences of behaviour and the recall of temporal patterns of auditory and other types of stimulation which, in their turn, provide the basis for the recognition of sound signals as releasers and as 'language'. Secondly, they must co-ordinate proprioceptive and re-afferent sensations with one another and with muscle control systems. On this depends much of orientation and the acquisition of motor skill. Thirdly, they must co-ordinate their perception of simultaneous rather than temporal patterns of exteroceptive or ex-afferent stimulation whereby they come to recognise 'form' in all its innumerable manifestations—including discrimination within the visual, tonal and chemical modalities.

All of these faculties, of course, overlap and intergrade. They may be partly and some of them perhaps wholly controlled by innate proclivities, faculties or instincts. Innumerable examples in all these classes are also acquired, developed and adjusted by what is broadly called learning.

It is the general conclusion of students of animal learning that younger individuals are better learners than are old ones—at least in tests where, as in maze learning, superabundant activity is an advantage. Older animals may perform better in tasks requiring directed activity (see Chapter IX of this book). Work on the rat, summarised by Munn (1950), suggests that learning ability for experimental problems is at its greatest between one and three months. Mason, Blazek, and Harlow (1956) have shown, as a result of discrimination, delayed-response and patterned-string tests with infant rhesus monkeys, that the first type of problem is readily mastered at 150 days of age. But the formation of learning sets (inter-problem learning) did not appear until later, and even then with a rate of improvement far below that of adult rhesus monkeys.

Imprinting as such has already been described and discussed in the previous chapter and in the introduction to this section of the book. It is of particular relevance to us here in that it appears to supply the best examples among animals of restricted periods of learning. It is particularly evident in life histories such that the animal has to learn one or two particularly urgent and important recognitions or responses very quickly. And so we can see the adaptive reason for the fact that the learning which we call imprinting is often restricted to a brief critical period and, within more or less wide limits, to a particular kind of pattern as object.

It is convenient to consider imprinting under three headings:

(*a*) Imprinted recognition of and response to specific patterns of visual stimulation;

(*b*) Imprinted recognition of and response to specific frequency patterns of auditory stimulation (fine periodicities or tonal patterns); and

(*c*) Imprinted recognition of and response to specific gross periodicities (rhythms) of stimulation.

In each case we shall pay particular attention to the extent to which the learning process is restricted to a particular period in the life cycle. We shall also consider any evidence that may be available to show what factors and types of factor—whether internal or external, excitatory or inhibitory—influence and control the onset and the termination of such sensitive periods.

(a) *Imprinted recognition of and response to specific patterns of visual stimulation*

Studies on the imprinted recognition of and response to specific patterns of visual stimulation are concerned very largely with birds, since birds are of course primarily visual animals and the following response of young birds has provided the 'classical' situation for the investigation of imprinting. With ducklings, Ramsay and Hess (1954) found that the critical age for imprinting was 13–16 hours; approximately half the 19 ducklings imprinted in this age group were completely imprinted to the 'right' object in all tests. Only three of the remaining ducklings gave perfect scores. Fear responses did not appear until 24 hours. Beyond this age only one showed any imprinting, and beyond 28 hours no imprinting at all could be detected, although four older ducklings (28 and 38 hours) were partially imprinted by association with well-imprinted ducklings during the imprinting-runs. Figure 10 shows the percentage of perfect scores for mallards imprinted in various age groups, and Figure 11 the percentage of positive responses made by ducklings and chicks in a test series.

Hinde, Thorpe and Vince (1956), working on the coot (*Fulica atra*) and moorhen (*Gallinula chloropus*), showed that it is not the learning capacity itself but the elicitation of the following-response which is restricted in time. It was further found that following one model does not prevent the later following of other models. The matter has been followed up by Boyd (1956) using the mallard (*Anas platyrhynchos*). Some of his birds which had been trained on duck models were afterwards found to have a preference for man, though they had never previously had the opportunity to follow man. This result could conceivably have been due to conditioning in the food-box when being hand-fed, but such a possibility was eliminated by Hinde, Thorpe and

Vince who found that if a moorhen does show a following tendency at all, it will subsequently follow a model which is as different from that on which it has been trained as a yellow football bladder is from a black wooden model of a moorhen, or a movable canvas-covered 'hide' 6 ft. × 2 ft. × 2 ft. is from a man. Thus it is hard to find evidence for the existence of a sensitive period for imprinting of the following response on a *particular* model in these species under the special conditions of the experiment. At the onset of the imprinting period the following-response is stronger than the tendency, which is also

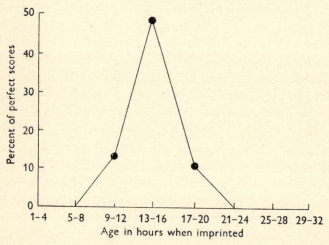

Fig. 10.   Percentage of 'perfect' scores for Mallards imprinted in various age groups. After Ramsay & Hess (1954).

present, to flee from strange objects. The result is that, in the course of the following experiments, the young bird gets used to a number of strange objects and so its fleeing tendency is weakened by habituation. Later, the fleeing drive becomes stronger and more difficult to habituate and consequently, with age, a strange object becomes less likely to elicit following. This is probably the basis of the conclusion of William James (1890) who, considering the experiments of Spalding, said 'when objects of a certain class elicit from an animal a certain sort of reaction it often happens that the animal becomes partial to the first specimen of the class on which it has reacted and will not afterwards react to any other specimen'. Hinde, Thorpe and Vince found that in the coot practically all the waning of the following-response can be attributed to the growth of the fleeing drive. That this, however, is not the full story in mallards

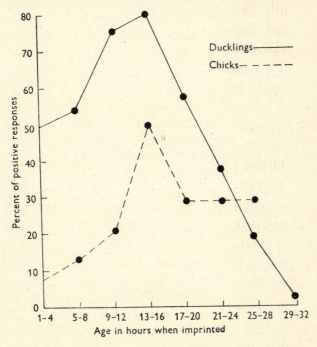

Fig. 11. Percentage of positive responses made by ducklings and chicks in test series. After Ramsay and Hess (1954)

is shown by the work of Weidmann (1956) who found that the growing escape tendency cannot, in this species, be the only factor in the cessation of the following period. Weidmann found that following stops in older non-imprinted ducklings even in the absence of any signs of fear at acoustic stimulation. Another proof for the existence here of some other factor than a competing fleeing drive, delimiting a sensitive period for imprinting, lies in his observation that non-imprinted ducklings which subsequently have become tame, only follow when they are hungry, and, moreover, show no sign of missing the parent or fellow ducklings when they are left alone.

While it is evident that the following-response is in some way self-reinforcing, in other words that no classical reinforcement theory can account for the learning which goes on during the following process (indeed it would not be easy to find a more striking example of non-reward learning), there is, nevertheless, ample evidence from studies on the commencement of the following response that the reinforcing effect of the whole situation is very complex. Thus Collias (1952), working with young chicks, found that the first 'cheep' calls are the result of cooling and loss of contact with the egg-shell. During the immediate post-hatching period (i.e. less than one hour) the frequency of calls, while diminished by contact, warmth and parental clucking, is in no way influenced by sight of a moving object. When the chick is more than an hour old a nearby moving object will inhibit these calls and much later, when the chick becomes 'imprinted', only a familiar moving object will inhibit the 'cheep'; a strange object will not. More recently Jaynes (1956), also studying the chick, described approaching- and following-responses to cardboard models which move irregularly about a ten-foot alley-way. He shows that the 'response' is not merely following but is a matrix of responses, including attention, vocalisation, approaching and following; which appear in that order. He finds that the reaction manifests itself suddenly within the first few minutes of exposure and develops progressively in the first half-hour—presumably as a result of the gradual combination and integration of the various elements involved in following. Further gradual improvement in performance is manifested over the first four days and is probably in part due to a general process of maturation. Thus although the response starts quickly, 'practice' certainly plays an important role and, as with the examples above mentioned, a decrement is shown when a strange object is introduced in place of the object on which the birds were trained. Hess (1955) showed that imprinting strength increases with the distance travelled by the animal in following the imprinting model during the imprinting process. His results indicate that it was the distance travelled rather than the time taken in the travelling which was the effective factor. Moreover, it would seem that there are two separate aspects of early experience of primary importance: Firstly, of course, there must be an initial tendency to orient and move towards an object, and, following on this, we have the second aspect, which is the amount of effort expended in the attempt to reach it.

It has been shown that, with mallards, the onset of the imprinting period may be delayed by denying the animal all opportunity to follow but that, if the first experience is delayed too long, i.e. more than five days or so, a fear of the object becomes dominant. That in nature there must, with some species, be a considerable delay in the onset of the process is shown by the fact that, where a large brood of ducklings hatches over a period of two or three days, no experience in following will be obtained by the oldest duckling until the youngest is ready to follow. This is particularly clear in the Shelduck (*Tadorna tadorna*) which breeds in rabbit burrows and in which the young can have no visual experience at all until they all sally forth on their first excursion behind the parent.

To sum up this work on birds, we can say tentatively that the imprinting period is initiated primarily by the maturation of an internally motivated, appetitive behaviour system which is in readiness, at the time of hatching or very soon after, to express itself in the following response. The time during which this internally controlled tendency can find its *first* expression in action is limited to a matter of a few hours or, at most, days, by internal factors; but during this time the response is ready to appear as soon as the circumstances permit it to do so. The termination of the ability to make a first response is, as we have seen, very largely influenced by the development of competing fleeing responses. On the other hand, of course, once the following behaviour has had a chance to establish itself, it will continue towards those models which have become associated with the achievement of the consummatory situation resulting from successful following and may be yet further generalised to others. Nevertheless, the conclusion that an internal change directly reduces the appetitive following drive with age itself is very strong.

With mammals, following and imprinting, as a result of visual stimulation, are much less familiar than with birds but enough examples among ungulates have been put on record (see Thorpe, 1956, p. 397) to convince us that there are genuine examples in this group comparable with those found in birds. It seems likely that the importance of odour for individual and species recognition in mammals, coupled with the ability of the animal to perceive its own body odour, may result in a 'self-conditioning' and so militate against abnormal 'fixations' to alien species. This would explain its apparent rarity in mammals as compared with birds. The only mammal in which the existence of a critical learning period has been at all carefully investigated is the domestic dog. Scott, Frederickson and Fuller (1951) distinguished five distinct and natural periods in the development of the dog which can be separated on the basis of the appearance of patterns of social behaviour and social relationships. They are (1) neonatal, (2) transitional, (3) socialisation, (4) juvenile and (5) adult. Theoretically, one would expect there to be important critical learning phases at the onset of periods (1), (3) and (5). In period (1) there is practically no capacity for learning, but periods (3) and (5) do show something like the expected learning phase. Thus, at approximately three weeks of age, which may be regarded as near the beginning of period (3), Fuller, Easler and Banks (1950) found that there is a marked change in the reaction of puppies to conditioning experiences, the change occurring almost overnight: from being very difficult or impossible to condition, they change suddenly to become reasonably good conditioning subjects. That there is a change at the onset of period (5) (not in learning *per se*, but involving the maturation of a particular response pattern) is suggested by the observation of Tinbergen (1942) on the eskimo dogs of East Greenland. Before sexual maturity, the young dogs do not defend territories and fail to learn the territories of others, with the result that they are continually attacked as they wander through the settlement. When they have become sexually mature they not only begin to defend territories themselves but at the same time they quickly learn the territories of others, and within a week their learning of such boundaries is almost complete.

Scott (1945) has pointed out that lambs isolated from their mothers for ten days and then returned to the flock neither followed the flock nor became strongly associated with it, even after ten years of experience. But here there is a difficulty in evaluating the results, since one must allow for the fact that a ewe will reject any lamb which is not her own by violent butting, so that the first experience of the isolated lamb returned to the flock is a rejection, which is probably reinforced many times. One cannot, therefore, really compare such sheep with dogs, which rarely reject their own or any other puppies during the nursing period and in which it is very easy to secure adoption.

(b) *The imprinted recognition of and response to a specific frequency pattern of auditory stimulation*

Here again birds have provided the best experimental evidence. The general conclusion of recent work is that call notes are primarily inborn and although, as we have seen, the call may provide an important cue for orientation for following in both chicks and ducklings, it is of course in the acquisition of full song that learning plays such an important part in birds, and here we do find an imitative approximation of the song of the mature bird to that of those individuals of the adult of the species to which it is exposed during the first breeding season. Most evidence for this comes from the chaffinch (*Fringilla coelebs*) and will be found in Thorpe (1956, 1958*a* and 1958*b*) and Hinde (1958*a*). A chaffinch which has been completely isolated from hearing the song of its species from the age of three or four days onwards is able to produce only a very restricted song, lacking almost all the fine details so characteristic of the normal song of the species. The song of the isolated birds is, in fact, a chaffinch song only in so far as it consists of a burst of notes of about the right number, about the right total duration and approximately the right pitch. If, as a fledgling of two or three weeks old or in early juvenile life, but before it is itself able to produce any kind of full song, the bird is allowed to hear the song of its species, it nevertheless shows in the following spring that it has, as a result of this exposure, learnt to divide its song into the three characteristic phases and to add the terminal flourish which is such a feature of the vocal performance of this bird. So evidently a good deal of song learning can take place in the early life. In nature, however, the fine detail is not acquired until the bird has been singing its subsong (a somewhat irregular and indefinite vocal performance with no communicatory function, particularly characteristic of the early spring performance of first-year birds) for some weeks and has then commenced its first effort to establish a territory. This it does by singing in competition with neighbouring birds, most of them experienced adults already in possession of a territory, and it is during this period that the chaffinch very quickly learns the fine details of perhaps two, three or more individual songs proceeding from its neighbours in adjacent territories. These two, three or more songs are thus learnt rapidly in the course of at most a few weeks' singing experience as a result of the competitive attempts to establish a territory. There is, of course, no more reward, in the classical sense, in this situation than in the experiments on the following-response which we have been discussing above; it is unrewarded learning; and by at latest the thirteenth month of life in the normal bird the learning period is over. The chaffinch has by that time acquired its song or songs—if an isolated bird, it will have only a simple inborn song; if a wild bird which has successfuly established itself in an area well populated with chaffinches it will have anything from one to four or even more elaborate songs. But whatever the case may be, no further song learning is possible and its song or songs at the end of its first year of life remain its song or songs for the rest of its life, no matter how varied or extensive its subsequent experience of bird song may be. Thus here we have a critical learning period, very indefinite at onset but culminating in a short period of intense experience and rapid learning brought suddenly to a close; presumably by internal factors acting directly on the song learning ability. For there is no evidence whatever that any conflicting drive is responsible in this case for terminating the learning period, nor that the song-learning 'capacity' of the bird has been satiated. Thus the experiments on the song learning of the chaffinch serve to demonstrate the existence of an inborn ability to produce songs of the right duration (approximately 2½ secs.). There must probably also be an inborn recognition of the tonal quality of the chaffinch's voice as providing an appropriate model for the imitative learning of the song and also almost certainly an inborn tendency to sing the song at intervals of from 10 to 20 seconds. Beyond this it seems as if in the chaffinch almost all is learnt; although the readiness with which the bird learns, as the result of experience, to divide its song into three more or less well-defined sections, and still further the readiness with which it learns a simple flourish as appropriate to the termination of a song, suggests that there may be a very imperfectly inherited tendency to respond to and perform these features of the normal song—a tendency so slightly developed that it can never issue in action without the stimulus provided by the singing of another member of the species. The limitations of this blueprint of the song are shown by experiments in which previously isolated inexperienced birds are kept together during the song-learning period, but still out of earshot of experienced birds. When this is done these 'Kaspar Hauser' birds will stimulate one another to sing songs of increasing elaboration, but the fine details of the song thus acquired may show little or no resemblance to the fine details of the normal performance of the species. But the songs of such isolated groups of naïve birds do show a slight tendency to be broken up into two or three phases, although there is no evidence of any ability to produce a terminal flourish.

It is the rhythmic element in such song learning which is perhaps the most significant point for our present purposes, and this leads us naturally to the third type of example to be considered, namely the imprinted recognition of and response to specific rhythms or periodicities of stimulation.

(c) *The imprinted recognition of and response to specific rhythms or periodicities of stimulation*

There are various observations and simple experiments described in the older literature which suggest that the diurnal periodicities of behaviour and physiological activity, so characteristic of human beings, are in some degree at least independent of the immediate environmental influences. Such periodicities include the diurnal temperature rhythms, the diurnal rhythms of visceral activity as well as diurnal variations in activity of the voluntary muscular system and of what may loosely be called 'performance'.

These conclusions have recently received confirmation from the work of Mills *et al.* (1952 and 1954) and Lewis *et al.* (1956) who find that the human 24-hour cycle of renal output of Na, K and Cl and of water persists independently of external stimulation. Disruption of the cycle by administration of alkaline potassium salts fails to produce any lasting derangement—the old 24-hour rhythm soon returns, and this fact suggests that some extra-renal influence must be responsible. When human subjects were exposed to prolonged periods of life on a 22-hour day only one individual out of eight showed any progressive improvement in adaptation to the 22-hour routine. With the remainder, renal function lagged behind as if in an attempt to maintain an 'inherent' 24-hour rhythm. One subject showed an unimpaired rhythm of electrolyte excretion even after 5 weeks' treatment. These experiments give interesting evidence for both intrinsic and imposed rhythms co-existing in the same subject at the same time, and Lewis, Lobban and Shaw (1956) make the tentative suggestion that an intrinsic cyclic influence in the hypothalamus may be responsible. It is of course very probable that these cycles in humans, however persistent, may have been conditioned by lifelong experience of the 24-hour day. But that this may not be all is suggested by the fact that a 24-hour cycle of mitotic rate is found in new-born mice. If this was conditioned by previous experience, it was experience *in utero*.

Interest in the subject has recently been renewed as a result of the discovery, in the last decade, of the exact time sense which seems to be necessary as a basis for the elaborate orientation mechanism found especially in bees and in migrating and homing birds, and to account for the tidal periodicities of many organisms dwelling in coastal waters. As will be seen, these results in general suggested, as did those of Brouwer (1928) for human beings, that the rhythmic behaviour of organisms is a combination of an internally driven and co-ordinated periodicity in some way keyed into or set by environmental factors which, although sometimes obvious enough, are more often very obscure.

Some of the more important studies of this subject up to 1954 have been summarised by Thorpe (1956). In the last three years, however, the studies of F. A. Brown, Jnr., and his co-workers, and of Pittendrigh (1954) have reached a point at which the whole subject emerges as one of exceptional interest and consequently it seems appropriate to reconsider the matter in a more thorough manner than was possible before.

One of the earliest records of persistent diurnal rhythms in animals was that of the regular retinal pigment migration in the eyes of the Silver 'Y' moth, *Plusia gamma* (Kiesel, 1894), and a number of other similar examples among insects and Crustacea kept and even raised from the larval stage in constant darkness in the laboratory may be found scattered in the literature of the next fifty years (see Stein-Beling, 1935, and Brown and Webb, 1948). Welsh (1941) maintained the lobster *Cambarus* in both constant darkness and constant temperature. When the animals were divided into two groups, one kept at 70° F. and the other at 21° F., no difference in the rhythms was detected, except that their persistence was rather shorter at the higher temperature. Workers on this subject naturally looked in the first place for the external factors amongst those to which the animals were already known to be readily responsive (e.g. light, temperature and humidity), extending their investigations to less likely stimuli as these obvious ones proved inadequate. The endogenous component demonstrated by such experiments was in due course supposed to involve some endocrine mechanism (e.g. the eyestalk hormone for Crustacean pigment migration) but the basis of the rhythm itself remained largely unexplored. The striking independence of temperature in the rhythm of *Cambarus*, disclosed by the work of Welsh mentioned above, led Brown and Webb to reinvestigate the subject in the Fiddler Crabs of the genus *Uca*. They concluded that at least in these animals the rhythm was in some degree sensitive to temperature changes and could, therefore, be presumed to be of metabolic origin. Following on this, Brown and Webb (1949) concluded that there must be in *Uca* two centres of rhythmicity, each one capable of having its rhythm altered independently of the other—the centre displaying the greater persistence of rhythms to some extent controlling the other. This conclusion was based on the discovery that, although the diurnal rhythm of the chromatophores (the cells responsible for body colour and colour change) can be inhibited by constant illumination, there remains a basic rhythm (assumed by Welsh to be maintained in the 'brain' or the fourth optic ganglion) controlling activity of the sinus gland which displays itself in the return to normal as soon as the inhibiting stimulus of constant light has been removed. Various experiments pointed to the conclusion that when a rhythm is altered, it is altered abruptly by a single light change or by a single period of illumination during a sensitive period—thus recalling an imprinting effect. Brown, Fingerman and Hines (1954), as a result of further experiments on the same animal, in which they produced a graded series as to the amount of shift, modified the theoretical mechanism of persistent shift. These experiments confirm and extend the conclusions of earlier ones and indicate that two operative factors, (a) strength of the stimulus in the form of dark to light change, and (b) the absolute brightness of the higher illumination, are influencing the persistence of the imposed rhythm. Moreover, it appears that the brighter the light the sooner the subsequent reversal of rhythm, and that, with higher illumination, reversal occurred on the first day. Once shifted, rhythms were sometimes found to be as stable in their new phase relations as in their original normal ones.

Harker (1953) provided remarkable evidence for similar 'imprinting' of rhythms with the nymphs of the mayfly *Ecdyonurus torrentis*. In this animal there is a temperature-insensitive 24-hour rhythm of activity, comprising six cycles of four hours each, which is so firmly established that it remains unchanged even after three months in continuous light. Although naïve nymphs which never had any experience of illumination changes show no rhythm whatever, nevertheless a single experience of a light-dark alternation during one 24-hour period is sufficient to set up a rhythm to all appearance as firmly established as that of the normal animal. She later showed (1954 and 1956) that in the cockroach (*Periplaneta americana*) diurnal rhythms of activity can be relayed from one insect to another if the bloodstream is shared, suggesting again that a secretion carried in either the blood or tissues is responsible for the transmission. More recently (Eidmann, 1956) it has been shown that the Stick Insect (*Carausius morosus*) does show signs of a rhythm even when hatched from the egg in continuous darkness.

While recent advances in knowledge of the widespread incidence of sun orientation mechanisms (see Thorpe, 1956) lead us to anticipate the discovery of internal rhythmical mechanisms in animals of very different kinds, it is very astonishing to find evidence of such mechanisms being insensitive to temperature over a wide range and so precise that, as was found in *Uca pugnax* by Brown, Fingerman, Sandeen and Webb (1953), a rhythm could persist for as long as two months (Brown, 1955) in constant darkness without there being any measurable drift away from the normal phase relations with the solar day and night—the internal clock neither gaining nor losing more than a few minutes during this time—i.e. an accuracy of the order of 1/50,000! Brown, Webb, Bennett and Sandeen (1954) show that in *Uca* the endogenous rhythm is temperature-insensitive between 6 and 30° C. Cooling to 3 or 4° C. stops the 'clock' for as long as the animal is exposed to this low temperature and this has been used as a convenient method for re-setting. Pittendrigh (1954) has also shown that, with the fruit-fly *Drosophila pseudobscura*, cultures of which show a remarkably consistent and predictable emergence time, there must be an endogenous clock which is temperature-independent. On this depends an inherent 24-hour rhythm. As in *Uca*, this clock is set by a second mechanism, in this case temperature-sensitive, which is itself apparently primed by the last dawn seen.

It seems then that the priming effect of a single exposure to a light stimulus is now well attested and such results again reinforce the conclusion that the rhythm is an inherent one. The experiments of Harker (1958) have shown that in cockroaches in which the rhythm has apparently been destroyed by reversed light-darkness treatment for three months, followed by continuous darkness, a resumption of the normal initial rhythm can be achieved even after three or four months in the dark. Harker's paper (1958) consists of a very valuable summary of the literature on this perplexing subject and her review has been heavily drawn upon in writing the rest of the present chapter.

The theory that the rhythms are based on the accumulation of toxic depressants or the elaboration and exhaustion of reserve products seems effectively ruled out by a number of considerations—not least that of a lack of temperature sensitivity. Theories of a hormonal or neurophysiological nature seem more plausible and, as has been shown above, there have been many attempts to locate the mechanism of the inherent 24-hour rhythm in invertebrates within a ganglion or group of neuro-secretory cells. Quite obviously both hormones and nerve tissues are involved in certain rhythms, but the discovery of 24-hour rhythms in Protozoa, in *in vitro* cultures of animal tissues, and in isolated plant tissues (e.g. carrot, potato and *Fucus*) shows that neither of these sources of control, as usually understood, are likely to be found to be universally necessary. It seems, in fact, that there may be a basic inherited 24-hour metabolic rhythm based on some 'a-metabolic' pace-maker present in the cells of all animals and perhaps all plants, and that this 'continues unchanged at a cellular level even when it is not evident in the behaviour or in the major physiological changes of the animal' (Harker).

Although endogenous stimulation seems now to be by far the most likely explanation of the observed facts, it is difficult finally to rule out every kind of exogenous factor, and so the search for the external stimuli capable of maintaining the highly persistent rhythms which have been revealed continues. For it is still not clear whether the temperature-independent clock can be supposed ever itself to possess the astounding precision which would seem to be required to account for the more extreme examples of long-maintained cyclic behaviour, or whether it is if not initially based upon, at least continually re-set by, stimuli not at present detected. But one by one plausible external stimuli seem to have been eliminated. Brown, Bennett and Ralph (1955) seem, it is true, to have demonstrated an effect on the *Uca* chromatophore rhythm of variation in exposure to cosmic rays and the intensity of cosmic ray showers is known to exhibit a diurnal periodicity. But it still seems extremely doubtful whether the normal diurnal alternation in intensity of cosmic ray showers could be the factor normally concerned in the precise maintenance of the 24-hour cycles of persistent daily rhythm in organisms; for there is no sudden peak or trough of intensity change in the 24-hourly cosmic ray cycle which could, it seems, provide a sufficiently precise marker for the highly exact rhythm reported for *Uca* and other organisms. If cosmic rays are providing the clue, then their perception must involve some very precise mechanism for summation over a period of hours. The mere ability to distinguish increasing intensity of radiation from constant or decreasing intensity, after the manner in which, as has been shown (Corbet, 1955), dragonfly larvae can respond to changing day length, would not seem to meet the case. Moreover, Wahl (1932) showed that at least the daily rhythm of bees could be maintained in a salt mine where presumably the animals were effectively shielded from cosmic ray stimulation, although it seems unlikely that, in this case, the behaviour of any temperature-insensitive clock was really being tested. We must conclude then that the theory that sensitivity to cosmic rays provides the key to the problem seems highly implausible, though not perhaps finally excluded. The regular daily changes in barometric pressure, on which the cosmic ray fluctuations themselves depend, might seem to be more easily detectable by organisms than changes in cosmic ray intensity are likely to be. But this is probably incorrect. The daily pressure fluctuations are similar to the cosmic ray fluctuations in that they equally lack sharpness, and it may be even harder to visualise a mechanism for summation of barometric pressure than of hard radiation. Moreover, it has now been shown that constant pressure does not inhibit the rhythm of O_2 consumption in *Uca*, and evidence has been provided (Brown, Bennett, Webb and Ralph, 1956) that *Ostrea virginia* and *Venus mercenaria* show a 27-day cycle of activity presumably controlled by radiation fluctuation with the 27-day rotation period of the sun. The baffling problem posed by this and by some of the reports of two inherent lunar periodicities is outside the scope of the present chapter. It is discussed by Harker (1958). Persistent daily rhythms in oxygen consumption are, of course, well known in plants as well as in animals (Brown, Freeland and Ralph, 1955) and it is hard to see how these could be controlled by barometric pressure changes.

In conclusion we can perhaps summarise very briefly this rapidly growing field of work on the time element in imprinting by saying that the evidence for a temperature-insensitive endogenous rhythmic mechanism is now very strong, but we cannot yet say categorically that the rhythm of this clock arises in complete independence of any similar periodicity of external stimulation. It is, however, clear that such clocks can be and often are set with great consistency by the first experience of a particular stimulus alternation. These clocks, it seems, can then control a variety of mechanisms, nervous, endocrine and 'metabolic'. The rhythmical behaviour which the animals exhibit as a result is the expression of (*a*) the temperature-insensitive internal clock, (*b*) the setting mechanism, and (*c*) the temperature-sensitive metabolic rhythms or clocks which may be inherent or later imprinted on the animal. Thus we can make an extremely provisional and tentative classification of internal rhythmic mechanisms into 'hard' and 'soft'. '*Hard clocks*' are presumed, for the time being, to be of inherent periodicity. They are primarily of solar (24-hour) periodicity, though there doubtless exist hard clocks of much shorter periodicity, down to minutes or even seconds (though it is now doubtful how far such clocks may be thermo-stable). Hard clocks are temperature-insensitive and so presumably 'non-metabolic'. They can be set by a single stimulus and, once set, can continue for long periods (sometimes apparently indefinitely) without re-setting. They may have an accuracy of the order of 1/50,000. '*Soft clocks*': While many soft clocks may also be of inherent periodicity, it is very usual for their rhythms as well as their setting to be determined by experience—i.e. learned. They are temperature-sensitive, and may be of any periodicity. They are relatively inaccurate and presumably 'metabolic'.

Where two hard clocks occur together in the same animal,

they may together give a combined rhythm. Thus it has been suggested that in Fiddler Crabs there are clocks for both lunar and solar day. This is thought to result from the coincidence of the two clocks in a tidal rhythm of 15 days. Where a soft and hard clock occur together, the harder of the two will, as in *Uca*, tend to control the rhythm of the other. The existence of two such centres of rhythmicity could, of course, provide for almost any period of activity.

The term 'clock' has been employed as a convenient everyday word to cover our ignorance. Pittendrigh and Bruce (see Harker, 1958) suggest that rather than a 'clock' the concept of a 'self-sustaining oscillator' provides us with a more useful model. Such an oscillator can be thought of as having a natural period which can be initiated by any single perturbation of, say, light or temperature of the required magnitude. To use a crude analogy—the watch starts when it is shaken. Such a theory of self-sustaining oscillators has led its authors to a number of complex conclusions and subsidiary hypotheses, some of which seem to fit observed facts remarkably well. It is maintained, for instance, that work on *Euglena* has revealed rhythms such as would arise if one oscillator were entrained by another and if the period of the entraining oscillator were exactly or nearly a whole sub-multiple of that of the entrained oscillator. This recalls the 'magnet effect' of von Holst and some of Pringle's suggestions as to the fundamental mechanisms of animal learning (see Thorpe, 1956, Chapter VIII). The consideration of such details is beyond the scope of the present discussion, but the general theoretical ideas on which they are based are of profound interest to ethologists and students of animal learning since they suggest an absolutely basic physiological mechanism which might be responsible for some of the most striking examples of 'imprinting' to certain periodicities of stimulation and could perhaps be extended to apply to other more complex examples also.

This brings us back to a final resumé of the imprinting concept. Thus, in many cases of 'imprinting', the animal is keyed by the constitution (whether of its sense organs, its central nervous system, or both together) to respond to particular kinds of stimuli or the impact of particular specific perceptions in the matter of its rhythmic behaviour. The intensity of stimulation itself may be important and this, of course, may again be related to the threshold of stimulation to which the sense organs themselves are sensitive. So besides having very convincing evidence for inherited rhythmic mechanisms, we also have overwhelming evidence for selective sensitivity both to simple stimulation and to more or less complex perceptions, as in the concept of the I.R.M., the effect of which is to direct and canalise the learning ability. The period during which the animal is sensitive to changes in such stimulation may be at one extreme co-terminous with its life and at the other restricted to an extremely short time span, usually early in its life cycle. This time span may be brought to a close in three different ways: (1) By internally controlled cessation of sensitivity to the stimuli concerned; (2) by a learning-performance (an imprinting) itself so rigid and unalterable that the particular phase of appetitive behaviour concerned can be said to have achieved its final consummatory situation and so become satiated; (3) the learning period for one response may be terminated by increasing sensitivity to other stimuli which compete with and so suppress it.

Summary

1. In children onset of specific learning is closely related to maturation of the nervous system, sense organs and effectors. The close of the specific learning period for a simple type of task may result not from a waning of the ability to acquire the simpler performance but because acquisition of more complex abilities, perceptions and skills, and the tendency actively to experiment with and explore the environment, render the subject less willing to restrict his attention. But there are rather isolated observations which suggest that there exist specific brain mechanisms ready to be activated only during a particular

period of the life span and that if they are not properly activated at the right time subsequent activation is impossible, resulting in permanent disabilities in later life.

To turn to the waning of learning ability in later life, there is extremely little evidence for pre-senile deterioration of the brain mechanisms mediating any of the specific learning abilities in adult human beings.

2. We can tentatively say that imprinting in nidifugous birds is initiated primarily by the maturation of an internally motivated, appetitive behaviour system which is in readiness, at the time of hatching or very soon after, to express itself in the following response. The time during which this internally controlled tendency can find its *first* expression in action is limited to a matter of a few hours, or at most, days, by internal factors; but during this time the response is ready to appear as soon as circumstances permit it to do so. The termination of the ability to make a first response is very largely influenced by the development of competing fleeing responses. Once the following behaviour has had a chance to establish itself, it will continue towards those models which have become associated with the achievement of the consummatory situation resulting from successful following and may be yet further generalised to others. Nevertheless, the conclusion that an internal change directly reduces the appetitive following drive with age itself is very strong.

3. The permanent waning, at about 13 months of age, of the ability of the chaffinch to learn new songs does, however, seem to represent the close of a period of specific learning ability. For there is no evidence whatever that any conflicting drive is responsible in this case for terminating the learning period, nor that the song learning 'capacity' of the bird has been satiated.

4. The ability of many animals to learn to respond to and to reproduce certain specific periodicities of stimulation is considered. The evidence for a temperature-insensitive endogenous rhythmic mechanism in many animals is now very

strong, but we cannot yet say categorically that the rhythm of this clock arises in complete independence of any similar periodicity of external stimulation. It is, however, clear that such clocks can be and often are set with great consistency by the *first experience* of a particular stimulus alternation, or even by a single stimulus, and so we have an extreme example of a sensitive period. These clocks can then control endocrine, nervous and other mechanisms, and the rhythmical behaviour which the animals exhibit as a result is the expression of (*a*) the temperature-insensitive internal clock, (*b*) the setting mechanisms, and (*c*) the temperature-sensitive metabolic rhythms or clocks, which may either be inherent or later imprinted on the animal. Thus we can make an extremely provisional and tentative classification of internal rhythmic mechanisms into 'hard' and 'soft'. '*Hard clocks*' are presumed, for the time being, to be inherent, and most of those so far studied are of solar (24-hour) periodicity—though future research seems likely to reveal many new examples of hard clocks of much shorter periodicity, perhaps down to minutes or seconds. Such hard clocks as have been investigated are temperature-insensitive or temperature-compensated over a wide range and thus called 'non-metabolic'. They can be set by a single stimulus and, once set, can continue for long periods (sometimes apparently indefinitely) without re-setting. They may have an accuracy of the order of 1/50,000. '*Soft clocks*': While many soft clocks are undoubtedly of inherent periodicity, it is very usual for their rhythms as well as their setting to be determined by experience—i.e. learned. They are in some degree temperature-sensitive, may be of any periodicity, are relatively inaccurate and are presumably metabolic. Where two hard clocks occur together in the same animal they may together give a combined rhythm. Where a soft and hard clock occur together, the harder of the two will tend to control the rhythm of the other. Such 24-hour clocks are now known to occur in isolated tissues of both plants and animals. They are perhaps evidence of some a-metabolic pacemaker present in the cells of all animals. They suggest the existence of an absolutely basic physiological mechanism which might be involved in some of the most striking examples of imprinting, of priming and of the 'one trial learning' of certain periodicities of stimulation. They can perhaps be extended to apply to other more complex examples also.

IX. DEVELOPMENTAL CHANGES IN LEARNING CAPACITY

1. *Introduction*

The way in which an animal's learning capacity develops with age is not yet understood. It is difficult to decide from the work published in this field whether there are developmental changes in learning separate from the effects of growth or of changes in neuro-muscular co-ordination, experience and motivation; or whether indeed the problem has ever been satisfactorily formulated. An early attack on this question was made in 1903 with J. B. Watson's work on 'the development of the psyche in the white rat', and in 1929 the question was to some extent diverted by C. P. Stone's monographs on 'the age-factor in animal learning'. Stone came to the conclusion that age differences appearing in animal learning experiments could be attributed solely to motivational change.

More recently work on children and some on birds, and also work on the effects of early experience on mammalian behaviour, suggest that although Stone's conclusions are important, there are developmental changes in the behaviour underlying learning which cannot be understood in motivational terms. The purpose of this chapter is to examine some of this work and discuss its significance as well as some of the experimental difficulties inherent in it.

The work on birds will be considered first.

2. *Some developmental changes in birds*

In all the work described in this section an attempt was made to keep motivation, in Stone's sense of a hunger 'drive', at a low level by allowing the birds to feed normally throughout the testing.

In the first experiment a clear-cut difference was found

between juvenile and adult finches in a simple learning situation. This situation has been described in detail by Thorpe (1956). The birds were fed at intervals with a certain type of food on a particular perch. Later the food was suspended from this perch on the end of a string which had to be pulled up and held before the bait could be eaten, as the string was suspended inside a glass cylinder (Plate I). In this situation some juveniles were successful within the test period of twelve half-hour trials but all adults failed. Records of the birds' behaviour while they were being tested showed that success was related to the length of time spent responding in the experimental situation: the young birds were more continuously engaged in pulling at the string and so on, and spent less time in each trial in such irrelevant activity as preening, feeding, sleeping and bathing than the adults. In respect to the experimental situation the young birds thus appeared to be more responsive than adults of the same species. In addition the records showed that whereas the response of all birds waned during any unsuccessful trial while the birds obtained no reinforcement by feeding on the bait, this diminution in the response was maintained from day to day (during 24-hour intervals between trials) only in the adults. Two suggestions were derived from this experiment. The first was that juveniles are more responsive than adult birds, and the second that the waning of an unreinforced response, which represents a distinct and rather delicate form of adaptation (extinction, or internal inhibition according to the Pavlov theory, Pavlov, 1927) is more stable in adults than in juveniles (Vince, 1958). Subsequent experience has suggested that the process of development is more complex than was suspected at this stage; nevertheless if responsiveness is greater in the juveniles and internal inhibition becomes more stable in the adults, we may well expect the manner in which birds adapt to their surroundings to vary with age.

The remaining experiments were carried out in an attempt to understand the development in birds of these two aspects of

PLATE I

A. 'String-pulling' situation.

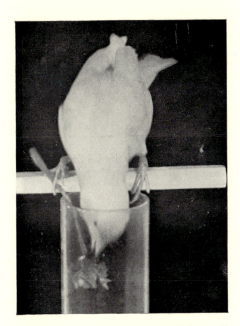

B. A canary reaches for the bait while holding
a pulled-in loop of string with one foot.

PLATE II

A hand-reared great tit removes white lid from dish containing food.
Differentiation situation.

behaviour: level of activity, or positive responsiveness, and internal inhibition, or the capacity to learn not to respond in situations where a response is not reinforced.

The second point will be dealt with first.

(a) *The development of internal inhibition in birds*

The test situation used in the next two experiments was the same and was very simple. The results of the previous experiments suggested that internal inhibition is weak in young birds. According to Pavlov's theory internal inhibition is the basis not only of extinction but also of a differentiation: the capacity to respond differently to similar stimuli some of which are, and some of which are not, reinforced. The birds were trained to remove a white cardboard lid from a dish containing food (Plate II). When the lid-removing response had become stable, they were presented in random order either with the dish containing food and covered with a white lid, or with the same dish, empty, and covered with a black lid. A bird responding adaptively in this situation continued to feed by pulling off white lids, and after pulling off the black lid a few times, gradually ceased to respond to this negative stimulus, indicating as it did an empty dish. In this situation therefore the time taken by the bird to remove the black lid (negative latency), or the number of correct negative trials when the black lid was not removed at all, was a measure of internal inhibition.

In the next experiment (Vince, 1959) twenty positive and twenty negative trials were given in random order to five groups of finches; a group of adult canaries and a group of juvenile canaries; also three groups of greenfinches—wild adults, hand-reared juveniles and aviary-reared juveniles.

The results of this experiment were, at first sight, confusing. The aviary-reared juvenile greenfinches achieved by far the highest proportion of correct negative responses: they gave up removing the black lid after very few trials and did not touch it again. The other juveniles gave results which were on the whole weaker than adults of the same species. This latter difference was not, however, entirely clear-cut. The hand-reared greenfinches and juvenile canaries actually made correct negative responses sooner than the adults and their performance was superior in the first ten trials, but it appeared that the task was easier for the juveniles as their responses to the black lid were weaker initially, giving longer latencies. Also in the second half of the experiment the adults of both species put up the best performance: they were slow to give up responding to the black lid but once they had done so they were able to ignore it more

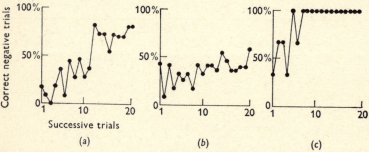

Fig. 12. Mean number of correct negative trials in differentiation experiment with finches (from Vince, 1959). (a) Adult canaries and adult greenfinches (b) juvenile canaries and hand-reared juvenile greenfinches (c) Aviary-reared juvenile greenfinches.

consistently than the juveniles, whose early differentiations tended to break down (Figure 12).

Although the results obtained from wild adult and hand-reared greenfinches and from adult and juvenile canaries confirm the view that there is an improvement with age, those obtained from the aviary-reared juveniles do not, and it is clear that these results are not consistent with any hypothesis that internal inhibition develops as a function of age alone; in greenfinches the best and worst performances were obtained from the two juvenile groups, the hand-reared and aviary-reared birds. The results can, however, be understood if a second variable is considered, that of individual experience, or conditions of rearing. The canaries had all been reared and kept under comparable, rather restricted, conditions and in this, as in the previous experiment, there was evidence of an improvement with age in this species. If variety of early experience is a factor affecting development, then the aviary-reared juvenile greenfinches, given their richer experience, may be supposed to represent a later stage of development than their hand-reared contemporaries, and if this is so we have again an indication of improvement which would normally occur with age. However, the performance of the adult wild greenfinches, although superior to that of the hand-reared birds, was inferior to that of the aviary-reared birds. It seems likely that the adults will have been through the stages of development represented first by the hand-reared and secondly by the aviary-reared birds. These results therefore led to the suggestion that the process of development of internal inhibition, if considered as a function of age, must be represented by a curve which rises to its maximum in the juveniles and later falls slightly; and also that a richer or more varied experience may well change the slope of the curve, giving a sharper rise, possibly to a higher level.

These suggestions have received some confirmation in experiments carried out on the great tit (Vince, 1960). Here the birds were originally all wild, but some were taken as nestlings, reared by hand and subsequently kept and tested indoors. The others were taken as juveniles, having been reared in the wild, and were tested under the same conditions as the first group. The hand-reared birds were tested at ages ranging from $4\frac{1}{2}$–73 weeks from the date of fledging. Only two groups of wild-reared birds, those whose ages can be assessed with some confidence, will be considered here. These were tested at approximately 8 and 20 weeks after fledging. These birds were all given the same task as the finches in the previous experiment, that of differentiating between a dish containing food and covered with a white lid, and an empty dish covered with a black lid. But they were given up to 40 instead of 20 positive and negative trials.

The level of performance achieved by each individual tested is indicated in Figure 13. Here the individual performances were assessed by counting the number of negative trials before the bird achieved the criterion of two successive correct trials, when the black lid was not touched at all. Individual records giving the latency in successive positive and negative trials and showing (*a*) a good performance and (*b*) a poor performance are given in Figure 14. Here short negative latencies indi-

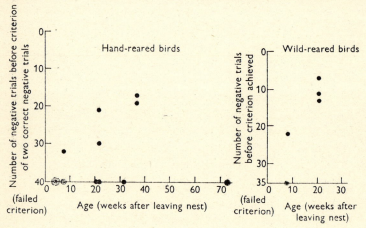

Fig. 13. Performance of individual hand-reared and wild-reared great tits in differentiation experiment (from Vince, 1960).

cate that the bird failed to make the differentiation, continuing to pull off black as well as white lids, while longer negative latencies indicate that the bird was able, to some extent at least, to ignore the negative stimulus.

The individual scores given in Figure 13 suggest that, although there are large individual differences, there was in fact some improvement with age in the capacity to learn not to respond, followed by a decrease in this capacity in the hand-reared adults. Although the wild-reared birds were few, there are indications that a similar curve might be expected from them and, as predicted, their performance is considerably superior to that of the hand-reared birds. There are, however, great difficulties in comparing the performance of animals reared under different conditions. These are discussed elsewhere in relation to the present work (Vince, 1960) and in more general terms have been set out by Beach and Jaynes (1954) and by King (1958).

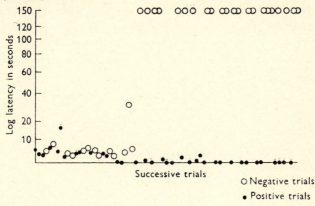

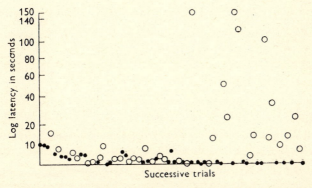

Fig. 14. Individual records from differentiation experiment. Top record wild-reared great tit, approximately 20 weeks after leaving nest; lower record, wild-reared great tit, approximately 8 weeks after leaving nest (from Vince, 1960).

(b) *The development of positive responsiveness in birds*

The first experiment described suggested that juvenile finches were more responsive than adult finches; the question which now arises is whether there are changes in responsiveness or level of activity which, if plotted against age, will provide a curve similar to that of internal inhibition.

Such curves can certainly be obtained. Figure 15 shows the changes with age in responsiveness in one particular situation in a group of six hand-reared great tits from a single brood (Vince, 1960). At intervals during the first year each bird was presented for fifteen minutes with a small brightly coloured

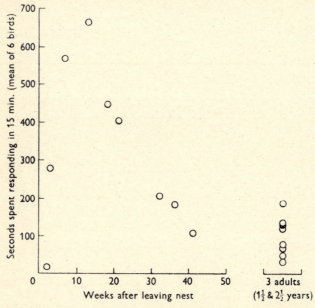

Fig. 15. The relation between age and responsiveness in hand-reared great tits (from Vince, 1960).

object and the length of time spent in pecking and pulling this about was recorded. Each point in Figure 15 gives a mean score in seconds for the whole group and for comparison the means obtained from three older birds are also shown. In this situation responsiveness increased from a very low level when measured at two weeks after leaving the nest, to a maximum between ten and twenty weeks. It then fell off rather rapidly and soon after thirty weeks reached the level achieved by a group of adults similarly tested. No data of this kind are yet available for the wild-reared birds.

If, therefore, we take birds of a particular species, rear and keep them under fairly homogeneous conditions, and test them at different ages we find quite complex developmental changes occurring during their first year. Responsiveness, in the positive sense, at first appears to be very low; it then rises rapidly to its maximum level at about thirteen weeks, and drops again to a

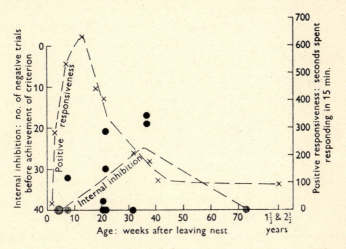

Fig. 16. The relation between changes in positive and negative responsiveness in hand-reared great tits (from Vince, 1960).

certain extent. Similarly, but apparently at a slower rate, there is an improvement in the capacity not to respond inappropriately or uselessly (internal inhibition). The relation between these two processes appears in Figure 16 where for the sake of clarity the trend of each curve is indicated by broken lines, although curves have not been fitted. The relative height of these two curves is, of course, of no significance, but the data suggest that internal inhibition is still being strengthened at an age when positive responsiveness is well past its peak.

This work therefore suggests the existence of developmental

changes in bird beeaviour of the type which must have some bearing on learning tasks. In view of the inhibitory as well as excitatory changes these can hardly be accounted for in terms of a single factor, such as learning capacity or motivation. In addition, although these changes can be related with age, the difference between birds reared under different conditions suggests that they cannot be accounted for exclusively in this way: individual experience is at least an important element in this type of development.

The question to be discussed below is whether these results bear any relation to, or throw any light on, existing work on the development of learning capacity in other species, and in man.

3. *The relation between age and learning ability in the rat*

In an early experiment Watson (1903) found that in a simple puzzle box or a simple maze young rats (25–37 days) entered the goal box in a shorter average time than older animals (63–70 days) and than adults of one year. The youngest animals were the best, but there was some suggestion of improvement after the earliest age of testing. In a more complicated puzzle box the adults were superior. Watson concluded that the younger rats had the advantage in a task depending on activity alone while the older animals gained when direction of activity was the important factor involved in solving the problem.

In another maze-learning experiment Hubbert (1915) carried out a more thorough investigation on rats aged between 25 and 300 days. The rats were scored for (1) the number of trials required to reach the criterion of learning, (2) total time required for learning to this criterion, (3) distance traversed, and (4) speed of movement. The younger rats (25–65 days) moved faster and learned in fewer trials while scores (2) and (3) showed an optimum age at 65 days, when the total time spent in learning and the distance traversed was less than in other groups. In addition Hubbert came to the conclusion that learning in the younger rats was most rapid during the early trials, while for the older rats it was more rapid during the later trials.

A suggestion made by both Watson and Hubbert that there may be an improvement with age during the period immediately following weaning is supported by Biel (1940) who used a maze and rats aged between 17 and 30 days; and by Liu (1928). Liu, in maze-learning involving seven groups of rats between the ages of 30 and 250 days, found an optimum performance in the 75-day group, followed by a deterioration up to 250 days. The same trend was exhibited according to all his criteria: number of trials required to reach the criterion of learning, time, and number of errors. This peak at 75 days is slightly less than the age at which activity in revolving and stationary cages, and also nest-building activity, reaches its maximum in the laboratory rat (Slonaker, 1912, and Richter, 1922). It may be worth noting here that Cruze (1938) has demonstrated an improvement with age in mazes and a problem box in chickens tested from 24 hours to 3 weeks after hatching.

Maier (1932b) substituted for the maze a problem involving the ability to combine two isolated experiences and found the performance of rats aged between 50 and 90 days *inferior* to that of rats aged between 120 and 300 days. Maier does not report any deterioration in the oldest rats, similar to that found by Hubbert, Liu and Watson, in maze learning. His findings may support Watson's contention that a task requiring directed rather than superabundant activity gives the advantage to older animals.

Stone (1929a and b) carried out an extended series of experiments on the age factor in learning in the rat. These experiments are frequently quoted as decisive disproof of developmental changes in learning (see, for example, Munn, 1950). Stone's conclusions were that the approximately maximum learning rate for mazes, problem boxes and a simple light discrimination is reached at 30–70 days and that this does not decline during the first two years of life. The differences obtained by others he regarded as spurious, due to faulty control of motivation, faulty techniques which 'provide pitfalls for the slapdash impulsive young or cater to the home cage habits which are inevitably fixed more strongly in adults than the young', and to time scores obtained from unhealthy adults.

Stone varied motivation (hunger 'drive'), age and intensiveness of training with his various learning tasks. As a result he found that younger rats learned in fewer trials than the older ones (1) when food deprivation was not too severe, (2) when an old habit had to be broken up and the opposite habit learned, (3) when early trials only were considered, (4) when training was more intensive, and (5) when the task was very difficult. Under conditions of severe food deprivation or relearning, when two experiments were run simultaneously and when later trials only were considered, there were no marked differences with rats of different ages. In a single experiment in an easy maze the older rats were superior. On the whole it seems that the older rats learned more slowly and, once established, their habits were more fixed, but when hard pressed they could reach a level of performance equal to that of the younger ones.

There are two ways of looking at Stone's work.

If Stone's conclusions are correct, then there is some kind of absolute 'learning capacity' which remains approximately constant from about 70 days onwards and is revealed under particular conditions in the older animals. On the other hand, the data provided by Stone give evidence of complex behavioural change with age which he does not attempt to account for. Some of the differences, for example that younger animals were superior in early trials only, and were more flexible, are consistent with those described in birds.

However, it may not be desirable to make too detailed a comparison between the rat and the bird work. In the rat experiments mazes, problem and discrimination boxes provide situations where learning can be observed, but whose complexity makes it difficult to analyse the processes involved in the learning. Such a criticism is of course irrelevant to the experimenters concerned because they were in fact looking for something absolute in the nature of 'learning capacity' which varied with age.

These experiments on rats may be summarised as follows: learning, as measured by scores obtained in mazes and puzzle boxes, becomes more rapid up to the age of about 70 days and later becomes less rapid. The performance of older rats can, however, be raised to the level of the juveniles by increasing motivation or hunger 'drive'. In addition to these main findings it would appear that older and younger rats learn in slightly different ways. There are also experiments where a different type of situation was used which gave results of the opposite kind: the performance of the adults was superior to that of the juveniles. This lack of consistency suggests that behavioural changes occurring with development may be more complex or more delicate than can be understood either in terms of 'learning capacity' or of 'motivation'.

4. *Some effects of experience on learning ability*

There was evidence in the work on birds discussed in section 2 that individual experience was a factor affecting the process of development. Individual experience was largely ignored in the rat experiments described in section 3. More recently considerable interest has been shown in the effect of experience, and especially early experience, on learning ability and behaviour in general. The work in this field has been reviewed and discussed by Beach and Jaynes (1954) and by King (1958).

Beach and Jaynes point out that there are special difficulties in this work: 'anatomical differences, nutritional requirements, sensory sensibility, motor development, and previous experience are closely interwoven variables with age and cannot usually be controlled independently of each other.' They consider that early experience may influence behaviour (i) by preventing the acquisition of other types of behaviour which could compete with a habit formed in response to a particular situation; (ii) because motivation may be more intense in the young; and

(iii) because certain types of early experience influence later behaviour by structuring the individual's perceptual capacities. They conclude also that there may be critical periods in development but that here much of the evidence may be of doubtful reliability.

Some of this work, and some published since, may be considered in more detail.

In experiments on the rat, Forgays and Forgays (1952), Bingham and Griffiths (1952) and Hymovitch (1952) have shown that animals reared in a varied environment are superior in various types of test to those reared under restricted conditions. The tests used were mazes, the Lashley jumping apparatus, and the Hebb-Williams closed-field test of rat intelligence, the last consisting of a series of problem-solving situations. Hymovitch reported no significant differences in performance in an alley maze. Forgays and Forgays found that the more varied the early environment, the greater its effect on problem-solving at maturity.

In another experiment reported in the same paper Hymovitch varied the age at which the 'free environment' was experienced. He reports that rats reared from 30–75 days of age in a free environment and later (85–130 days) in 'stove-pipe' cages were superior in problem-solving to animals reared at the early stage in the 'stove-pipe' cages and later in the free environment. However, they were no worse than animals reared early and late in the free environment; and again, rats reared under conditions of continued restriction up to 130 days were no worse than those given early experience of the restricted environment. He concludes that the effect of the late experience is negligible. Similarly Forgus (1956) has shown that rats which have had perceptual experience with specific two-dimensional forms are superior in discrimination and form generalisation to rats which have not had such experience. He found also that rats which have this experience from 0–41 days are superior in form discrimination and generalisation to rats which have the experience from 41–66 days.

Thompson and Heron (1954a and b) have considered the effects of early restriction on activity and the problem-solving capacity of dogs. They have interpreted 'problem-solving' rather widely, and some of their experimental situations are outstandingly simple. They show (1954a) that the exploratory activity of dogs reared under restricted conditions was subsequently greater than that of those reared normally, even when both groups had been living under identical, non-restricted conditions for a year or more, and younger dogs were more active than older dogs. In other experiments (1954b), they found that the animals which had been reared under restricted conditions tended to perseverate, e.g. they had more difficulty than normals in breaking the habit of running straight to food when a chicken wire barrier was put between them and it, and they tended to go to the corner where they had been trained to feed even when they were shown food in a different corner. They were also virtually incapable of delayed reactions. Again, the effects of restriction were detectable after a year.

Forgays and Forgays, Hymovitch, and Thompson and Heron interpret their results largely in terms of visual experience and perceptual learning. Hymovitch has reported that rats reared in small-mesh cages where the animals could see out, and where the cages were moved every day, were superior in problem solving to rats reared in enclosed activity wheels where the opportunity for physical exercise was great, although free movement was restricted, and this particular although somewhat limited comparison supports their interpretation. Forgus (1955) has stressed that the early experience can be a hindrance or an aid to problem solving, depending on the nature of the experience and its relationship to the problem.

All this work is concerned with the rather specific problem of the effect of early restriction on later behaviour, rather than with the developmental problem as such. The animals were reared under specific conditions and tested later. Nevertheless it is of great importance in this discussion as it brings the factor of experience into the developmental problem in no uncertain way. There are, in addition, suggestions of a link between the effects of experience and the effects of age. For example, Thompson and Heron (1954a) have shown that restricted dogs and young dogs were more continuously active in an exploratory situation than normal and older dogs, respectively. This tendency to go on responding occurred in young as contrasted with older birds in the 'string-pulling' situation (section 2). The tendency of restricted dogs to perseverate in situations requiring a change of response (presumably involving inhibition of the direct or previously trained response), as well as their failure in the delayed response test, suggests that internal inhibition had not developed normally in the restricted animals, and is thus also consistent with the work on birds described in section 2. But it is not, at the moment, possible to do more than guess at what specifically may be missing from the environment of the restricted animals to produce this result.

5. Developmental work, mainly concerned with children

Developmental work in which the method used is most comparable with that in the bird work described in section 2 has for the most part been carried out on children. This raises obvious difficulties when we come to generalise from one to the other: differences of complexity, of life-span, of the presence or absence of speech, and of experience (which in the human child is almost completely beyond experimental control) have to be borne in mind.

There are two main lines of work which are relevant here, each with its characteristic approach, but as the conclusions are similar they will be discussed together.

Luria (1932) has dealt with the development of some very simple reactions in small children. His observations were made in situations which adults would normally regard as elementary: making rhythmic pressures in a pneumatic apparatus, making simple reactions such as pressing a button on receiving a signal, and making 'delayed' movements, the last being a slow controlled up and down movement, the kymographic trace of which has a 'regular cupola form' in the adult. In testing children aged from 2 to 7 years he found that the capacity to make these responses develops slowly. A child of 2 to 3 years was capable of making the right kind of pressure, but in successive responses the rhythm rapidly broke down. Children of 4 to 5 could make simple reactions to a signal, but each signal tended to produce many such movements so that gaps between signals were often filled with extra responses; sometimes these extra responses could be inhibited even in children younger than 3 or 4, but this might disrupt the habit so that the occurrence of a signal inhibited the reaction. As a rule Luria found the 'delayed' movement to be impossible for a child up to 7 years; in the younger children such movements were indistinguishable from rapid responses, and in older ones signs of slowing down might appear, but late, or spasmodically, giving an appearance of conflict in the trace.

Luria maintains that the excitatory aspect of behaviour presents no problem for the small child. The difficulty lies in controlling the 'intense masses of excitation' provoked by a signal. Every small child is capable of direct activity, and in order to understand development it is necessary to study the regulation and control of behaviour.

In a more detailed account and a somewhat more clear-cut situation Paramonova (1955) reports that a 3-year-old child can learn to press a key in response to a signal such as a red lamp; however, he usually gives the same response to other signals, such as a yellow or a green light. At this age it is beyond a child's powers to refrain from making this generalised response, even though the instructions to do so are well understood. Signs of restraint may appear, but belatedly, or inappropriately as a series of negative signals may well destroy the positive response. These difficulties are usually not overcome until 5 or 6 years. Similarly Panchenkova (1956) has shown that a differentiation is established quite slowly in rats aged between 18 and 32 days and breaks down rather easily, whereas the differentiation is formed more rapidly in 4-month-old rats, and remains stable.

Support for the view that behaviour depending on the inhibition or control of response is weak in the very young child and develops quite slowly is also given by the work of Lambercier and Rey (1935) and by Rey (1954). Lambercier and Rey were concerned with the evolution of a manual skill involving (i) a detour and (ii) the use of an intermediate agent. In younger children (4 to 5 years) they found that a direct response to a reward which was visible but out of reach was so strong as to prevent exploration of the problem. Not until $5\frac{1}{2}$–7 years did the children pause and appear to reflect before acting. Rey (1954) gave to children of 5–13 years, and also a group of adults, the task of drawing a line as slowly as possible across a piece of paper. The younger children were unable to make very slow movements and he concluded from his results not only that the capacity for restraint was weakest in the youngest children, but also that it was dissipated most rapidly in them in the course of making the trace. The adult level of performance was not reached until 11 or 12 years.

Certainly it would be rash to push too far any parallel between this work on children and the animal work. Nevertheless it would be difficult not to agree with Luria that regulation and control, the inhibitory aspect of behaviour, is of great importance in development; also, it appears to be weakest in the very young and to develop gradually. Apart from that, the course of development is not necessarily parallel: for instance, it is only in the bird experiments that an increase with age in the capacity not to respond appeared to be followed by a decrease in this capacity.

6. Discussion

The evidence suggests that we know very little about the development of behaviour in animals, and in man; nevertheless it appears that there are indeed behavioural changes related with age and that these changes may have considerable bearing on what is understood as 'learning capacity'. It is also apparent that these developmental changes do not occur automatically with the passage of time, but depend to some extent at least on individual experience—that is, on environment or earlier activity. In addition, these changes may be exaggerated, or concealed, by the motivating conditions.

It is true that if the work discussed is considered in the light of traditional psychological concepts, the results appear confusing, or even contradictory; they also depend too closely on particular experimental conditions for it to be possible to say whether 'learning' changes in any specific way with age. The suggestion made in this chapter is that the confusion may be produced by the concepts, which are too crudely descriptive to allow a precise analysis to be made. The work on birds presents the same difficulty if it is considered in this way. For example, if the results obtained from finches in the differentiation experiments are discussed in terms such as 'intelligence', 'learning capacity', or even 'speed of learning', the interpretation will vary according to the criterion of success chosen. In the first half of the experiment the juveniles were superior and later the adults were superior. The behaviour in fact developed, and the differentiation was established in a different way at different ages.

It may well be the case that a deeper understanding of developmental change underlying what we understand broadly as learning capacity will depend on an analysis which is not only more detailed, but which is of a rather different kind from that exemplified by the concept of learning. It is clear that with the progress of adaptation—the accumulation of experience—behaviour develops as a whole. Some aspects of this development have emerged in the foregoing discussion. Firstly there is the question of responsiveness. This appears to increase with age in the juvenile hand-reared great tit, and then to decrease. Secondly there are changes in a different type of responsiveness—internal inhibition—and here again there appears to be a rise, and also probably a slight subsequent fall with age. Here the rise, as measured in a group of hand-reared great tits, occurs fairly slowly and continues for some weeks after the peak of positive responsiveness is past. Changes in these two aspects of behaviour might well give rise to inconsistent results if 'learning' is regarded as a unitary factor and different types of 'learning' tasks compared indiscriminately. It would appear, for example, that in tasks depending mainly on the level of activity the younger juveniles are likely to be superior, while older juveniles will gain in tasks depending mainly on the rather more delicate type of adaptation which involves not responding when a stimulus ceases to be reinforced.

Thirdly there is the question of motivation in Stone's sense of hunger 'drive'. The work on birds was carried out under conditions which were normal for the individual tested; they were not starved, and some measure of food deprivation might well affect the results. Despite the difficulties involved in equating the effects of this kind of deprivation at different ages, an investigation of the kind made by Stone should prove to be of considerable value here, particularly if the effects of extreme hunger on internal inhibition were included in it. It seems unlikely, however, that the two curves given in Figure 16 could be accounted for in terms of motivation alone, and indeed it might be preferable to treat motivation in the more physiological sense of excitability, the level of which changes continuously with external as well as internal factors and which is an essential part of any response, rather than as a category on its own which can be manipulated separately from behaviour coming under the category of 'learning'.

Fourthly, there are the effects on behaviour of early experience or environment. Thompson and Heron state that two-year-old dogs reared under restricted conditions behave in some respects more like young dogs than like normal adults. Some of their experiments and relevant experiments on birds may be interpreted as suggesting that internal inhibition develops more rapidly and more completely in a varied environment, presumably as a result of greater activity, stimulation, opportunity for adaptation, and so on. It could indeed be the effect of adaptive behaviour alone which is responsible for the negative changes associated with age, but we do not know enough even to guess whether the positive changes can be considered in this way. However, it is impossible to grow older without in some way adapting to and being changed by external conditions, however varied or homogeneous these may be.

We have, therefore, provisionally four aspects of behaviour connected with the problem of development of behaviour. The question which now arises is whether the traditional descriptive categories such as 'learning', 'motivation' and the like have been discarded, only to be replaced by four other separate categories, 'excitability', 'inhibition' and so on.

For the sake of clarity, positive responsiveness and internal inhibition have been treated separately, but it is not meant to imply by this that they are two separate processes, or are manifested by entirely distinct elements in behaviour. It is clear that the two curves indicated in Figure 16 could represent changes in different processes, or could be simply two different measures of the same process. In the work on birds the measurement of internal inhibition appeared to be a relatively simple matter, based as it was on the analysis made in existing work on conditioning; the concept of 'positive responsiveness' or 'activity' in general is, however, much less satisfactory and it might be rather difficult to say exactly what is being measured in the data given in Figure 15. The simplest view of this peak in responsiveness during the juvenile stage might well be that of Luria, who suggested that the child's cortex, being functionally weak, is unable to repress the 'large masses of excitation' provoked by a signal; the peak, it would seem, is due to the weakness of internal inhibition at this stage. It is also conceivable that the development of internal inhibition could not occur without the preliminary rise in excitability. In any case it is not likely that excitation alone can be regulated in behaviour, as the intensity of a response will be measured at all stages by inhibition of the inductive, reciprocal kind, as well as by internal inhibition. Indeed, the downward slope of the curve of 'positive responsiveness' became steeper when the stimulus was changed from one which was indifferent for the birds to one which had previously a conditioned significance, thus introducing the effect of extinction (Vince, 1960), and it is possible that families of similar curves could be obtained by varying the experimental conditions in this and other ways. Again, when measuring internal inhibition in the extinction and differentiation situations, the behaviour varies not only with the strength of internal inhibition but also with the level of excitability. It would be undesirable, therefore, to consider these two aspects of behaviour as distinct 'factors'.

Further, if motivation is considered simply as excitability it becomes not an added 'drive', but an integral part of the behaviour. Moreover, excitability, as all the conditioning work shows, is intimately bound up with the environment or with the minute-to-minute as well as cumulative effects of individual experience. The minutely controlled environmental changes in Pavlov's work, for example, with their positive and negative effects on the animal's behaviour, may be regarded as a most detailed study in 'motivation'; or, to look at it another way, in this work the maintenance of an individual's equilibrium with a changing environment is expressed in terms of excitation and inhibition. If behaviour is regarded in this way, as a ceaseless long-term as well as short-term process of adaptation to the individual's environment, it is possible to regard the four 'aspects' of behaviour described above almost as a single process, but still without abandoning the attempt at analysis.

As an example of the cumulative effects of experience, the view has been put forward that the development of internal inhibition may depend largely on experience. This is supported by Pavlov's demonstration (Pavlov, 1927) that the rate of establishment of inhibition is improved by practice. If we consider behaviour as central nervous activity, this adds weight to a further implication of the work discussed above: that age and/or experience produce more than an increase in the individual's 'store' of sensory impressions, habits and the like, they result in actual change and development of the behavioural mechanisms. In comparing the behaviour of a very young animal or one reared under very restricted conditions with a normal adult of the same species, we may well be dealing with two mechanisms which function rather differently.

It is true that this approach produces experimental difficulties. For example, the adaptiveness of behaviour makes it difficult to separate the effects of age from those of environment or early experience. If testing is carried out at different ages, the results may be misleading because animals tested in this way will inevitably have acquired different amounts or kinds of experience. It is simpler to control age and vary experience, but still difficult to avoid or measure 'transfer' effects produced by adaptation to a particular environment in the early stages. One way of tackling this age/environment difficulty would be to consider development as adaptation to a particular environment, and to compare the details of this process with those manifested by adaptation to a different environment in a comparable group of animals.

7. Conclusion

The view put forward is that there are behavioural changes during development which can give rise to complex, indeed confusing, age differences in traditional 'learning' tasks. It is suggested that the concept of 'learning' may be too crude, or too general, to allow an adequate investigation into these changes, and a different approach is described. Such an investigation must ultimately look for changes in the mechanisms of behaviour, which changes are likely to a considerable extent to depend on the established state of adaptation of the animal: its previous activity or experience. An approach which is concerned with the effects of experience leads, of course, to the consideration of development as an individual process.

X. HUMAN PERCEPTION AND
ANIMAL LEARNING

1. *Introduction*

In recent years most investigators of animal learning have tended to treat as irrelevant the results of experiments on human perception. This division between two types of data has not always been so sharp. Some of the most eminent early students of animal problem-solving were convinced of the importance of perceptual factors, whose nature could be revealed by experiments on man as well as on animals. Köhler's (1925) classic study on apes provides a sufficient example. The animal experiments to which such a view gave rise were mostly intended to show the importance of the momentary stimulus situation in producing or preventing learning, while those who confined their attention to animal learning as an isolated problem were more concerned with such variables as the frequency of past responses in similar situations. A typical experiment of the first type is the demonstration that an ape has difficulty in reaching a banana outside its cage when the only available tool is a branch of a tree within the cage.

Let us describe this experiment as a gestaltist might, retaining the right to criticise the description in the next paragraph. By analogy with human perception, one might say that the branch is seen as part of the tree, and not as a reaching-stick. To achieve the latter perception requires a restructuring of the field analogous to the human experience felt when looking at an ambiguous figure such as a Necker cube. When such an insightful restructuring is made unnecessary, as for instance by placing a stick at the front of the cage pointing towards the banana, learning is immediate. In such a case the perceptual field contains within itself tendencies to closure which will produce a correct solution.

248

This example shows not only the strength but the weakness of the perceptual approach to animal learning. It is likely that the language used in the last paragraph will produce an uneasy feeling in many readers: certainly it requires a deliberate effort from the writer. Words such as insight, restructuring, and closure are very difficult to define in such a way that they mean the same to all those who use them. By derivation they are names given to perceptual experiences, and it is doubtful whether these experiences are really identical in all people placed in the same situation. Not all of us cry 'Eureka!' and leap from our baths when solving a problem in hydrostatics: at some point the internal processes must become different in different individuals. And in addition to the ambiguity of terms developed in this way, there is a danger of missing factors which are essential to learning but which do not appear in verbal reports of experience. For example, Maier (1931) compared the problem-solving performance of two groups of human beings, treated identically save that to one group the experimenter made a casual and apparently accidental gesture (setting a hanging rope swinging). This group did better at solving the problem, but the experimenter's gesture did not appear in the reports of their experiences while reaching a solution. Other similar experiments are numerous, and show the danger of using purely phenomenological concepts in attacking learning. Finally, there is a logical difficulty which may be indicated by introducing the language of information flow, which will be used repeatedly in this chapter. Behavioural experiments normally involve inserting certain information at the input end of a system whose nature we do not understand, and observing the resulting output. There may be more than one output channel: in human beings, one may observe a verbal report after a stimulus, and also a bodily response. In any system, if there is more than one output terminal the information at one of them must have passed through a part of the system which was not used by the others.

It is only features of the shared parts of the system which will be equally reflected in all the output channels. Consequently there is no guarantee that principles appearing in spoken accounts of experience will also appear in bodily response to the situation. Because a soldier accurately reports having seen an enemy, it does not follow that he shot at him, and still less that he hit him. The presence of speech in human beings is an additional means of drawing inferences about events within the nervous system; but it may well require an additional mechanism in that system, and the advantage of the extra source of data is probably offset by the extra complexity of the problems which human behaviour poses.

Concepts drawn from perceptual experience have, then, the disadvantages of ambiguity, incompleteness and lack of any necessary connection with non-verbal behaviour. These are probably the major reasons for their rarity in the modern literature on animal learning. But although they are valid objections to a particular language, they are not objections to studying the effects of the stimulus situation on behaviour. Experiments in which the momentary input to the nervous system is varied are obviously as necessary as those in which attention is concentrated on the output; and, as we shall see, there are reasons for thinking that simpler and more general principles of behaviour may appear as functions of input than as functions of output. This type of experiment, when applied to human beings, would normally be called a perceptual one even though the response used for measurement is not verbal. Human subjects have the advantage for such work that suitable outputs for each possible input may be rapidly established by instructions, and a good deal of time may thus be saved in research. Although any findings on man will need confirmation on other species, this applies whatever the species first studied: there is a regrettable tendency amongst animal psychologists to assume that findings on the rat do not need verification with other species.

The present chapter will, therefore, discuss experiments on both animals and human beings, and will attempt to show the importance of stimulus variables in types of behaviour which are often approached only through response measurement. Two fields will be considered: that of reactive inhibition, and that of the continuity-discontinuity controversy in learning.

2. *Reactive inhibition*

Reactive inhibition is a postulate in the Hullian system (Hull, 1943), but it is accepted by many who differ from Hull in other respects. In simplified form, the hypothesis is that every response produces a decline in the probability of that same response recurring. This drop in probability is temporary rather than permanent; so that when a response has ceased due to accumulation of reactive inhibition, it will reappear after an interval of time.

This postulate is introduced in the Hullian system to explain temporary experimental extinction of conditioned responses. The conception is that a conditioned response followed by reinforcement (food or other drive-reducing conditions) acquires an increased probability which will normally balance or exceed the drop in probability due to reactive inhibition. As soon as reinforcement is omitted the influence of reactive inhibition is unopposed, and the conditioned response will cease after it has been elicited a number of times. The disappearance will only be temporary, and it is indeed true that an extinguished response will spontaneously recover in time. Similarly the rate of extinction will be greater if less intervening time is allowed between each response; and this is also true.

Reactive inhibition is also of use in accounting for 'spontaneous alternation'. The latter is a feature of the observed behaviour of rats in mazes. If two paths of equal length both lead to food, the animal does not normally continue to take the same path every time it is placed in the maze. It will take one path on one trial and the other on the next. One can easily see that this is very reasonable to an upholder of reactive inhibition, since the most recent response is associated with the greatest inhibition and therefore is at a disadvantage compared with the less recent response. Similarly, reactive inhibition may be applied to explain the variation in behaviour which character-ises trial-and-error learning. The cat placed in a puzzle-box makes a response which is not reinforced by escape; the response is not repeated, because of reactive inhibition, but is replaced by some other response and so on until the animal succeeds in escaping. Again, discrimination learning, in which the animal learns to make a response to one kind of stimulus but not to another, is explained in similar fashion. It is likely, however, that any theory which gives an explanation for ex-tinction, including spontaneous recovery and the effects of varying the interval between conditioning trials, will also account for these other phenomena. The essence of reactive inhibition is the explanation of extinction.

Recent experiments on animals, however, do not support the postulate in the form in which we have stated it. As normally held, it predicts that extinction will be greater when more responses have occurred. In contrast to such a view, it seems rather that extinction is a function of the stimuli which have been presented: that is, that it is a perceptual effect in the sense in which we have agreed to use that term. At first sight it may seem that this is a distinction without a difference, since in the normal conditioning situation the stimulus is always presented when the response occurs. But by considering rather abnormal situations we can distinguish between the effects of repeated stimulation and those of repeated responses. Let us examine some such cases.

A simple first example is that of stimulation of low intensity. Many years ago J. B. Watson conditioned a child to show fear response to a woolly animal. In order to extinguish this response, he did not repeatedly frighten the child by presenting the animal so as to accumulate reactive inhibition. On the con-trary, he introduced the animal only at some distance so that the fear response did not occur. Gradually the animal was brought closer, and eventually the child was cured of its fear without needing to show the response. This example is a familiar one to most psychologists, but its implications had been strangely neglected until recently, when Kimble and Kendall (1953) performed a closely similar experiment on rats. They compared a group who extinguished a response by repeatedly making it, with another group who were repeatedly given faint stimuli, and therefore did not actually respond. On a test stimulus of normal intensity, the latter group showed more extinction than the former. That is, extinction occurred even though no responses had taken place, but only repeated stimuli.

A second method of dissociating stimulus and response is to remove some parts of the situation, so that response becomes impossible. For example, a rat may be placed in a Skinner box which normally contains a bar which the animal has learned to press. If the bar is absent, the response cannot occur although most of the stimuli from the normal situation are still present. It has been shown by Hurwitz (1955) that such treatment produces extinction of the bar-pressing response. Once again the extinction depends on stimulation rather than response. There are also experiments by Deese (1951) and by Seward and Levy (1949) in which rats are shown a food-box which they normally reach after running a maze. The maze-running response is depressed by the stimulus of the empty food-box— which is a further demonstration of the possibility of extinguishing a response without eliciting it. (Hurwitz's situation has the advantage of avoiding a possible criticism that secondary reward was operative in the maze-running task.)

A third means of distinguishing input and output is to take the case of spontaneous alternation, mentioned above. In this case the response is for the animal to turn right or left, while the stimuli concerned are those from the corresponding alleys. If the maze is cross-shaped rather than T-shaped, and if the rat is started from opposite arms on alternate runs, then any alternation of response will reveal itself by the animal continuing to turn down the same path. A right turn when travelling north brings one to the same destination as a left turn when travelling south. In fact under such circumstances the rat tends to go alternately east and west, that is, to make a series of responses of the same kind but to alternate the stimuli which arouse the response. This work is due to Glanzer (1953a and b) and Montgomery (1952b).

There is thus some doubt about the postulate of reactive inhibition. Instead of saying that each response produces a decline in the probability of that response, it might be more suitable to say that each stimulus produces a decline in the probability of any response to that same stimulus. An alternative way of putting this statement is that the incoming information from the sense organs is filtered at an early stage in the nervous system, that only part of the information proceeds further, and that the filtering is not random but appears most likely to discard information on channels from which other information has recently been passed. The most plausible reason for the existence of such a filter is that the capacity of later stages in the nervous system is limited—that is, it cannot transmit information at more than a certain rate. This is a familiar situation in the telephone engineering case for which the language of information theory was originally designed.

Before proceeding to further experimental evidence, it may be as well to note possible objections which might be raised to the above argument. Perhaps most important is the suggestion that extinction is in these experiments due to counter-conditioning of some response incompatible with the one studied; in particular, that conditioned inhibition is present, which is recognised by the Hullian system as an addition to reactive inhibition. To some extent, the objection is invalid. Conditioned inhibition is supposed to obey the same rules as those which govern the learning of any other response: to Hullians, this means that its acquisition must be reinforced by reduction of a drive such as hunger or sex. Normally, reactive inhibition itself is supposed to act as the drive which reinforces conditioned inhibition. If the former does not occur, the latter can only be supposed to arise by abandoning the postulate that drive reduction is always necessary for learning. There is an alternative of postulating secondary reward, curiosity drive, or some similar device, but operationally such concepts are indistinguishable from abandoning the reinforcement postulate. The same difficulty arises with any other response which is supposed to be counter-conditioned.

From another point of view, however, the counter-conditioning view is merely the same one which we put, in different terms. If counter-conditioning means only that some fresh process now takes place in the nervous system when the animal is placed in the conditioning situation, and that this process is incompatible with the process which formerly followed the conditioned stimulus, then it is indistinguishable from the conception that the input information is filtered and that a repeated stimulus is unlikely to pass the filter in the future. This also applies to another possible objection, that we have not been sufficiently sophisticated in taking bar-pressing or turning left as the responses which produce reactive inhibition. On the contrary, this objection would urge, it is the visual response, internal and unobservable, which is subject to such inhibition; and the experiments cited do not destroy such a view of reactive inhibition.

Such a re-interpretation of Hullian theory is in line with the approach of Berlyne (1951), and seems to be perfectly valid. One major reason for preferring our description, in terms of filtering of information, is simply that of semantic hygiene. When the terms 'stimulus' and 'response' are normally used for observed quantities, it is highly confusing to speak of counter-conditioning unobservable responses, or of reactive inhibition accumulated by such responses. If anybody has the language of the Hullian system strongly ingrained, however, he may legitimately adopt such a description, provided he remembers that the 'responses' in this case are neural processes occurring in pathways near the sense-organs, and that extinction is thus a function of observed input rather than observed output.

The last objection that might be raised is that the experiments so far cited are species-bound, dominated by the white rat. Let us therefore turn to experiments on human behaviour. Prolonged work by human beings is a considerable practical problem, and has therefore been the subject of much study. Many of the American investigators felt that the postulate of reactive inhibition would be a valuable guide and framework in studying human work, and so up to a point it is. Human performance in skilled tasks does deteriorate with time and recover after rest pauses. But some skills show this effect much more than others; Gagné (1953) reviewed a number of studies and pointed out that skills in which the stimulus-complex was repetitive or cyclic agreed better with the reactive inhibition view than did skills in which this was not the case. For instance, studies on the Pursuit Rotor, in which the subject has to follow a spot moving regularly on a gramophone turntable, show results which fit the reactive inhibition account. Studies on complex co-ordination tasks in which, for example, a subject may have to match a light pattern by manipulating aircraft controls, give much less marked deterioration in performance. Here the stimulus part of the task does not repeat itself. This suggested to Gagné that the relative roles of stimulus and response repetition should be examined more closely. Recently Adams (1955) has performed an experiment which seems to meet this need. He used the Pursuit Rotor, but investigated the effect of rest-pauses in which the operator had to watch somebody else perform the task and press a button whenever the second operator was off-target. Thus the subjects were receiving the same stimuli as in the task and were having to take decisions about those stimuli. But they did not have to make the responses normally made in the task. If a period of performance on the main task was followed by a period spent on this similar task, a subsequent third period on the main task showed poor performance by comparison with performance after a rest away from the main task. Adams tested control groups to show that standing or button-pressing during the intervals between sessions on the main task did not produce decrements; in short, it seemed that performance of the main task was being impaired by 'fatigue' due solely to the stimulus side of the performance.

This line of research seems to have come to the conclusion, then, that human performance deteriorates under repeated stimulation rather than repeated action, and the reactive inhibition postulate as simply formulated will not serve. Adams, it may be noted, adopts the plan suggested above and speaks of the visual response as producing reactive inhibition. But it is clear that these experiments, like those on animals, throw the emphasis on the observed stimulus rather than the observed response.

British investigations of prolonged work have been less guided by Hullian theory. But it has emerged empirically that it is very difficult to impair performance on lengthy tasks needing light manual work, and comparatively easy to show ill-effects from periods of watching radar screens or inspecting finished products on an assembly line (Wyatt and Langdon, 1932; Mackworth, 1950). In such work the man is making few overt responses, but is exposed to repeated stimulation. The level of efficiency in these tasks is subject to a number of influences, but it seems to be possible to connect together results from a number of experiments by a theory of the type we have already suggested. In other words, one may suppose that the early stages of the human nervous system have a limited capacity, that therefore only part of the information reaching the sense-organs passes further along the system, and that passage of information from one source lowers the probability of subsequent passage of other information from the same source. It follows from such a view that certain kinds of tasks will be more liable to ill-effects from prolonged work, while others will be almost unaffected. On the whole these predictions are fulfilled (Broadbent, 1953a and b).

If one may summarise in popular terms the views which studies of prolonged work by human beings suggest, tiredness in the muscles has little importance for many industrial tasks. The difficulty of long work spells is rather that attention wanders, that the worker does not notice parts of the situation to which he would have attended when fresh; but he may be able to keep up a series of actions without difficulty. Here we have again the language of every-day perceptual experience, but the evidence on which these views are based is essentially similar to the evidence which we have cited from animals. It is repeated stimulation which produces inefficiency, rather than repeated response.

We have now completed the main argument of this section, that the phenomena frequently explained by reactive inhibition are in essence perceptual, so far as that term is operationally meaningful. But before passing on to the next section it is only proper to mar the simplicity of the argument by introducing some complications. In the first place, we have only been considering learned responses in mammals. Hinde (1954) has shown that when instinctive behaviour is considered, changes in responsiveness may sometimes be functions of stimulation and sometimes of response. Thus, for example, if a chaffinch has been exposed to one kind of predator (say, an owl) and has responded repeatedly until the frequency of response has declined, the response to another kind of predator (say, a stoat) is also diminished. Here we have a case which seems to fit the orthodox reactive inhibition view, and not our 'perceptual' alternative. The explanation we may give for this is that both owl and stoat are predators and information about them passes the filter because they are predators. Exposure to one predator-stimulus lowers the probability of any response to another predator-stimulus: the fact that the response is in both cases the same is accidental.

This point has been made clear by a recent experiment by Schroder and Rotter (1954), who taught animals to turn right in a black maze and left in a white maze. If both habits were established using food reward, extinguishing one extinguished the other. If one habit was established using water reward and the other using thirst, extinguishing one did not extinguish the other.

Thus Hinde's experiments need not cause us to abandon our view that it is repeated presentation of a class of stimuli which produces extinction, though we must remember that the classification of stimuli may be by relevance to a motive and not by any physical characteristic of the stimuli.

Finally, it must be admitted that extinction is not capable of one simple explanation. As has been said, Hull required two forms of inhibition to account for it; and it is quite conceivable that more different varieties must be distinguished. In the terms we have been using, it is clear that a stimulation-produced drop in the probability of information passing the filter will not account for permanent extinction, but only for that part of the phenomenon which disappears with time. Permanent extinction represents a change in the output which corresponds to a particular input, and this is a mechanism distinct from filtering of incoming information. There may also be other possibilities.

What then remains of our general argument after these troublesome complications have been considered? Certainly it is true both of man and of the rat that the probability of a particular response may be reduced without a series of occurrences of that response. One way of describing the effect of extinction is that some of the information passing through the nervous system is lost; presenting the input, the stimulus of the conditioned response, produces no detectable change in output. The fact that repeated monotonous stimulation, rather than response, can produce this condition suggests that the locus of the loss of information is rather early in the nervous system. If this is so, then the establishment of particular input-output relations may take place at a point in the nervous system after the filtering of incoming information. In S–R terms, the building up of particular stimulus-response connection may be dependent on the stimulus being free from inhibition; but these terms do not readily express the point at issue.

3. Continuity–discontinuity controversy

The continuity controversy is an issue in discrimination learning over which there is much diversity of opinion. The continuity position is, broadly, that the probability of a particular response appearing to a particular stimulus depends on the total number of previous reinforced occurrences of that stimulus-response combination. The non-continuity position points to an undoubted fact about discrimination behaviour, namely that rats learning to run a maze react systematically to certain stimuli which may not be those treated as correct by the experimenter. Thus the rat may for a time enter all lighted alleys and ignore all dark ones, although many of the lighted alleys turn out to be blinds rather than correct paths through the maze. After a number of such responses, the rat may begin to turn regularly to the right, although this also is not always rewarded. In mentalistic language, the animal forms a hypothesis about the correct solution and acts on the hypothesis. Only when one hypothesis has proved false, does it try another, and the suggestion is therefore that the success or failure of any particular response is relevant only to the keeping or discarding of the hypothesis then in operation. If the animal is reacting to light, it might be that it does not notice the other features of the alleys it enters, and so never establishes an association between the sight of a triangle at the entrance to an alley and the food-box at the far end of the alley. Therefore the probability of a response to the correct stimulus will be increased only by reinforcement occurring after the correct hypothesis has been set up. The upholders of continuity accept the regular systematic reactions of the rat to particular cues: they merely assert that this systematic behaviour is irrelevant to the acquisition of the correct response. It arises, they believe, from a temporary excess in the number of reinforcements following entry of, say, the lighted alleys as opposed to the dark ones. Since half the lighted alleys are correct by chance, it will take a number of trials before the light-turning habit is overtaken by some other. But whatever the habit which is momentarily dominant, all the reinforcements received count in the establishment of the relative strength of other habits; so sooner or later the stimulus approved by the experimenter will become the one to which approach has most often been rewarded.

The experiments which are carried out to test these theories illustrate the differences between them. In one type an animal is given two different stimuli and reinforced for approaching one of them. Before he has learned the discrimination the roles of the two stimuli are reversed, the former rewarded stimulus now being unrewarded or even punished. Does the reversed preliminary training impair the learning of this second discrimination? On a discontinuity theory it might be argued that it would not, for the rat will not have noticed that, say, it was reinforced for going to a triangle, as long as it was acting on the hypothesis that right turns are correct. Some workers do in fact claim that reversed pre-training does not impair learning. Others have done experiments in which such training does impair learning as compared with a control group. Blum and Blum (1949) have reviewed these experiments and have made the important contribution of pointing out that the two types of finding come from slightly different experiments. Results supporting the discontinuity view come from training using highly similar stimuli, with training trials separated by short time intervals, and with punishment for incorrect responses. The other school use more different stimuli, space their trials, and use reward alone. The Blums argue that the former situation is more inhibitory than the latter, that in consequence the shift from one type of reinforcement to another may act as a 'disinhibitor', and so conceal a true continuity learning by a spurious improvement following the reversal of reinforcement. The less inhibitory situation, on the other hand, reflects the true continuity of learning.

Ingenious as this solution is, it presents certain difficulties. 'Disinhibition' appears empirically as the improvement in response which may follow some irrelevant stimulus. To regard it as following a change in the particular stimulus-response combination to be reinforced is making a rather large step. Furthermore, disinhibition is very little studied and its theoretical interpretation is one of the greatest weaknesses of Hullian theory. The Hullian view of it may be fairly paraphrased as asserting that any combination of stimuli may have quite different properties from its parts, and that therefore an irrelevant stimulus along with that for an extinguished response changes all previous predictions. It is difficult to devise any experimental test of such an imprecise view, but at least it is clear that it classes disinhibition as a perceptual process, a function of the stimulus pattern. Much more detail is required before one can distinguish an appeal to disinhibition from the more direct perceptual approach of the discontinuity theorists.

It is clear that the discontinuity view stems from a parallel with human perception. Our experience is selective, picking out some features of the world and ignoring others. The very word 'hypothesis' is taken from a human activity, and may well have offended some of those who fear anthropomorphism in dealing with animals. None the less, the approach can clearly be put in non-phenomenological terms similar to those used in the last section. It is in effect asserting that the input information is filtered at a stage before the mechanism which determines the output for each input. Some of the incoming information passes the filter and therefore may produce changes in the input-output relationships. Other simultaneously presented information does not pass the filter and so cannot produce such changes. Although an outside observer cannot decide exactly which items have not passed the filter, he can tell that some items have certainly done so because the information contained in them is preserved in the output. Thus when a rat is responding systematically by approaching light and avoiding dark, it is clear that the light-dark information is passing the filter. Information on the shape of the stimulus objects being presented may or may not be passing the filter; but the probability is lower than it is for light-dark and, on the average, changes in the output to such signals are less likely. The continuity position, on the other hand, denies that filtering of information takes place before the system which varies input-output relationships.

The resemblance between the discontinuity view and that

advanced in the last section is very clear. In the latter case we were able to show that both human and animal researches led to the same conclusion about prolonged performance. The human experiments had the advantage of greater flexibility, in that observable responses can be rapidly attached to a large number of stimuli, but they merely reinforce the finding that the stimulus rather than the response is the important variant. Let us now consider typical results from human experiments on this question of selective perception: they will lead us to certain curious gaps in the animal experiments.

In the first place, let us consider a case in which the filtering of incoming information is particularly clear. If a man receives two different speech messages, one on one ear and one on the other, he can be asked to repeat continuously the message reaching one ear as fast as it arrives. Such a task means the transmission of information through the man at a high rate, varying with the statistical interdependence of successive words, but none the less among the higher rates of which human beings have been shown to be capable (Miller, 1951; Quastler, 1956). In general any communication channel has a limiting capacity, and if the channel is already accepting a sequence of signals at a rate approaching that capacity then other independent messages cannot be transmitted over the same channel without interference with the first sequence. In fact it was shown by Cherry (1953) that a listener performing this task on one ear is able to say very little about the message reaching his other ear, even after the experiment is over. The information from that ear has been largely discarded, and the filter has passed chiefly information from one ear. Chiefly, but not entirely. It would not be true to say that the man 'attends to one stimulus and ignores the other'. Some of the information even from the neglected ear is transmitted through the man. Thus he may be unable to report whether the language used was English or German, or whether the words were spoken forwards or backwards. But he can say whether the voice was a man's or a woman's, and he will report replacement of the voice by a pure tone. Some of the aspects in which the presented stimulus differs from other possible ones are reported, while other aspects are not.

There is a further reason for preferring the language of information flow to that of stimulus and response: the latter terms are ambiguous, and may lead one to ignore the quantitative nature of selective perception. Another experiment which shows this is that of Broadbent (1956a), in which the interference produced on a speech-task, by a buzzer-signal requiring a hand-response, varied with the probability of buzzer. If the buzzer was only one of two possible signals (that is, if it conveyed more information), it produced more interference in a simultaneous task. Other related experiments are reviewed briefly by Broadbent (1956b): it seems that two simultaneous stimuli may both receive an adequate response if they convey little information, but not if they convey much information.

The importance of these findings for the continuity controversy is that the stimuli presented for discrimination by animals do not seem to have been analysed from an informational point of view, and usually have conveyed little information. The animal has been asked to discriminate between only two possibilities: quite apart from the other statistical weaknesses of this procedure (Cane, 1956) it means that, in mentalistic language, the animal may have been able to attend to other features of the situation as well as to the one which was controlling response. For a true test of the discontinuity view it would be desirable to present a number of alleys between which the animal can choose on each trial; and also for these alleys to differ in numerous ways, only one of which is chosen by the experimenter as correct. In some of the reported experiments the two alleys seem to have differed only in position and illumination, the lighted alleys being the correct ones in preliminary and incorrect in later training or vice versa. The filter might well pass position and illumination information simultaneously and so support a theory that all stimuli reaching the sense-organ have an equal chance of establishing S–R connections. Yet a more varied situation would have led to an opposite conclusion, because the filter could not pass an unlimited number of differences between alleys; and the experiments which support the discontinuity view are indeed usually on alleys differing in several respects.

An experiment by Lawrence and Mason (1955) shows this relationship directly: the animals were in this case faced with alleys that differed in spatial position and in either one or two other ways. If there were two other dimensions rather than one, learning was slower and 'hypothesis' behaviour (systematic reaction to an incorrect cue) more often present.

To return to the human case, it is known that even when stimuli of high information content are presented simultaneously they may be dealt with by various means. If we present three digits, say 736, to one ear and at the same time three other digits, say 215, to the other ear, all six digits may be reported by the listener (Broadbent, 1954). But they cannot be repeated in the order 723165: normally they will be in the order 736215 or 215736. The information reaching the sense-organs is indeed filtered, just as in Cherry's experiment, and only that on one ear passed; but afterwards the information on the other ear may also be passed. The same phenomenon will appear with other sensory combinations (Broadbent, 1956c), and it is not simply peculiar to the binaural case. Thus if we wish to be sure that some stimuli will not produce a response, special measures must be taken, such as presenting stimuli for only a brief period. The experiments cited shed some light on the time intervals involved. If the series of digits is slowed down to the point where a two-second interval separates the beginning of each digit, then the listener can alternate digits from two series in his response: that is, he can produce the response 723165.

The work of Neirmark and Saltzman (1953) shows the importance of this point in human learning. Their subjects were presented with numbers visually; one group were instructed to learn the numbers, one to mark certain numbers with a circle, and one to circle and also learn. All groups were afterwards given recognition tests to measure their learning. When stimuli were presented rapidly, the two 'circle' groups learned equally little and were worse than the third group who had merely to learn. When stimuli were presented slowly, the two groups with instructions to learn did equally well, and better than that with no such instructions. In other words, additional time was required to treat the stimuli in two different ways rather than one.

To put the matter in everyday language, attention wanders; and if an object is present for an unlimited time many features of it will secure attention. It is only with a briefly presented object that some of the major aspects will fail to be noticed. In information theory language, the limit on a system is its rate of handling information and not the information it will transmit in unlimited time. But this point is largely ignored by animal experiments on the continuity of discrimination learning: the animal can observe the stimulus-object for an indefinite period.

Experiments on human perception, then, suggest that work on continuity of learning in animals has not produced the kind of results needed for establishing a theory. In the human case, it is quite possible for more than one stimulus to receive a response, provided the information transmitted is small and/or the time allowed is long. Equally in animals it is entirely reasonable that stimuli other than that governing response may contribute to habit strength: the filter may sometimes pass such extra information. To demonstrate continuity of learning in some cases does not mean that there is no filter. Until animal experiments are performed with the results on man in mind, the matter remains unsettled.

Is there even a balance of probability in either direction? To the writer it does seem likely that the mechanism which attaches outputs to inputs, stimulus to response, acts only on filtered information. Some experiments, as has been said, do show discontinuity in learning, and it is quite reasonable that others should not. Unless there is some flaw in the technique of these experiments, they seem explicable only by a filter theory or by the Blums' disinhibition suggestion. As has been said, it is not clear whether the latter is meaningfully different from the former. Furthermore, even the experiments which nominally support the continuity view show certain features which suggest the opposite. Although in these experiments animals which have had reversed pre-training are worse than those without it, they are not so bad as they ought to be. If a rat has received 20 trials in which stimulus A was reinforced, and is then taught to avoid A and go to B, it should—on a naïve continuity view—take at least 20 trials to reach the same level as animals without the preliminary training: on the average it will make at least 10 more errors before learning the new problem. Yet McCulloch and Pratt (1934), for example, gave 348 preliminary trials to some of their animals and found only about 60 extra errors. This kind of finding is easily explained by a filter theory, but requires more effort from sophisticated continuity theorists. Spence (1940) suggest that pre-training equalises tendencies to respond to other cues, such as position, and that this makes the simple assumption given above invalid. This explanation, like that of the Blums, seems almost equivalent to admitting the opponent's case. The equalisation of response to other cues can only affect the building up of the correct S–R association if response to those cues interferes with the correct stimulus-response sequence. Unless there is such interference, habit-strength for the correct S–R combination should build up regardless of other aspects of behaviour. And it is difficult to distinguish a view which postulates interfering responses from the doctrine of selective learning; in fact the writer would regard Spence as saying essentially the same as that urged in this chapter, in a rather different terminology.

In addition, the reported differences between experiments which support continuity and those which do not are not difficult to interpret on a filter theory. Thus the supporters of discontinuity typically use rather difficult discriminations, as the Blums point out when urging their disinhibitory solution. From our point of view, presenting a difficult discrimination seems, in the human case, to be in some way equivalent to demanding a high rate of transmission of information. Long reaction times may be produced either by increasing the information per stimulus (Hick, 1952a) or by making the stimuli hard to discriminate (Crossman, 1955); interference in a second task may be increased either by increasing the information in the first task (Broadbent, 1956a) or by requiring difficult discriminations (Broadbent, 1956d). Using difficult discriminations is especially likely to reveal the role of the filter.

There seems, therefore, to be a slight balance of evidence in favour of discontinuity of learning, but the matter cannot be said to be established. It would be desirable for animal experiments to study quantitatively the relation between amount of relevant and irrelevant information in a discrimination, and the degree of effect of reversed pre-training. Furthermore, the time of exposure of the stimulus should also be controlled and varied.

Why should such effort be made? Is there any interest in the presence or absence of a filter selecting part of the incoming information? In fact, such a filter provides a possible escape from a pressing dilemma of current learning theory. From one point of view, the problem of learning is that not everything is learned. The simple Watsonian view, that any repeatedly occurring stimulus-response combination is learned, is clearly inadequate since following some responses by presenting food or a mate gives those responses an enormous advantage. The Law of Effect, or reinforcement theory, developed by Thorndike and taken up by Hull appeared to offer a simple and scientific solution: responses followed by drive-reduction become more probable than others. But with the years many situations have been found which seem to contradict this Law. Animals given preliminary experience of a maze show sudden improvement in performance after food reward is introduced, showing that their experience without reward had produced some learning; if two stimuli are repeatedly presented together and one of them afterwards attached to a response by reward, the other will show a slight tendency to give that response; resistance to extinction of a response may be greater if it is learned by rewarding only occasionally than if it is always rewarded; a response of withdrawing the leg is learned worse if a shock is given after every trial than if the shock is omitted when the response occurs. All these instances can be fitted into reinforcement theory by postulating secondary reinforcement, acquired drives, and similar devices. The essence of such explanation is that other events, such as stimuli previously present during reward, may take the place of rewards. But this throws us back on our original problem. Why are some responses not learned, since secondary drives and reinforcers are so powerful? Conceivably a highly quantitative theory of secondary reward might be produced to answer this question, but so far it has not appeared.

A filter theory, however, avoids the difficulty. Suppose we say that if information on the occurrence of two events passes the filter, then the conditional probability of one event given the other will be stored within the nervous system. In such a theory learning would not depend on reinforcement, and the various difficulties of the Law of Effect will cease to trouble us. But not all conditioned probabilities will be learned: only those of events which pass the filter. As was said in the last section, it is possible that the drive-state of the animal may influence this selection, just as hunger may cause men to see ambiguous shapes as resembling food (Levine, Chein and Murphy, 1942). To give a direct physiological example, Hernandez-Péon, Scherrer and Jouvet (1956) recently recorded activity in the cochlear nucleus from an unanaesthetised cat, and found the usual potential changes after click sounds. But when the cat was shown a mouse the effect of the clicks disappeared. It is obvious that the conditional probabilities of events involving the mouse are more likely to be learned by the cat than those of events involving the click sounds, because the latter information does not travel far into the nervous system. Yet in the absence of a mouse the relation of clicks to other stimuli might well be learned.

If this view were true, one might expect that the incidence of failures of the Law of Effect would vary with the drive-state of the animal. A hungry animal would notice food to the exclusion of all else: his filter would not pass information concerning water. An animal which was neither hungry, thirsty nor under any other strong drive would take in a less biased sample of information from his surroundings. Consequently if such an animal was run in a maze containing both food and water, and was afterwards made hungry or thirsty, his behaviour would show that he had learned the position of food or water. If the preliminary experience was given to a hungry rat, he would learn the position of the food but not of the water. This is in fact the trend of the evidence (Thistlethwaite, 1951). Selective perception in animals, filtering of incoming information, therefore offers a valuable alternative to reinforcement theory and it is to be hoped that the necessary detailed study of animal behaviour, taking account of results in human perception, will be carried out.

Let us return to mentalistic language to summarise this chapter, because of its familiarity. Human perception is selective, we attend to some things and ignore others. Attention wanders so that a monotonous series of signals will rapidly produce inefficiency. But by and large we learn only what we have perceived. Crude and demonstrably inaccurate as these statements are, it seems possible for them to be refined in such a way as to avoid the weaknesses of phenomenology, and when this is done it is likely that animal behaviour fits them as well as human behaviour does. In the after-effects of monotonous stimulation there is a clear parallel between man and rat; in the continuity of learning there is no parallel because the animal experiments do not consider points which the human ones show to be important. To put formally our mentalistic statement: it is possible that the capacity of a mammalian nervous system is too small to transmit all the information reaching the sense organs, and that therefore the incoming information is filtered before the 'coding' mechanism which allocates a particular output to each input. The probability of information passing this filter depends both on properties of the stimulus and on those of the organism. In particular it decreases with recent passage of information on the same channel. It also decreases with low physical intensity of the stimulus; and different classes of stimuli differ in their probability of passing the filter. That is, pain stimuli have a higher probability than touch, and high-frequency noises than low-frequency noises in man. In addition, when the animal is in a drive-state, stimuli of the kinds usually called reinforcing are more likely to pass the filter. In this sense, it is indeed likely that a connection exists between human perception and animal learning.

Left	Right	Effect of choosing Left
O	×	Wrong
×	O	Right
×	O	Wrong
O	×	Right
×	O	Wrong
×	O	Wrong
O	×	Right
O	×	Right
×	O	Wrong
×	O	Wrong

Fig. 17. A reversed pre-training experiment for the reader, supporting the continuity theory. The page should be covered so that only one line at a time is exposed. Choosing the left-hand figure in each line is 'hypothesis' behaviour and its consequences are shown for each choice. For the first two trials the cross is rewarded, and thereafter the circle. The reversed pre-training is likelier to hamper the reader in reaching the finally correct solution, thus showing that some learning does take place of cues irrelevant to the hypothesis being acted upon.

Left	Right	Effect of choosing Left
□ ○ I A	△ II × B	Wrong
△ II × A	□ I ○ B	Right
□ I × B	△ II ○ A	Wrong
△ II ○ B	□ I × A	Right
□ II × A	△ I ○ B	Wrong
△ I × B	□ II ○ A	Wrong
△ I ○ A	□ II × B	Right
□ II ○ B	△ I × A	Right
△ II × B	□ I ○ A	Wrong
□ I × A	△ II ○ B	Wrong

Fig. 18. A reversed pre-training experiment more likely to support the discontinuity theory. General procedure as for Figure 17. As in that case, the cross is rewarded for the first two trials and the circle thereafter. Reversed pre-training is likely to show less effect that in Figure 17 because there is so much information presented that the reader will only be able to remember the success or failure of his hypothesis. The argument of the text is that most animal experiments supporting continuity are of the Figure 17 type.

XI. THE PLACE OF REINFORCEMENT IN THE EXPLANATION OF BEHAVIOUR

1. *Introduction*

The term 'reinforcement' has perhaps as great a variety of meanings as has any of the technical terms to be found in experimental psychology. It is the principal purpose of this chapter to discuss one of these and to enquire how far, in any sense, it is true that learning and some other aspects of behaviour require the notion of reinforcement for their explanation.

The account of reinforcement which will be discussed is that provided by those theories which identify reinforcement with 'drive-reduction'. There can be no doubt that the analysis of reinforcement and of the conditions governing behaviour which has been developed in these terms has seemed to apply with surprising generality to the behaviour of animals. Much experimental work has been stimulated by predictions drawn from this theory, but it is now apparent that several serious objections may be made to the theory. These include (1) the existence of behaviour which cannot be explained in terms of the operation of drives unless special, and for various reasons implausible, assumptions are made concerning the factors which increase and decrease the strength of such drives. Avoidance learning and exploratory behaviour are instances of this category which present difficulty. (2) There seems to be direct evidence that not all reinforcement can be analysed in terms of drive-reduction. The preference for saccharine, the results of experiments in which food is administered by gastric fistulae, and other related work will be considered in this connection. (3) One of the key functions of the concept of reinforcement is its use in the explanation of learning in general. The identification of reinforcement with drive-reduction fails to provide an explanation for several kinds of learning. Latent learning and irrelevant incentive learning will be examined under this heading. (4) Finally, secondary reinforcement is an essential concept in drive-reduction explanations of behaviour. It will be argued that the theoretical implications of this concept are obscure and that experimental evidence suggests that this is a major defect in the drive-reduction analysis of reinforcement. These objections, and the experimental evidence on which they are based, seem to suggest an alternative analysis of reinforcement. It may be useful here to outline very briefly some of the main features of the drive-reduction explanation of reinforcement and behaviour, and to contrast these with some of the principles which now seem to be suggested by experimental evidence.

The basic contentions of drive-reduction explanations of behaviour, as presented for instance by Hull (1943, 1952), are roughly as follows. Behaviour is to be regarded as composed partly of innate, or reflex, responses to stimuli and partly as acquired responses to such stimuli. It is with the latter that we are here principally concerned. It is supposed that, during the acquisition of such responses, some kind of 'connection' is established between sensory and motor mechanisms. No attempt is made (indeed it is frequently explicitly avoided) to specify any further the nature of such connections; all that is claimed is that, as a consequence, stimuli which did not previously evoke a given response will subsequently come to do so. It remains, however, to state the conditions under which (a) such connections will be formed, and (b) under what conditions, once formed, they will be utilised in subsequent behaviour. The answer to the first question involves the concept of reinforcement. It is claimed that when a stimulus and response occur in temporal contiguity, and when this combination is followed by reinforcement, then such a connection between the stimulus and response will be formed. It is further suggested that reinforcement consists, basically, in the correction of some primary state of physiological need, such as hunger, or upon the alteration of the activity of some correlate of such a state in the nervous system. This last is usually referred to as the 'drive-stimulus'. Hunger sets up such an internal stimulus, and reinforcement consists in its reduction as a consequence of, for instance, the consumption of food. It must be supposed that there is a limited but unknown number of such drives, each of which sets up its own distinctive drive-stimulus. Now it is clearly not the case that, once such a connection has been established between a stimulus and a response, an animal will always make the response when the stimulus is presented; for, if we train an animal to make some response to obtain food, it will not be found that it will continue to make the response in this situation once its hunger has been satisfied. Hence, in answer to the second question (b) above, it is supposed that all behaviour, excepting isolated reflex behaviour, requires for its production and maintenance the activity of some one of these drives. In general, the drive mechanisms perform at least three functions. They set goals for behaviour and thus initiate activity. They register the fact that these goals have been attained and thus terminate activity. And they act as reinforcing factors which determine which responses shall become associated with cues during learning. It is, however, found that this set of assumptions is not sufficient to explain any considerable range of behaviour. A necessary additional postulate is that of secondary reinforcement. It is assumed that any environmental, or internal, stimulus which occurs shortly before some degree of drive-reduction takes place will also come to have reinforcing properties. An additional, though related, source of reinforcement is thus provided.

Now this account breaks down at least in the ways listed above. The experimental evidence seems to suggest that we should modify our explanations of behaviour in at least the following ways. First, animals seem to learn, to a very large extent, what cues to seek rather than what responses to make. The explanation of such learning requires no postulate of reinforcement—in the sense of some event which has a retroactive effect upon what the animal has experienced or done shortly before. When an animal has learned that one cue is usually followed by another, the appearance of the latter after the former is always 'reinforcing', in the only sense that is relevant here, in that it confirms the learned association of the cues. There is a sense in which secondary *reinforcement* can be regarded as nothing more than the learning of such associations based upon temporal contiguity alone. Secondly, in order to provide a framework within which any specific instance of behaviour may be explained, reference must be made to the mechanisms which select the final goal to be sought—e.g. food—and register the fact that it has been found. The first part of this task may be accomplished by something partly similar to the mechanisms which drive theorists have had in mind; the second part, the recording of the arrival of the goal, requires a mechanism similar to that which records the arrival of any other cue but with the additional function of terminating, or 'switching off', behaviour. Thirdly, although it appears that such behaviour may be controlled by mechanisms, in which the selection of goals is determined by variations in the physiological condition of the organism, there are also other instances of behaviour for which the goal-selecting and recording mechanisms have to be conceived rather differently. Finally, we should distinguish between the mechanisms which are involved in learning and in goal-selection and those which control, very roughly, how much effort an animal will expend in the task of getting to the goal. This last distinction appears to some degree in most theories, including Hull's explanations of behaviour.

We shall return to these points after considering the evidence which appears to be incompatible with the drive-reduction hypothesis. Before, however, proceeding to examine the experimental evidence upon which these claims are based, a preliminary point should be discussed. It may be argued that, since all speculations above drive-mechanisms are necessarily vague and difficult to relate conclusively to experimental evidence, we should be well advised to content ourselves with the limited aim of listing and classifying those factors which may be observed to act as inducements to the performance of various types of behaviour. That is, we should compile empirically a list of conditions which could be called 'reinforcing'; and in classifying them in this way we should *imply* that they would serve as inducements to performance. The reasons for which they have this property could be ignored. There seem to be at least two reasons why this programme is unsatisfactory. First, of course, we do in fact wish to know the nature of the mechanisms which control behaviour by setting the goals which animals seek at any particular time. A case could be made for saying that, in the first place, such a specification can only be derived by means of inferences from the observation of the conditions under which, and the ways in which, goal-seeking behaviour occurs. That is, the purpose of any such classification would be in part to provide evidence for just such speculations. Secondly, most explanations of behaviour inevitably have reference to the goals which the animal is presumed to be seeking. Quite apart from any direct interest in motivational mechanisms, it is difficult to proceed at all with, for instance, the explanation of learning without making certain assumptions about the goal which the animal is attempting to attain. In fact, two kinds of assumption are usually involved here. We must know whether animals, in a given experiment, are searching for, for instance, food rather than water; and apart from this, and given that we know what incentive is governing the behaviour, the *kind* of explanation of learning which we shall give is closely related to the kind of specification which we have in mind for the goal-selecting and recording mechanisms. In the absence of knowledge of the first kind, we may be led into error by supposing that learning has occurred when in fact the motivational condition of the animal has changed, or vice versa. It will sometimes be difficult to determine which of these has occurred if the *only* means by which we may define the goal for which an animal is seeking is the observation of its behaviour in a given instance. We must have developed at least some specifications of the kinds of goals which may be operating in the circumstances. But in so doing we must avoid the difficulty inherent in, for instance, McDougall's theoretical work, of the undue multiplication of such goals which are not independent. In connection with the second kind of assumption, it should be noticed that important reasons for accepting or rejecting a given explanation of learning may be drawn from hypotheses about the nature of motivational mechanisms. Thus a drive account of motivational mechanisms rather suggests that what occurs during learning is the connection of stimuli with responses. If, on the other hand, we suppose that the goal for which a motivated animal seeks is certain special stimuli, we may be led to regard learning more in terms of association between cues.

We may now consider the drive-reduction hypothesis of reinforcement in relation to certain types of experimental evidence.

2. *The factors controlling drives and the kinds of drive required*

Some slight modification or addition to the hypothesis is required in order to explain the case of avoidance-learning, for it seems to be a necessary requirement here that the drive— whether it be called fear, pain or anxiety—should be switched on by factors external to the organism. In the case of other obvious candidates for the class of primary drives, such as hunger, thirst, sex or some sub-classes of these, the drive is directly related to internal conditions and waxes and wanes with these. But in the case of fear the factor controlling the drive is clearly external in that it depends upon some painful environmental stimulation. Moreover, the drive cannot be held to fluctuate in the same kind of way. The concept of 'satiation', for instance, cannot be used in connection with a fear drive. It is further necessary to assume that the external stimuli which control the drive in this way include not only those which may be supposed to do so in virtue of innate connections such as, for instance, those stimuli resulting from electric shock, but also a larger group which have come to do so as a result of learning. In many ways these are major modifications of the hypothesis, but for present purposes the hypothesis can be stated, without too much alteration, in one of two forms. In the one case it will be supposed that environmental stimuli, as a consequence either of inborn or of learned connections, set up a characteristic physiological condition analogous to that of hunger or thirst, together with a distinctive 'drive-stimulus', and that, when the animal escapes from these stimuli, this condition will diminish more or less gradually. An example of an elaboration of this kind of view may be found in Mowrer and Lamoreaux's (1942, 1946) earlier explanation for simple avoidance behaviour. Reinforcement here will consist in the reduction of this general physiological condition and of the drive-stimulus. Alternatively, a second possible hypothesis is that the external stimuli act directly to set up this drive-stimulus, the general physiological characteristics of fear being concomitant phenomena rather than an essential step in the causal sequence. Again reinforcement will consist in the reduction of this drive-stimulus.

It is not entirely clear which of these two hypotheses Hull had in mind in his proposals concerning 'secondary drives'. In so far as he supposed secondary drives to occur not only in the case of fear but also in that of other drives such as hunger, it seems likely that the latter form of the hypothesis would have commended itself to him. But, however this may be, it is clear that the notion of drive-reduction plays a central part in these explanations of avoidance learning, for the tendency of previously neutral environmental stimuli to evoke the drive is, on all these views, held to be dependent upon the conjunction of the reception of the stimuli and the onset of the drive being succeeded by a reduction in this drive; that is, the animal must escape from the punishment in order to learn that certain environmental signals are danger signals. It should be noted, however, that another approach may be adopted, namely that of 'drive-induction'. This idea has been proposed by, among others, Mowrer (1956), the essential difference between this and the preceding hypotheses being that it is no longer proposed that a learned connection between a neutral stimulus and a drive, or drive-stimulus, depends for its formation upon subsequent reinforcement by drive-reduction. The temporal contiguity of the stimuli and the onset of the drive is here regarded as a sufficient condition for the stimuli to acquire the properties of danger-signals. Unfortunately the experimental evidence on this point seems to be somewhat scanty, but two experiments (Mowrer and Aiken, 1954, and Davitz, 1955) suggest that an environmental stimulus is learned as a danger-signal more quickly when it immediately precedes the beginning of punishment, than when it precedes the end of punishment; and this finding tends to support the drive-induction hypothesis. The latter experiment goes at least some way to eliminate the confusing effects of temporary escape while the shock is being delivered and of possible adaptation to the punishment. If this interpretation of the facts is correct, we have here a case in which we must say either that not all learning is dependent upon reinforcement, or that reinforcement does not always consist in drive-reduction.

Now a partially similar but more important difficulty arising from the drive-reduction hypothesis concerns the explanation of exploratory behaviour in animals. It seems clear that when, for instance, a rat is neither hungry nor thirsty, its movements about its environment are not random. Montgomery's work demonstrates that under these conditions the animal's behaviour is systematic in that it investigates each part of its environment in a comparatively orderly sequence. A similar tendency seems to be imposed upon the behaviour of hungry animals when, for instance, food may be found at either of two places. Here we find alternation of behaviour, and indeed this seems to be a characteristic of most behaviour in circumstances in which the rat must make a choice between alternatives. It also appears that for the most part the animal does not alternate responses, such as turning first left, then right, then left, etc., but rather alternates the places it seeks. This is especially clear in the experiments performed by Montgomery (1951, 1952*a* and *b*), Walker, Dember and their associates (1955), and in an experiment by Galanter (1955). These experiments taken together appear to eliminate reactive inhibition as a possible explanation of this kind of behaviour, for this hypothesis must predict that responses, and not the places or stimuli sought, will be alternated. The behaviour in question gives every appearance of being exploratory in nature, and it may be so described for the moment without thereby committing ourselves to any particular explanation for it. The question now is whether the drive-reduction theory of reinforcement can provide a reasonable explanation for this kind of behaviour.

The first question which must be asked is what drive, in these instances, we must suppose to be governing the animal's performance, the reduction of which will constitute reinforcement. As noted earlier in this chapter, there is a definite risk involved in the use of 'reinforcement' in a vague sense, in that it may lead us erroneously to multiply the number of basic goals which some animal seeks. In the context of the present discussion, of course, this means in turn that the number and variety of drives will be so multiplied. It is therefore important to determine first of all whether 'exploratory behaviour' cannot be regarded as, for example, a search for food (even though the animal is not sufficiently hungry to eat), for other animals, for a mate, and so on.

There seem to be several reasons why any hypothesis of this kind must be rejected. In the first place, it is difficult to see how the alternation results can be explained in this way, for here the animal is certainly hungry, but we must suppose that some further tendency is superimposed upon the food-seeking behaviour since it is rewarded whichever choice it makes. If the animal were searching for some object in addition to food, such as water or a mate (and this the general evidence suggests is unlikely, at least in rats), we should expect it to learn in due course that this other goal was not available in the experimental apparatus. The alternation should, therefore, tend to diminish, but this effect is not found, yet exploratory behaviour in other conditions often does diminish with repeated trials. The latter result seems to depend rather closely upon the details of the environmental conditions. Several experiments, for instance those of Montgomery and of Berlyne (1955), agree in demonstrating a decrease in exploration on successive trials on the same day. Montgomery, however, finds no significant decrease in exploration of the same environment from day to day, whereas Berlyne does find such a decrement. All these results, however, apply to the case of satiated animals and not to the case of hungry animals running to alternative food rewards. It is in any case not entirely clear why, if such animals were searching for some goal object in addition to food, they should be led to alternate their choices in the observed manner.

A second reason for rejecting the hypothesis that exploratory behaviour reflects the operation of some other of the 'primary' drives may be found in the fact that, at least under some conditions, hungry animals explore less than satiated ones. This result has been found by Montgomery (1953b) and by Chapman and Levy (1957). The former found that rats traversed fewer maze sections, and in a less regular manner, when hungry than when satiated. The latter found, first, that animals will traverse a straight alley leading to a goal box, in which walls and floor differ from those of the alley, faster when satiated than when hungry; and, secondly, that when the stimuli in this end box are changed this led to a marked acceleration of running by the satiated animals, but only to a very slight increase of speed on the part of the hungry animals. On the other hand, Alderstein and Fehrer (1955) and Fehrer (1956) have found contrary results. In the first case a much more complicated maze than that used in Montgomery's experiment was employed; and in the second case it was demonstrated that hungry animals are more prone than are satiated ones to leave a familiar environment (in this case a box) for an adjacent strange box in which they had previously never been. On the other hand, Montgomery (1955) has shown that a conflict may be established between the tendency to explore and what appears to be fear of novel stimuli. It seems likely, therefore, that we must suppose that there are two distinct types of behaviour here, exploratory behaviour and food-seeking behaviour; and that whereas hungry animals are more active and more ready to leave a familiar place in the face of potentially frightening stimuli, they are not so prone to explore their environment in that they will not continue to investigate novel stimuli which initial trials have shown not to be associated with food. It should be mentioned here that Wike and Casey (1954) have shown that even thirsty animals, satiated for food, will consistently run to food even though they neither eat nor hoard it. But it is not entirely clear at this stage whether these rats ran to the food in virtue of its character as food or whether the food merely serves as a comparatively novel stimulus, in an otherwise comparatively homogeneous environment, which consequently evokes exploratory behaviour of the kind under discussion.

This account brings us to a third reason for rejecting the hypothesis that exploratory behaviour reflects the activity of some primary drive. For it has been shown by Montgomery (1954) that animals will learn a maze, and by Montgomery and Segall (1955) a black-white discrimination task, in order to reach a complex or spacious environment. It is clear, therefore, that exploratory behaviour cannot be explained solely by reference to the stimuli that are impinging on the animal at the choice-point. Rather, the crucial stimuli in these experiments appear to act as goals for a sequence of behaviour in a manner similar to that in which other stimuli such as food act as goals. Now it could only be maintained both that stimuli other than food, water etc. act in this way and also that the behaviour was due to some primary drive, if we suppose that it is the complexity of the stimuli which is the important factor; that is, we should suppose that animals sought complex stimulus situations because they could not be certain that—in such a region—

food, water, etc. was not to be found. Now although it appears that complexity and spaciousness of environment do serve as goals for exploratory behaviour, so also does *novelty* of stimulation. Thus Berlyne and Slater (1957), Kivy, Earl and Walker (1956) and Dember (1956) have shown, in various ways, that a novel stimulus, or a familiar stimulus in a novel context, will lead to exploratory behaviour. In Berlyne's experiment, rats were given a period of pre-exposure to a card with a pattern drawn on it. They were subsequently allowed to choose between investigating a goal-box containing this and another containing a similar card but with a different pattern upon it. It was found that a measure of 'sniffing time' was significantly greater when the latter was being investigated. Kivy, Earl and Walker (1956) and Dember (1956) placed rats in the starting alley of a maze the goal arms of which were, in the first case, both painted the same colour, and, in the second case, different colours. The animals could see into those arms but could not enter them. In Kivy, Earl and Walker's experiment one of the arms was then changed to a different colour and it was found that, upon being tested, the animals entered this arm. In Dember's experiment the arms were altered so that they were both the same colour and in this case the animals again entered the arm the colour of which had been changed. It is difficult to see how in these experiments, as also in the earlier experiments of Walker and his associates, this kind of environmental change could be such as to lead an animal which was searching for food, water, etc., to explore in the observed manner. It appears, at least in these cases, to be the *novelty* of the environmental conditions which governs behaviour.

There seem to be good reasons, then, to reject the idea that exploratory behaviour is based upon one or more of the primary drives. This naturally leads at once to the hypothesis that in addition to these there is at least one other, variously termed an 'exploratory drive', 'curiosity', and so forth. But before examining this idea it should be stressed that the problem is not that of explaining why the animal is active at all when apparently no drive is operating, but rather that of explaining why it is active in the ways observed. It has in fact been shown by Montgomery (1953a) that exploratory behaviour of the kind under discussion is independent of previous opportunities to exercise in such devices as the activity-wheel.

Quite apart from the danger of undue multiplication of basic drives, there seem to be certain difficulties inherent in the concept of an exploratory drive which deserve consideration. The object of such a postulate is to maintain the drive-reduction hypothesis as a principle governing all behaviour. It is necessary, therefore, that the drive concerned should be formally similar to the other primary drives: that is, there is no purpose in raising the hypothesis in verbal form if the meaning of 'drive' is greatly altered. We may therefore ask what factors increase and reduce this drive, what constitutes deprivation and satiation, in this instance. It has already been seen that some modification of the concept of drive has had to be made to accommodate avoidance behaviour, in that it is difficult to see what constitutes satiation and deprivation. These modifications were not too drastic, however, since the factors which increased and decreased the fear drive could be specified; but it is not entirely clear how even this step is to be taken in the case of the exploratory drive. The earlier specifications of this drive involved that the drive was activated by external stimuli, as in the case of the fear drive. Here, however, the characteristic which determines which stimuli will so activate the drive is novelty. It was supposed, further, that as the animal investigated the stimulus it would become increasingly familiar, and hence the drive would be reduced; and this also is somewhat similar to the reduction of the fear drive as the animal removes itself from the region of the 'danger signal'. However, it is clear that no such simple formulation will predict the observed behaviour, for consider the case of an animal exploring a T-maze for the first time. Let us suppose that it has reached the choice-point and that the two arms of the maze are, at this stage, equally novel. Other factors, or chance, will presumably determine in which direction the animal will turn upon this first trial. Let us suppose further that it enters the left alley, remains there for a short time, and is then removed. Now on the present account of the exploratory drive we must suppose, first, that the drive is already active when the animal reaches the choice point, or that it becomes active when novel stimuli are observed; and, secondly, that exploration of the left-hand alley will lead to the stimuli in that alley becoming more familiar, and hence to some reduction of the exploratory drive. We have here, then, a left-hand turn followed by a drive-reduction. Hence, on ordinary reinforcement principles, there should on future trials be a tendency to turn left rather than to turn right. If a prediction is to be made, therefore, it follows that we must expect the animal to repeat its previous choice, whereas the behaviour observed, and that which the notion of the exploratory drive was devised to explain, is the alternation of choices. It seems probable that a similar difficulty will arise in the explanation of many examples of exploratory behaviour.

This difficulty has been recognised by several workers in this field. Montgomery (1954), for instance, has proposed that in the case of exploratory behaviour an *increase* in drive strength must be regarded as the reinforcing factor. But this proposal involves so radical an alteration of the drive-reduction hypothesis as to deny it the principal advantage it offered of providing at least some degree of theoretical meaning for the notion of reinforcement. We shall find ourselves once again in the situation in which drives or goals may be multiplied unduly, since it will certainly be possible to regard all aspects of behaviour as governed either by the increase, or by the decrease, of some drive or other. Reinforcement would again become an entirely empirical concept, leading to the difficulties involved in this use of the term.

An alternative approach to the problem may be to suppose that familiarity, or lack of novel stimulation, initiates the drive, and that novel stimulation reduces it. In this case we should have something similar to deprivation when an animal is contained for long periods in a familiar environment; and we might also expect to find something in the nature of satiation in the exploration of a very novel environment. There appears to be little experimental evidence on the first point. Butler (1957) has found that when chimpanzees are contained in a cage which does not permit any perception of the outside environment—in this case the animal-keeping room—the frequency with which they will subsequently perform an instrumental response, in order to look through a window in the cage, increases with increasing periods of detention. The major part of the increase, however, occurs in the early part of the curve, only a slight rise in the rate of response being observed with periods of detention greater than four hours. These results are clearly open to the interpretation that for very short periods of detention the animal has still much to discover about the cage in which it finds itself and that, when the window becomes available, there will be some conflict between the tendency to explore the cage and the tendency to investigate the rest of the environment. After longer periods, however, the animal will have become thoroughly familiar with the cage, and so this conflict will not arise; and we should expect that thereafter there would be only a slight rise in the rate of response with further increase in the period of detention. This interpretation does not involve *deprivation* of novel stimuli in any sense of the term similar to that in which an animal may, for instance, be deprived of food. In any event, it is not clear that this approach to the problem provides a satisfactory solution. For if novel stimulation is to have the function of reducing the drive, then we should expect to find something similar to satiation, in that an animal exploring a novel environment which is effectively unlimited in extent should show a gradual decrease in exploratory activity, quite apart from that due to the increasing familiarity of certain sections of the environment. There seems to be no direct evidence bearing upon this point, but general observation of exploratory behaviour, at least in rats, suggests that it is distinctly improbable that this prediction is correct.

In summary, we have, at least in rats, a type of behaviour, designated for descriptive purposes exploratory behaviour, all instances of which cannot be explained in terms of reactive inhibition, or of some kind of stimulus inhibition operating at the choice-point, or by reference to the general degree of activity or exercise. Further, there does not seem to be any way of explaining the behaviour in terms of reinforcement, when by this term is meant some kind of drive-reduction. Attempts to modify this formula to account for the evidence render the use of the terms 'drive' and 'reinforcement' so vague as to be of doubtful utility. On the other hand, the behaviour is in many cases quite plainly goal-seeking, and this is therefore a case in which the drive-reduction hypothesis of reinforcement ought to apply. Some alternative mechanism must be sought, and to this we may return below.

There remains at least one further example of behaviour which deserves mention in this context. It has been shown by various workers—for example, Marx, Henderson and Roberts (1955), and Kish (1955)—that animals, in these instances rats and mice respectively, will learn and perform an instrumental response in order to switch on a light, provided that this is neither too dim nor too bright. It seems to be demonstrated that although the rate of response under these conditions is increased when the animals are hungry, the light cannot plausibly be regarded as having a secondary reinforcing function with respect to any of the primary drives. We must therefore suppose either that this behaviour is exploratory in nature, and should be grouped with that discussed above, or that light serves as a primary reinforcement in its own right. The available evidence is not adequate to decide between these hypotheses. It has, for instance, been reported that the rate of responding does not decline with time as might be expected if the behaviour were similar to other types of exploratory activity, but this finding may be due to the complexity of the situation. If it becomes necessary to suppose that light can serve as a primary reinforcing agent, difficulties analogous to those arising in connection with exploratory behaviour will be found here also.

3. *Drive reduction as the goal of behaviour*

Thus far we have discussed cases of behaviour which, for various reasons, seem not to be capable of explanation in terms of the drive-reduction hypothesis. There are, however, experimental studies which bear more directly upon the hypothesis in that they suggest that even in those cases in which behaviour is motivated by one of the primary drives such as hunger, reinforcement does not consist simply in the reduction of this condition. It may be useful to list some of the principal findings derived from experiments on the conditions governing motivation. First, non-nutritive substances such as saccharine have been found to have reinforcing properties. Sheffield and Roby (1950) have demonstrated that rats, both hungry and satiated for food, prefer certain saccharine solutions to water, and also that they will learn an instrumental response in order to obtain a saccharine reward. These animals showed no reduction of preference during a period of ten days; Sheffield and Roby conclude, therefore, that the behaviour cannot be explained by supposing that a sweet taste has strong secondary reinforcing properties. Smith and Capretta (1956), on the other hand, claim that extinction of secondary reinforcing properties of taste would not be expected within the context of these experiments, since the animals, upon the occasion of feeding each day, would tend to associate the taste of the saccharine with food once again. They compared the time required by two groups of animals to drink a given quantity of saccharine solution. One group was always given the solution two hours after feeding, and the other group received the solution twenty-one hours after feeding. The former group came in due course to drink the solution more quickly. This result they interpret as evidence of an extinction of secondary-reinforcing properties in the case of the twenty-one hour hungry group, based upon the fact that they are only presented with the saccharine solution when their stomachs are comparatively empty. They also find that the two-hour hungry animals will learn a maze to obtain a saccharine reward with fewer errors than will the other group. There seems to be some inconsistency between these findings and some results obtained in an experiment in which I found that, over much longer periods of time, there was no tendency to extinguish a preference for saccharine solutions as compared with water, and that this preference was particularly marked when the animals were trained to eat their whole daily ration of food within one hour. In this case the saccharine was never present at the same time as the food, whereas when the animals were allowed to eat the same quantity of food in their own time, so that saccharine was present together with food for a considerable part of each day, the preference, although still detectable, was very much smaller. It is possible, however, that secondary reinforcement might still be maintained under these conditions, or that Smith and Capretta's results should be explained in terms of variations of liquid consumption due to the food deprivation arrangements used in their experiment. Further investigation seems to be required.

Secondly, Sheffield, Roby and Campbell (1954) have shown that hungry rats will drink more dextrose solution (and run to it faster) than they will water, that both measures are further increased for a saccharine solution, and are further increased again for a combination of dextrose and saccharine. Smith and Duffy (1957a) find that both sugar and saccharine solutions are consumed more than is water both by hungry rats and by animals satiated for food. They also find that, while animals who have access to a sugar solution eat less food during a twenty-four-hour period than do animals for whom only water is available, saccharine solutions have no such effect. In these cases, therefore, both taste and nutritional factors seem to be influencing the behaviour. In a second experiment, Smith and Duffy (1957b) have further shown that animals will learn a maze to obtain a very small quantity of sugar solution at about the same rate when they are satiated as when they are hungry; hungry animals, on the other hand, learn very much faster when the amount of reward is increased. These results are taken as evidence in favour of the hypothesis that two entirely distinct motivational mechanisms are involved; for one of these taste is the reinforcing factor, while the reduction of hunger is the reinforcing agent for the second.

Thirdly, it has been shown by Kohn (1951) that pre-feeding by means of a stomach fistula significantly reduces the rate at which rats will subsequently press a panel in order to obtain food, but that this reduction of response is less than that obtained by the ordinary consumption of the same quantity of food. Berkun, Kessen and Miller (1952) have found a similar effect upon the amount of food that an animal will eat after such pre-feeding; and Miller and Kessen (1952) have further shown that hungry rats will learn a discrimination task in order to obtain milk by means of a stomach fistula. Here again, however, the learning is not so efficient as is that obtained when the reward is the same quantity of milk consumed in the normal way. Miller, Sampliner and Woodrow (1957) have found similar results in the case of thirst.

Fourthly, Coppock and Chambers (1954) and subsequently Chambers (1956) have shown that rats and rabbits will learn an instrumental response when the reward provided is an intravenous injection of glucose, and that the rate of this learning is increased with greater degrees of hunger.

The above experiments, taken together, seem to establish the following points. First, the proposal advanced by Sheffield and others, and similar to the view of Tinbergen, that the reinforcing factor with respect to the primary drives is the performance of a consummatory response cannot account for a large part of the available evidence; for it is difficult to see that a consummatory response is in any way involved in the case of the fistula experiments, or in those of Coppock and Chambers. Second, no simple drive-reduction hypothesis will meet the facts, for here the saccharine studies seem to be inexplicable;

and, further, some explanation is required for the apparently greater reinforcing effects of ordinary feeding as opposed to fistula feeding. Third, it cannot be supposed that taste alone is the reinforcing factor, unless it can be shown that the fistula experiments and those involving intravenous injections produce changes in the mechanism of taste, and do so with great rapidity. It appears that this might be true in the case of intravenous injections; but it is not clear that food injected by means of a stomach fistula could have any such effect. Fourth, the drive-reduction hypothesis can only be sustained in modified form if some definite assumptions are made concerning the manner in which secondary reinforcement operates. This latter concept forms an essential part of any explanation of learning in terms of reinforcement by drive-reduction, and will be considered again below. As far as the present point is concerned, it may be supposed that, for instance, a sweet taste has secondary reinforcing properties, and that it is this reinforcement which, in addition to that involved in drive-reduction, produces the greater efficiency of normal feeding as compared to fistula feeding. It has been seen that there is some conflict of evidence over the question of the susceptibility of saccharine preferences to extinction; and hence it is not clear whether a sweet taste should or should not be regarded as having only secondary reinforcing properties. But however this may be, if the hypothesis is to explain all the preceding results it must be supposed that secondary reinforcement operates by reducing the intensity of the drive at least for a short period. That is, when a neutral stimulus has constantly been associated with, for instance, the reduction of hunger, it must be supposed that it acquires the property of reducing the animal's hunger, at least temporarily. For not only is it the case that animals will learn a motor response (involving a muscle or limb movement) more rapidly when normal as opposed to fistula feeding is used as reinforcement; they also consume less food after normal feeding than after fistula feeding. Such an interpretation of the action of secondary reinforcement gives rise to difficulty in connection with the explanation of learning, and to this point we shall return. But, apart from this, if saccharine preferences are to be explained in this way, then the food consumption of animals given access to solutions of saccharine should be smaller than that of those given access only to water. There seems to be no evidence in favour of this prediction and certain studies, such as those of Smith and Duffy and others, bear against it.

The nearest approach to a comprehensive explanation for all these findings seems to be that adopted by Deutsch (1953). The central characteristic of this theory is that, both in the case of learned and of innate behaviour, the animal is regarded as searching for cues or stimuli. Within this framework, a distinction is made between 'drive' and 'need'. This is in some ways similar to the distinction between 'drive-stimulus' and 'drive' in the work of Hull and of others. It was mentioned above that Hull eventually adopted the view that it was the reduction of the 'drive-stimulus' which was the crucial reinforcing factor for behaviour. Similarly Deutsch proposes that it is reduction of the activity of this neural correlate of 'need' which brings a sequence of behaviour to an end and provides terminal reinforcement. The crucial difference between the theories which is relevant here is that on Hull's view this reduction is normally brought about by the correction of the need—e.g. food is digested and the need eliminated; while in Deutsch's theory this reduction is accomplished by the direct action of certain special stimuli. In order to explain the major part of the experimental results it is necessary to suppose that, at least in the case of hunger and thirst, there are two groups of stimuli which are effective in this way. Thus both taste and stimuli from the stomach, arising in turn from the consumption of food, must be supposed to have the property of terminating the sequence of behaviour and also of reducing the level of the drive. But in order to explain both the results obtained with fistula feeding and those derived from the experiments in which saccharine is used as a reward, it will be necessary to suppose further that there are some stimuli, notably certain taste

A. J. WATSON

stimuli, which terminate the sequence of behaviour but do not reduce the level of the drive. A similar distinction between the action of different groups of taste stimuli cannot be made within the framework of the traditional drive-reduction hypothesis of reinforcement.

Two of the preceding results remain to be explained. The first of these is the fact that animals which are not hungry prefer saccharine solution to ordinary water and that this preference is maintained. It should be noted that although under these conditions an indisputable preference is detectable, and that it is maintained over long periods of time, it is nevertheless a minor effect when compared to the degree of preference exhibited by hungry animals. It seems possible therefore that this finding should be considered together with Wike and Casey's results mentioned above. In both cases it remains to be established whether the behaviour is directly related to food-seeking behaviour or whether it should be regarded more as an example of exploration. The latter type of behaviour has already been seen to present serious problems for the drive-reduction hypothesis; if the behaviour is of the former kind it is not clear why food-seeking behaviour should be displayed by satiated animals. The second question raised by these experiments is why both hungry and satiated animals should learn at approximately the same rate for a very small nutritive reward, while hungry animals learn much faster than satiated rats for a larger reward. Various partial explanations are possible here; but in general it seems likely that this is an instance of a variation of incentive, in the sense of that term in which incentives were varied in Crespi's (1942) experiments.

4. *Drive-reduction and learning*

It might reasonably have been expected that, from the very considerable body of experimental work upon latent and incidental learning, some conclusion could be drawn about the alleged dependence of learning upon reinforcement in the sense of drive reduction. In fact this evidence proves to be conflicting and, for the most part, it is not clear what interpretation should be placed upon these results. There are, however, a few findings which appear to be inconsistent with the drive-reduction hypothesis of reinforcement. The classical findings of Blodgett, Reynolds and others may be explained without violation of the drive-reduction hypothesis, provided that it is assumed, first that rate of learning depends only upon the number of reinforcements obtained, and not, for instance, the amount of drive-reduction involved; secondly, that there is an additional incentive factor governing performance; and thirdly that some slight reinforcement was present during the 'latent' period of these experiments, when the animal reached the end of the maze. The first of these assumptions may be true, and there is direct evidence in these experimental results for the truth of the third assumption. In an experiment by Buxton, however, precautions were taken against the unintentional presence of some reinforcement during the unrewarded phase of the learning by means of varying the places at which the animals were introduced into and removed from the maze. Again, in an experiment by Herb (1940), while there is evidence, as in the case of the earlier experiments, of the presence of reinforcement during the unrewarded period, food was subsequently placed in the maze in such positions that this previous reinforcement should have hindered rather than helped the animals in learning the position of the food. In fact, both experiments gave results showing that some learning had occurred during the unrewarded period. In these cases, then, it appears that learning occurs independently of reinforcement in the sense of drive-reduction.

If we now turn to the studies of irrelevant incentive learning. i.e. those cases in which an animal is required to learn how to obtain an incentive for which it is not motivated whilst searching for one for which it is motivated, it appears that such learning may be found under certain conditions. It is necessary either that the animal should not be too strongly motivated for the relevant incentive—usually water—or that there should be considerable spatial separation of the relevant and irrelevant incentives, or that the prevailing motivation should not be reduced, or that some combination of these conditions should obtain. An explanation of these results consistent with the drive-reduction hypothesis of reinforcement has been proposed in terms of a 'fractional anticipatory goal reaction'. This hypothesis, however, is unsatisfactory, since the presence of any such reaction appears to be unverifiable, and, if the hypothesis is to be interpreted strictly, it has been pointed out by Deutsch (1956a) that it is not in fact adequate to explain the behaviour observed. Since there appears to be no other way in which this type of learning may be explained in terms of reinforcement in this sense, further exceptions to the general rule of the dependence of learning upon reinforcement must be made here.

5. *Secondary reinforcement and the function of incentives*

A concept essential to the explanation both of latent learning and of other forms of learning is that of secondary reinforcement. This term is here interpreted as meaning the acquisition, by previously neutral stimuli, of reinforcing properties as a consequence of the organism's experience of their temporal conjunction with primary reinforcement. It should be noted that the mechanism of operation of this form of reinforcement is normally left vague; that is, a specification of the manner of operation of primary reinforcement is given in terms of the reduction of drive but no analogous statement is to be found concerning secondary reinforcement. Stimuli are simply said to 'acquire reinforcing properties'. Two questions arise. First, is it in fact the case that secondarily reinforcing stimuli act, as it were, retroactively to confirm previous connections in a manner analogous to that in which primary reinforcement is supposed to operate; and secondly, do they have this property by virtue of an ability to reduce the level of drive, if only temporarily, or do they act independently of any such effect?

It has already been seen from the studies of motivation that there is some evidence against the hypothesis that secondarily reinforcing stimuli act by means of reducing the drive if, for instance, the taste of saccharine is to be regarded as a secondarily reinforcing stimulus. Further difficulties would arise in the study of learning if such a view were maintained. It is found, for instance, in the case of lever-pressing for a food reward, that a rat will learn the response more rapidly if some stimulus such as a sound is introduced into the interval between the response and the reward; and similarly it is found that the presence of such a stimulus will retard the rate of extinction when food no longer follows the response. These effects are commonly attributed to secondary reinforcement. If, however, this is supposed to operate by means of drive-reduction, then we must suppose that in both these cases the animals will be less motivated, and so should consume less food than animals who have not experienced these stimuli. This prediction appears, *a priori*, to be implausible; and if it were to be supposed that such an effect would be undetectable because the drive reduction involved would be extremely transitory, the hypothesis would be unsatisfactory on the grounds of unverifiability. It would also, on this latter view, be difficult to explain the reduced consumption of food as a result of normal, as compared with fistula, feeding, as a consequence of secondary reinforcement. If, then, secondarily reinforcing stimuli are to be regarded as *reinforcing* in the same sense of that term as that used in 'primary reinforcement', it seems likely that they must be supposed to have this property independently of any effect upon motivation. And this therefore constitutes a major addition to the theoretical implications of the term 'reinforcement'.

If we set aside this point for the moment and turn to the first of the two questions above, it may be expected that the degree to which stimuli acquire secondary reinforcing properties will depend, among other things, upon the magnitude of the primary reinforcement with which they are associated. The experimental evidence upon this point is somewhat conflicting.

It appears that if different groups of animals are rewarded in a distinctive goal-box by differing amounts of food, and if this goal-box is subsequently placed in a maze without further primary reward, there is no difference in the rate at which the various groups learn to run to this part of the maze. The work of, for instance, Hopkins (1955) and Lawson (1953) supports this conclusion. On the other hand, it has also been shown that if the same group of animals are taught to run to two different goal-boxes in which they are rewarded by different amounts of food, and if they are then placed in a T-maze in which these goal-boxes are placed at the end of the arms, a preference is shown for that box in which the greater reward was obtained. These results are due to the work of, among others, D'Amato (1955), of Powell and Perkins (1957) and to a further experiment by Lawson (1957). Lawson suggests that stimuli will acquire differential secondary reinforcing strength only when an animal has experienced differing amounts of primary reinforcement. This point must now be considered more generally.

It is clear from the work of Crespi (1942) that variations in the amount of food given to hungry rats affects the speed at which they will run along a straight alley in order to obtain this reward. The main effect of variations in amount of reward seems to be upon the maximum running speed which will be attained, although some difference is also found in the rate at which the performance improves to this maximum. There does not appear, however, to be any very satisfactory evidence that the rate of maze learning is much affected by such variations in amount of reward, if the measure of performance is in terms of errors made in traversing the maze, rather than in terms of time required to do so. It may be supposed then that learning, in the senses of elimination of errors and of speed of running, are independent of each other, and that moderate variations of incentives influence the latter but not the former. If this were so, it would follow that information concerning the amounts of reward to be found in various places must be in some way stored and that this information governs the speed at which the animal runs; the elimination of errors, on the other hand, would depend upon information about the 'lay out' of the maze, including the position, but not the amount, of reward. In the case of the first group of experiments mentioned above no difference in the extent of preference for the secondarily reinforcing goal-box would then be expected, since this would depend only upon the use of the type of information which would be common to all groups. In the second group of experiments, however, the first type of information would also be relevant, in that the animals would be faced by a choice between proceeding to stimuli associated with differing amounts of reward. In general, then, maze learning might be divided into at least two factors. The learning of the way through the maze to any particular goal would be independent of the amount of incentive involved. But the speed of running, and also the selection of which goal to adopt, would be governed by, amongst other things, such variations of incentive. At least one reservation to such a simple scheme must, however, be made. It must be supposed that if the reward is very small indeed, although the route through the maze might be learned nearly as quickly as when a larger reward is involved, this will not appear fully in the animal's performance since it will tend to search elsewhere for other, more satisfactory, rewards. The results of Smith and Duffy's experiment referred to above, which showed that hungry and satiated animals learned a maze equally quickly for a very small reward but that hungry animals learned much more quickly for a larger reward, might be explained in this way.

It should be noted that this kind of explanation does not require the assumption that the elimination of errors in a maze depends upon either primary or secondary reinforcement in the sense of some kind of retroactive effect upon the learning occurring immediately prior to reward. The function of motivation becomes that of determining what goal should be sought; and reward has the function of signalling the fact that this goal has been achieved and of determining, first, how fast an animal will run to the goal, and, secondly, of selecting goals when more than one is available. In general it might be said that this kind of behaviour is regarded more in terms of 'cue-seeking' and less in terms of the connection of stimuli and response due to subsequent reinforcement. A detailed theory of this general kind has been developed by Deutsch (1953, 1956b). Such a theory, as mentioned above, also seems to be consistent with at least the majority of the findings of the studies upon motivation. The point of importance here is that both secondarily reinforcing stimuli and also primary reinforcement are regarded as performing two functions. The first of these is that of providing a cue used by the animal in learning its way about its environment. In this respect these stimuli do not differ from any other cues used by the animal; all such cues would be secondarily reinforcing. The terminal cue for which the animal seeks, however, is regarded as having the additional function of signalling the achievement of the goal and of determining the incentive value of the reward. This kind of approach to these problems clearly bears some affinity to Tolman's views. The point arising here, however, is that in so far as the concept of reinforcement is involved at all in this general kind of account, it bears very different theoretical implications from those implied by the drive-reduction hypothesis.

6. *Conclusion*

It has been argued that it is desirable to use such terms as 're-inforcement', 'drive' etc. in such a manner that they carry with them some theoretical implications concerning the nature of the mechanisms responsible for the behaviour which these terms are employed to describe or explain. One such interpretation is to be found in the 'drive-reduction' hypothesis. It appears that this interpretation leads to difficulty in the explanation of certain features of avoidance learning, in the explanation of the results of some of the studies on motivation, in the explanation of latent learning and in the discussion of the nature of secondary reinforcement. Although it is not the purpose of this chapter to attempt to establish the validity of any alternative account of reinforcement, it may be suggested that these difficulties may substantially be overcome if reinforcement is analysed in terms of two separate factors, namely the learning of association based only upon the experience of temporal contiguity of cues, and the function of reward as determining the value of incentives involved in any given environmental situation. To speak of a stimulus x as reinforcing may, therefore, imply either that it accompanies or follows some other stimulus, y, and that the presentation of it in conjunction with y will strengthen some stored correlate of this association; or it may be to imply that this stimulus is one of the special class which determine the degree of incentive to performance involved in the experimental situation, and the reception of which brings a given sequence of behaviour to an end.

There remains the problem of the explanation of exploratory behaviour. It has been argued that, while the evidence is not conclusive, it seems unlikely that this type of behaviour is dependent upon the operation of drives, and upon the reduction of these in a manner analogous to that in which, for instance, food-seeking behaviour has been held to be so dependent. It may be suggested, very tentatively, that in the light of the preceding analysis of reinforcement this behaviour may be explained by supposing that when the animal is satiated for the 'primary incentives', that cue is selected for the goal of behaviour which is least associated, according to one or more criteria, with other cues. In this case, reinforcement would only enter into the explanation of exploratory behaviour in that sense in which it refers to the temporal association of cues. But, clearly, much more experimental evidence is required before it would become profitable to develop any such analysis in more detail.

PART IV

THEORETICAL APPROACHES
TO BEHAVIOUR

INTRODUCTION

Advance in biology has always been in large degree dependent upon advances in other sciences, in particular physics and chemistry. But we have witnessed in recent years a surprising transformation of biological thinking which can only be ascribed to the impact of modern technology. Principles derived from control and communications engineering are being increasingly brought to bear upon biological problems and 'models' derived from these principles are proving fertile in the explanation of behaviour. Although opinions differ as to the propriety, and usefulness, of physical models in biology, it would certainly be unwise to dismiss the whole approach as a transitory freak of fashion. After all, if it is accepted that the principles of physics govern biological systems, 'models' constructed in accordance with these principles—whether they be purely conceptual or actually embodied in 'hardware' should give us valuable insight into their organisation. This, at all events, is the hope that unites scientists drawn from many disciplines under Norbert Weiner's banner of Cybernetics. Can their hope be translated into solid biological achievement?

This new influence has been particularly potent in experimental psychology at Cambridge, where it owes much to the brilliant insight of the late Kenneth Craik. Indeed, Craik (1947 and 1948) was one of the first to apply concepts derived from control engineering to human behaviour and to treat the human operator of a machine as himself an engineering system. Gregory carries some of Craik's ideas further when, in Chapter XII, he urges that organisms in general should be viewed as engineering systems and that their performance should be assessed with reference to the engineer's criterion of efficiency. He insists that the organism involves in its construction no principle that is not actually or potentially known to physics and argues that it should therefore be possible to erect, on the basis of known principles, 'models' of a biological system in accordance with the specifications of the engineer. But he makes it clear that such a 'model' is certainly not intended to provide a finished explanation of biological phenomena and may well be found not to correspond with biological fact. Its sole purpose, Gregory contends, is to suggest, in the light of the principles of machine construction, how an organism *might* adapt to its environment or how certain of its parts or systems might execute their proper functions. In short, 'models' serve no purpose other than the direction (or re-direction) of biological research in the light of the principles of physics.

Gregory applies these ideas in a most interesting way to some problems of neurology, in particular that of the cerebral localisation of function. He points out that many classical conceptions in neurology, e.g. that of discrete anatomical 'centres' each subserving a particular function, are difficult, if not impossible, to reconcile with the position of one who approaches the nervous system from the standpoint of the engineer. In this connection, he issues a timely warning against too ready acceptance of the methods of ablation and stimulation in the study of the brain. Until we understand its basic design, Gregory contends, interference with its parts cannot help us to understand how it really works. Although many will disagree with this view, the onus of proof is placed squarely on the shoulders of those, such as Weiskrantz and the present writer, who have advocated ablation as a legitimate method of attack upon the problem of brain function.

The positive gain to behaviour study of an 'engineering' standpoint is well brought out by Barlow, in Chapter XIII. He is concerned essentially with the 'coding' of sensory information in the central nervous system and the physiological mechanisms that may be supposed to bring it about. With Gregory, Barlow holds that the nervous system must utilise in its mode of construction principles acceptable to the communications engineer, and he is therefore constrained to formulate the problem in the first instance in terms of a machine. Having stated the basic principles in this way, he then proceeds to seek for actual physiological mechanisms which might, in fact, effect the necessary 'compression' of sensory information. Certain of these are almost certainly peripheral, and in this connection the well-known phenomena of sensory adaptation come to be seen in a new and exciting light. The phenomena of lateral inhibition in the retina, our knowledge of which owes much to Barlow's own work, are likewise seen to have a hitherto unsuspected bearing upon the 'compression' problem. Further, Barlow is not afraid of extending his ideas to central processes, in spite of the fact that direct physiological information bearing on the issue is unfortunately still very much more limited than one might wish. He even goes so far as to suggest that essentially the same type of 'compression' mechanism may govern reaction at the psychological level and bring about the economy of memory and thought. Although admittedly speculative, Barlow's ideas will be of great interest to the psychologist seeking a basis to his subject in the mechanism of the brain.

Chapter XIV, the last in this section is of a rather different kind. Violet Cane is less concerned with theory than with methodology, but her approach brings out well the gain to biology of a proper understanding of probability theory and modern statistical method. Like Gregory and Barlow, Violet Cane understands well the essentially statistical character of scientific observation and theory and this is perhaps the real link between the three chapters. Her chapter is concerned with the logical and methodological considerations involved in the study of behaviour, whether it be the deceptively simple field observations of the ethologist or the more contrived situations of the experimental psychologist. In both cases, she shows that the results depend to a degree seldom acknowledged upon the ways in which the relevant variables are chosen. As she rightly points out, the results of many experiments have been vitiated by the fact that too few variables have been chosen or that the measures used have not been truly independent. Very real theoretical difficulties can often be ascribed to unsuspected error in experimental design.

Violet Cane's chapter has the further merit in that it brings out the underlying unity of behaviour study in spite of the diverse forms which it has taken and the diverse disciplines in terms of which it has been organised. This unity is to be found in the belief in experiment common to all workers and in acceptance of the criteria, logical no less than empirical, which experiment must attain in order to be good experiment. Her analysis makes it clear that, from the standpoint of empirical science, there is no fundamental difference between 'observation' and 'experiment', provided only that the measures are well chosen and that proper control of the relevant variables can be attained. There is a sense, then, in which Violet Cane's chapter epitomises the unity of the discussion group which gave it birth. If for this reason alone, it may be regarded as the most appropriate for our closing chapter.

XII. THE BRAIN AS AN
ENGINEERING PROBLEM

'The brain serves to cool the blood'
                                    Aristotle
'The brain is like an oven, hot and dry,
Which bakes all sorts of fancies, low and high'
                              The Duchess of Newcastle

1. *Introduction*

Biologists generally refer to the activity of living organisms as
'behaviour'. When talking about machines, engineers tend to
use the word 'performance'. To interchange these words is to
raise a smile, perhaps an appreciative smile, but the speaker
risks being labelled quixotic. It does appear, however, that
the terms 'behaviour' and 'performance' are interchanged
much more now than in the past, the reason almost certainly
being the influence of cybernetic ideas, which have unified
certain aspects of biology and engineering. Some biologists
even go so far as to regard their subject as essentially a branch
of engineering, and some engineers use examples from biology,
such as living servo-systems, to illustrate their principles. The
activity of organisms is most often referred to as 'performance'
when their efficiency is being considered. Thus play-activity
is called 'behaviour', while a skilled worker's activity may be
called 'performance'. This change is interesting, for it brings
out the influence of the engineering way of thinking upon even
lay thought about human and animal activity.

It is worth stressing that physical principles have not always
been accepted as appropriate to biology. Aristotle did not make
any basic distinction between the living and the non-living,
but a sharp distinction was drawn by Kant in the *Kritik der
Urtheilskraft* (1790). Perhaps Kant was so influenced by the
patent inadequacy of Descartes' attempts to describe organisms
in terms of his Natural Philosophy that he was led to say that

307

the behaviour of living systems cannot be governed by causal principles applicable to the physical world. To Kant, living systems are somehow outside the dictates of the laws of nature, and this has been held by some biologists since—certainly as recently as E. S. Russell (1946), who regards 'directiveness' as a special property of living organisms. The influence of Kant's teaching upon biology has been profound and (to the cybernetically inclined) disastrous. Historically, it has led to the creation of special entities to distinguish the living from the non-living, such as Dresch's Entelechy, Bergson's *élan vitale* and the Emergent Properties of the Gestalt school of psychology.[1]

We do indeed think of inanimate matter as somehow different from animate matter. If we did not, these words would have no special meaning, for no distinction would be implied. The point is this: is it useful to describe, or to explain, this difference by postulating some *special factor* which is held to be present in animate and absent in inanimate matter? To biologists looking for general explanatory concepts, after the manner of the physical sciences, such postulated special factors must appear harmful. These factors do not enable us to relate phenomena; they do not provide any sort of picture; they do not enable predictions to be made. The trouble with Entelechy, *élan vitale* and the rest is that they do not help us to understand. Such terms give a sacrosanct air to life, which may be pleasing, but which tends to warn off further enquiry. The Gestaltist's plea for the special nature of 'organic unities' is effectively a warning against attempts at further analysis, the doctrine being that it is *in principle impossible* to analyse the whole in such a manner that its activity can be completely described by the causal relations between the parts. It is, however, just this sort of

---

[1] Köhler, in his book *Die physischen Gestalten* (1920), takes a different view from that of most Gestalt writers. He does not suppose that organisms are unique in this respect, but rather that Emergence is to be found in many physical systems. Some philosophers have also taken this view. It leads to the difficulty that 'emergence' is used so generally that it points to nothing special. This point is well discussed by Madden (1957) and is also considered by Gregory (1953).

analysis which is the goal of exploration in the physical sciences. Further, it is important to note the *in principle impossible* here: it is not the complexity of the task which is held to make analysis impossible, but rather the claim that the organic world is such that analysis into parts is doomed to failure, however complete our knowledge of it may be. Curiously, this is regarded by some as an exciting and interesting discovery about living systems. This is an attitude puzzling to those who believe that useful explanations in science should take the form of analysis into simpler elements. Now it *could* be that there is something irreducible about living systems which defies such analysis, but surely we have no right to claim this until the traditional types of explanation have failed for a very long time, and certainly not now while exciting advances are being made in the biological sciences. If we seek the types of explanation found in the physical sciences, *élan vitale*, or the concept of Emergence, will appear as doctrines of despair. To postulate such special unanalysable factors is to make a philosophy of pessimism. To say that *x* is an Emergent Property is to put *x* into the limbo of the unknown and shut the door upon it, while warning others against peeping through the keyhole.

To regard the brain as a problem in engineering is to look for possible solutions in terms of engineering principles to the questions set by biological enquiry. This chapter is concerned not with answers to specific questions—such questions perhaps as: How are memories stored? How does the eye guide the hand? What are dreams made of? But rather will it attempt to discuss some of the difficulties in taking over engineering methods into biology, and some implications of this approach for the study of the central nervous system.

An alternative to the Kantian doctrine is to say that living systems are *machines*. The cybernetic view is often put in this way, but it has objections. If we use the term 'machine' to include living organisms, it loses its major classificatory use. Further, the term 'machine' is very difficult to define in general terms. We might call a given system a machine though it has no predictable output, displays goal-seeking behaviour, and is in fact indistinguishable in its behaviour from at least simple living systems. If we mean merely that it is man-made, then the distinction is trivial. We cannot get away with an ostensive definition of 'machine' (pointing to all existing machines), for we must allow the possibility of future new kinds of machine, and these could not be included. If animate systems are called 'machines', at least two important things might be meant: (1) that their functioning could be described in terms of known physical principles, or (2) that their functioning could be described, if not in terms of principles known at present, at least in terms of principles which *could* be known to us. This is to say that living organisms are in fact so constituted that we could in principle understand them as engineers or physicists understand their systems. It appears that to call an animal a machine is to indicate that its manner of functioning is not *essentially* different from machines which might be designed or made by men. To deny that animals are machines is, it would appear, to suppose that they *are* essentially different. Those who take the former view feel that existing or possible machines performing similar functions may provide clues as to how animals work, and in particular how their central nervous systems are organised. Those who hold that animals are not machines refuse to accept that this could ever give the whole story. Both types of biologist might well agree that we should go as far as we can in looking for analogies, while being careful not to oversimplify or to accept similiarities in a naïve manner.

2. *The use of engineering criteria for deciding between models of brain function*

When a biologist or engineer considers what sort of system might be responsible for producing a given function, he may run up against one of two difficulties: (1) that there does not seem to be *any* known type of system capable of just the observed functions under the given conditions, or (2) that there is a *large number* of possible mechanisms, any of which might provide the required functions. We cannot say anything here about the first contingency, except of course that further observation, experiment or thought might suggest possible mechanisms, but we can say something about the second. It is worth thinking about this, for the principles available for deciding which of various alternative types of mechanism are appropriate are just the principles we need for verifying cybernetic hypotheses. Without such principles we can do no more than guess.

Consider an engineer in a position of doubt about how an unfamiliar machine works. We may take an actual example of a dramatic kind: consider the problem of discovering the manner of function of the control mechanism of an enemy's secret weapon, such as the V1 rockets during the last war. The engineer could make use of the following considerations. First, it was clear that the rocket had been made recently by men in Germany. This knowledge that they were man-made was clearly enormously important, though probably never explicitly stated. Martian rockets would offer many more alternatives, including the high probability of principles quite unknown to us. As it was, new principles were unlikely, though possible. Secondly, examination of rockets which failed to explode revealed many already familiar components such as motors, condensers, valves etc., and a great deal was already known about these. Thirdly, it would seem certain that the rockets must have been designed as efficiently as possible. Now how far does the biologist examining brain function share these assets?

(i) Since living organisms are not designed and made by men, any number of new principles might be expected, as in the imaginary case of Martian rockets. As an example, it is now believed that feed-back loops are important in organisms, but these were not known to the engineer until Clerk Maxwell's work in the last century, and there could always be further more or less fundamental principles involved which are so far unknown to engineers.

(ii) Examination of the brain reveals many identifiable 'components', such as Betz cells and amacrine cells, but the functional properties and circuit potentialities of cells are not as well understood as the functional properties of electronic or mechanical components—and even these have their surprises.

(iii) Efficiency is a difficult criterion to apply to biological systems for a logical reason: it cannot be assessed without some idea of purpose. It is, however, important to note that the notion of efficiency (and also that of purpose) does not imply specific design for a known end. Thus it might be said that a screw-driver makes a good paint scraper, though it was not designed for that purpose. For something to be said to be efficient, it must be efficient for a stated end though not necessarily for a designed end. Thus if it said that some postulated brain mechanism is more efficient than some other mechanism, we must know what end these mechanisms are supposed to serve, and we must know how to assess relative efficiency towards this end. We may ask, for example, 'how efficient is the eye?' and its efficiency may be measured. Thus its acuity and its sensitivity may be measured and expressed in appropriate units. The difficulty arises when we do not know what to measure through not knowing the functional significance of the structure or system involved. Clearly we could not talk about the efficiency of the eye if we did not know that it subserved vision. If a system is found to be highly efficient, in general but few possibilities are left open when it comes to guessing how it works—not many engineering tricks would be good enough.

When an engineer talks about efficiency he may mean a number of things; he may simply mean that it works well, or that its fuel consumption is low, or that the capital or running cost is low, or a number of other things. If the biologist is to make a reasonable guess at which type of mechanism is responsible for a given type of function, and he wants to use efficiency criteria, he must be clear which criteria it is appropriate to take over, and this raises a number of difficulties. Let us, for the fun of the thing, consider a few engineering efficiency criteria in the context of biology.

(a) *Thermal efficiency*

This may be used for power systems. The efficiency E of a heat engine is given by $E = W/JQ$, where W is the useful work done by the machine when a quantity of heat, JQ mechanical units, is supplied to it. Since no machine can create energy, W cannot be greater than JQ, so that no engine can have an efficiency greater than 100%.

Now, knowing the total thermal efficiency of a given machine, and knowing the expected efficiency of the type of system by which it is supposed to function, it is clear that if the actual efficiency is higher than predicted, then the hypothesis is false and some other explanation must be sought. If, on the other hand, it is too low some cause for the loss may always be postulated. It follows that where the predicted efficiency is high, more possible solutions are ruled out as too inefficient to be likely, and so the criterion is more useful. This criterion might be used in biology to test a hypothesis about, say, conversion of chemical energy into mechanical energy in muscle. It is hardly applicable to the brain because its thermal properties do not seem important to us, though they did to Aristotle when he regarded its function as cooling the blood.

(b) *Information efficiency*

Information rate may be defined by the rate of transmission of information defined by binary choices, or 'bits'. The Hartley-Shannon Law, which is basic here, states that

$$C = W \log_2 (1 + S/N),$$

where C is the channel capacity, W the band width, S the average signal power, N the average noise power. Some communication systems are more efficient than others. In particular, a change in the manner of coding the information might make a large difference to the efficiency of the system. Now this does appear to be directly applicable to neurological systems, which is one reason why quantitative estimates of information rate for human subjects, as made by Hick (1952a) and Crossman (1953) among others, are of great importance here. The most efficient type of coding is important once we think of the nervous system as handling information and as subject to the same limitations as a man-made system. This criterion has in fact been applied to test between different possible codes adopted by neurones in transmitting information. MacKay and McCulloch (1952) decided, tentatively, in favour of pulse interval modulation for peripheral nerve fibres on this basis.

(c) *Capital cost*

This is difficult to assess. We might at least say that where general, or some specific, nutriment is in short supply, cells may be 'expensive'. Further, weight may be at a premium, which will limit the permissible number of cells. Also, it might be the case that increase in the number of cells would impose an informational strain on the available information coded in the gametes.

(d) *Running cost*

Not much can be said about this, beyond the obvious point that if food is scarce it will be an 'expensive' commodity.

(e) *Simplicity*

This is difficult. The engineer favours the 'neat' solution to a problem, and he dislikes certain 'complicated' types of mechanism. This may be in part due to the aesthetic appeal of simplicity, but simple mechanisms perhaps also tend to be cheaper and more reliable, though not necessarily so. Carburettors for petrol engines have in fact become more and more elaborate, with gain in running economy and over-all reliability.

Have we any reason to suppose that we should find in nature the 'simplest' way of going about engineering problems? Certainly nature is handicapped by lack of many materials and techniques indispensable to the engineer. It is striking, for example, that flight with flapping wings is for an engineer more complicated, and in every way inferior to, flight with fixed wings, though the former is found in nature. But then nature has not got a suitable engine to provide forward drive independently of the wings, and nature has not got true bearings, or the wheel. This case is far removed from neurology, but neurology also provides examples. A familiar one is that of the retina, which is 'inside out'. The light has to pass through layers of blood vessels, ganglia and supporting cells before it reaches the receptors. This optically shocking arrangement appears to be dictated by embryological, or perhaps basically developmental, considerations. Considerations of this kind make the use of the criterion of simplicity difficult and dangerous to apply.

(f) *Length of life*

Some types of machine outlive others. There are many reasons for this—choice of materials, friction between moving parts and many more. This criterion for deciding between rival designs can hardly be applied at present to the living machine because of its self-repairing properties.

We conclude that some engineering design criteria can be applied to biological systems in order to ascertain which, among many possible types of mechanism, is the most likely to be operating in any particular case. Efficiency criteria, particularly thermal and information efficiency, seem to be the ones most readily applied in the biological context, but in some cases other criteria might also be used. If this way of linking behaviour study with neurology is adopted, then rather precise 'engineering-type' data will be required. It is unfortunate that it appears difficult to apply the other criteria commonly used by engineers. As a result, cybernetic writing easily becomes science fiction, where the supposed theories and mechanism may be limited in variety only by the imagination of their inventors. This is unfortunate for a vitally important approach to biology.

3. *Localisation of cerebral function*[1]

What is meant by saying that some feature of behaviour is localised in a part of the brain? It cannot mean that the behaviour itself is to be found in the brain, or that a region of the brain can be sufficient for any behaviour. The intended meaning is that some necessary, though not sufficient, condition for this behaviour is localised in a specific region of the brain.

The evidence for localisation is mainly from studies of ablation and stimulation of regions of the brain. If, for example, when a point on the occipital cortex is stimulated, flashes of light are reported by the patient, it is generally held that this region of the cortex must be important for vision. If an area in the left frontal lobe is damaged and speech is found to be disturbed, it may seem that we have found something causally necessary for speech. But have we?

This area may be *necessary* for speech (i.e. if it is removed, speech may disappear) but so also are a number of other parts of the organism, for example the vocal cords, the lungs and the mouth. There is nothing special about the brain here. It may be that the 'speech area' is concerned only with speech, but if so it is not unique in this respect either: if we except coughing, the vocal cords have no other function but to subserve vocalisation. Now we may say that the vocal cords are *causally necessary* for speech, and also that the 'speech area' is somehow *causally necessary*, but it is not clear in the second case just what the causal functions are, though we do understand the causal role of the vocal cords. There is an important point here: we may say that *A* is the cause of *B* if *A* is found inductively to be a necessary condition for *B*, and the evidence may be purely

---

[1] I would like to express my gratitude to Mr. A. J. Watson and Professor O. L. Zangwill for their help in discussion of this section.

inductive for this type of causal argument. No understanding of the mechanisms involved is required to assert the causal relation between A and B. But we may also say that A causes B on *deductive* grounds, when we understand (or think we understand) the mechanism by which A produces, or causes, B.

Once we distinguish these two types of argument from physical structure and function to causal relationship, we should ask which sort of causal argument is being used in discussions about brain function. Take the case of the speech area. It would appear that the reason why this region of the brain is held to be associated with speech is that speech is found to be defective or absent when the region is damaged. This is clearly an inductive argument, and it does not presuppose or imply any knowledge of how the speech area works, or what causal part it plays in the production of speech. Again, we know fairly clearly the causal role of the vocal cords, but not that of the 'speech area'.

Consider now the word 'function'. We may say that it is the *function* of the vocal cords to vibrate in certain ways, producing pulses of air which resonate in cavities ... we see the causal role of the vocal cords and we come to understand the mechanism of speech production. And now what about the word 'localisation'? What is it to say that a *function is localised*? The question is: How can we say that a function is localised until we know what the function (of a given bit of brain tissue) is? To say this we need to know in some detail how the system works. It seems that before we can talk usefully about localisation of function we must have some idea of *how* the system works.

It might be interesting to consider how an electronics engineer deals with, and represents, specific functions in a complex device. He uses three types of diagram to represent an electronic machine. These are (a) *blue prints*, showing the physical locations in space of the components, with their sizes and shapes. These drawings will give dimensions and describe the structure, so to say the cytoarchitecture, of the machine. (b) *Circuit diagrams,*

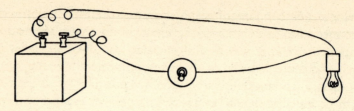

1. The 'blue print' type of diagram

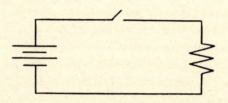

2. The 'circuit diagram' type

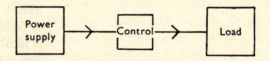

3. The 'block diagram' type

N.B. In a complex system each box
will include many components

Fig. 19

(*a*) Shows a simple *blue print* type of diagram. Pictorial representations of components are linked with paths of conductivity (wires), also shown pictorially. This may be compared with histological descriptions of the brain.

(*b*) Shows a *circuit diagram*. The symbols show conventional functional properties of the components. This may be compared with physiological descriptions of the brain.

(*c*) Shows a *block diagram*. The blocks show the functional units of the system, indicating the causal processes in terms of the flow of power or information. This may be compared with cybernetic descriptions of the brain.

318

in which each component is shown in diagrammatic form with its connections (usually paths of perfect conductivity) with other components having idealised properties. (c) *Block diagrams*, in which there are a number of boxes connected together with flow lines, each box being labelled with its 'function'. Thus one might be labelled 'Radio frequency amplifier', another 'mixer' and so on. Now what knowledge of the system is conveyed by (*a*), (*b*) and (*c*)? And what knowledge, therefore, must the man who designs the diagrams have of the system? (*a*) requires a knowledge of the look of the thing; (*b*) implies information about certain selected properties of the 'components' of the system. Thus, to see ⊣⊢ and marked, say, $0 \cdot 1 \ \mu\mathrm{f}$., conveys a great deal of information about what will happen to voltages passing through the component. For example, that direct current will be blocked, while high frequency will hardly be affected except for a phase change. But the general effect of these changes produced by a component will not be apparent except in terms of (*c*), the block or function diagram. The condenser may then turn out to be part of an oscillator, and then its purpose within that functional section of the device could be stated. For example it might provide feed-back between anode and grid of a valve. Now when a trained engineer looks at a circuit he can very often see almost at a glance what it will do and how it does it. With an unfamiliar circuit this might be difficult, but he could probably predict its function and performance given the circuit valves and a slide rule. To do this he will make use of a number of generalisations, or Laws, such as the conditions for obtaining oscillation, or linear amplification or whatever it is. The point is that (*c*) implies these general principles while (*b*) does not. Only a skilled engineer could design or work out a block diagram, but anyone who can recognise components by their appearance could draw a circuit diagram. The question for neurology is: Do we want diagrams of structure of the 'circuit' with component characteristics, or functional block diagrams? I would suggest that only the last describe how a system works. They are comparatively simple, but alone convey (and imply for their drawing) a knowledge of the function of the system.

The neurologist can discover the properties of *his* components, the neurones; he may discover how the components are wired up. Can he discover the causal mechanisms involved? Can he find out where they are situated in the brain?

There seems to be a widespread hope that, by ablation and stimulation of parts of the brain, functional regions may be discovered, these being logically the same as the boxes in a block diagram. Is this a reasonable hope?

4. *Ablation and stimulation as techniques for discovering functional regions of the brain*

Suppose we ablated or stimulated various parts of a complex man-made device, say a television receiving set. And suppose we had no prior knowledge of the manner of function of the type of device or machine involved. Could we by these means discover its manner of working?

In the first place, and most important, to remove a part of a machine, even a discrete component, is not in general to remove a necessary condition for some feature of the output. If a part is removed from a complex machine, we do not in general find that simple elements or units are now missing from the output. It should be noted here that the functional processes taking place in the components, or groups of components, of a machine are generally quite different from anything in the output. Thus we do not see the spark in a car engine represented in its output—we see wheels turning and the car moving: no spark. If a component is removed almost anything may happen: a radio set may emit piercing whistles or deep growls, a television set may produce curious patterns, a car engine may back-fire, or blow up or simply stop. To understand the *reason* for these 'behavioural' changes we must know at least the basic principles of radio, or television, or car engines, or whatever it is, and also some of the details of the particular design. Of course, if we already know about radio, or engines, then these abnormal manifestations may well lead to correct diagnosis of a fault: the difficulty is to reverse the procedure.

Consider a television set which has, of course, two quite distinct outputs—sound and vision. Some 'ablations', or 'extirpations', may quickly reveal which parts are *necessary* for each output, and also which parts are *common* to the two outputs. In the case of the brain, there is a large number of inputs and outputs: the limb movements, the face with its various expressions, the voice, and so on. It may be a fairly simple matter to discover regions of the brain which are necessary for these various outputs, and in general they will lie near the peripheral output of the system. The inputs, the senses and their projection areas, we might also expect to locate in this way without undue difficulty. What I suspect *is* difficult, indeed impossible, is to locate functional regions of the system. It seems to me that this conclusion is forced upon us by considering the possibility of isolating elements of a complex output in a single channel in the case of man-made machines. In a serial system the various identifiable elements of the output are not separately represented by discrete parts of the system. Damage to a part may indeed introduce quite new factors into the situation, and these could only be comprehensible when we are provided with a model indicating the function of the parts. If the brain consisted of a series of independent parallel elements with separate output terminals for each, like a piano, it might be possible to identify behavioural elements with particular parts of the system, as the various notes of the piano might be regarded as being 'localised' in the piano; but where output is the result of a number of causally necessary operations taking place in a series, then this is not possible. The removal, or the activation, of a single stage in a series might have almost any effect on the output of a machine, and so presumably also for the brain. To deduce the function of a part from the effect upon the output of removing or stimulating this part we must know at least in general terms how the machine works. The point here, perhaps, is not so much that the piano is a parallel rather than a serial system, but that it is a set of largely independent machines in one box. Where they do interact, as in the pedal systems, then one 'ablation' may affect all the notes. Parts of the brain could be independent.

The effects of removing or modifying, say, the line scan time-base of a television receiver would be incomprehensible if we did not know the engineering principles involved. Further, it seems unlikely that we should discover the necessary principles from scratch simply by pulling bits out of television sets, or stimulating bits with various voltages and wave forms. The data derived in this way might well lead to hypotheses once we knew something of the problem in engineering terms.

But we should, in some systems, be able to map projection areas and delimit pathways, and this is a good deal. Analogy with familiar physical systems strongly suggests that to go further these studies should be used to test rival hypotheses of brain function, rather than to attempt to isolate functional regions. This brings us back to the idea of physical model explanations, with ablation and stimulation studies as one way of trying to decide between rival models. We are left with the difficulties besetting this approach: in particular, the brain might work on some novel principle, and then its true manner of function would never come up for testing by any experimental technique. It would clearly require a most highly sophisticated set of techniques to discover a quite new principle in the living brain, but this is conceivable. Perhaps the principle of scanning, or heterodyning, could be discovered by these techniques, even in a jelly.

It is a common finding that with electronic equipment several very different faults may produce much the same 'symptom'. For example, anything which produces a change in the supply voltage will first affect the part of the system most susceptible to supply changes, and so anything affecting the supply will tend to produce the same fault. To aggravate the position, faults affecting the supply voltage are not limited to the power pack supplying the voltage to the various parts of the system, but may be in any of these parts, increasing or decreasing the load and so affecting all the other parts in greater or lesser degree. Thus the removal of any of several widely spaced resistors may cause a radio set to emit howls, but it does not follow that howls are immediately associated with these resistors, or indeed that the causal relation is anything but the most indirect. In particular, we should not say that the function of the resistors in the normal circuit is to inhibit howling. Neurophysiologists, when faced with a comparable situation, have postulated 'suppressor regions'.

Although the effect of a particular type of ablation may be specific and repeatable, it does not follow that the causal connection is simple, or even that the region of the brain affected would, if we knew more, be regarded as functionally important for the output—such as memory or speech—which is observed to be upset. It could be the case that some important part of the mechanism subserving the behaviour is upset by the damage although it is most indirectly related, and it is just this which makes the discovery of a fault in a complex machine so difficult.

We may consider one or two further points. Since learning is important in at least the mammalian nervous system, it is clear that where animals and men have had different past experiences their brains are likely to be in some ways different. What is 'stored' must at any rate vary between individuals of the same species. It is known that for man surgical removal of some areas of the brain, e.g. the frontal lobe, may pass almost unnoticed in some individuals, while in others it produces serious defect of function. This might perhaps be due to the different importance of specific causal mechanisms in individuals employing different 'strategies', or possibly to the unequal importance of various pieces of stored information. In any case we should expect, and do in fact find, individual differences. This is a complicating factor in interpreting ablation studies which would hardly concern an engineer using man-made machines, except indeed for certain electronic computers.

A further point that might be made is this: Suppose we ablate or stimulate some part of the brain, and lose or evoke something in behaviour, then it is not clear—even quite apart from previous considerations—that this region is the seat of the behaviour in question. Might it not lie along a 'trunk line' or 'association pathway'? A cut telephone line might affect communication over a wide area, principally behind the region of damage. This has at least two important implications: first, unless the region is known not to lie on a 'cable' the region cannot be identified with a brain 'centre' responsible for some aspect of behaviour, since the 'centre' responsible but cut off might lie anywhere from this region along the trunk line. This is further complicated by the consideration that it might be cut off in some conditions but not in others: it might conceivably depend upon whether the animal is motivated in a particular way, receiving information from a particular 'store', or countless other possibilities, whether this block will matter; and the same is true of damage to, or stimulation of, a 'centre', even if this word is taken as meaningful. In many machines it might be possible to remove large parts without any effect except under certain working conditions.

There are two points here: (1) damage might produce, so to say, a shadow within which brain function is lost to regions of the brain on the 'other side' of the damage. If the better analogy is a short-circuited power line, the effect may extend both ways along the cable. (2) The damage may be important only under certain critical circumstances. It does not matter that a car's trafficators are not functional until the driver wishes to turn a corner in traffic—or that his brakes do not work until he tries to stop.

This view of what we mean by 'function' is important in considering brain 'centres'. These are supposed loci for particular types of behaviour: thus Hess has a 'sleep centre' for the cat, in the hypothalamus. This idea of 'centres' has been taken over by Ethology and is particularly important in Tinbergen's writings. But we may well feel worried about the concept of functional centres when we do not know what is going on, in functional terms, in the region concerned. The above considerations apply here *mutatis mutandis*. Why, if stimulation of a given region produces sleep, should this region be regarded as a 'sleep centre'? To take a facetious example: if a bang on the back of the head produces stars and a headache, is this a 'centre' for stars and headaches?

In summary: (1) it might be argued that 'localisation of function' means that some feature of behaviour has certain vital (but not sufficient) causal mechanisms located in a given region of the brain. But before we know how, in general terms, the brain works we cannot say what these supposed causal mechanisms are, and thus it is very difficult to say what we mean by 'localisation of function'.

(2) Stimulation and ablation experiments may give direct information about pathways and projection areas, but their interpretation would seem to be extremely difficult, on logical grounds, where a mechanism is one of many inter-related systems, for then changes in the output will not in general be simply the loss of the contribution normally made by the extirpated area. The system may now show quite different properties.

(3) It would seem that ablation and stimulation data can only be interpreted given a model, or a 'block diagram', showing the functional organisation of the brain in causal, or engineering, terms. Such data may be useful in suggesting or testing possible theoretical models.

(4) These models are explanations in the engineering sense of 'explanation'.

5. *Conclusion*

It would be nice to say something more constructive about the use of engineering thinking in biology. Given that there are certain difficulties in taking over engineering ideas of design into biology, can we not still use engineering techniques and devices to make some better-than-random guesses about how the brain works?

We have throughout looked at the brain as an engineering problem in a general way: we have not considered any particular engineering techniques, or mechanisms, or machines which might throw light on biological function. We have mentioned radio sets and car engines when thinking about localisation of brain function, yet it is at least clear that brains are very different from these. We could certainly think of machines more like brains—and this might be worth doing. What about computers? Obviously we should expect more similarities between computers and brains than between car engines and brains, for the inputs and outputs are similar for the one though not for the other. Now we might go further and ask: what *sort* of computer is most like the brain? There are many different types of practical computer. As is well known, they are divided into two main classes: analogue and digital. Each has certain advantages. The former are usually simpler in construction, they are fast, and are generally subject to rather large random errors. The best-known example is the slide rule. Their inputs and outputs are usually continuously variable, though this is not always so: a slide rule might be made with click stops and still be called analogue. The essential point is that the input variables are represented by the magnitude of some physical variable, such as a length or a voltage. Digital computers, on the other hand, are generally slower, and their answers tend to be either correct or wildly wrong. They work in discrete steps, and according to some fixed rules or calculus. The functional units (essentially switches) of a digital computer take up certain discrete semi-stable states according to a code. For some purposes the analogue type would be chosen by the engineer, and for others the digital type. Thus we may now ask: which would be the most suitable type of computer for a brain, an analogue or a digital computer? Or perhaps a mixture?

To answer this question we may make a list of the relevant properties of the brain and try to decide which type of computer fits best (Gregory, 1953). Some of the difficulties we anticipated at the beginning: we found that engineering criteria are not easy to apply, and that some are indeed inappropriate. The basic efficiency criteria evidently may be applied, but they have their difficulties unless we know a good deal about the functional properties and efficiencies of the components of the brain. Thus it is not possible, for example, to say whether the brain works too fast to be a digital computer unless we know the rate at which the components can change their states, or count. If we also knew the minimum number of steps logically required to reach a given solution with the available data it would be possible to say whether the brain *could* work digitally.

Similar considerations apply to testing the hypothesis that the brain is an analogue machine. We may ask: is the brain too accurate for an analogue machine? We cannot answer this until we know how the 'templates' representing the variables work; we need to know more about the actual ironmongery available; 'ideal' considerations are not adequate here, we must know the properties of the components. If we invoke feed-back principles the brain might be an analogue device given rather variable templates—there are many such 'saving' possibilities. In fact this view that the brain is in essential respects analogue is perhaps borne out by the type of errors observed in control situations. The point is that engineering here supplies the hypotheses for testing, and also (up to a point) the manner of testing them, but to make these decisions it is important to know in detail the functional limitations of the components of the brain. It is also important to have 'engineering' performance data. Much experimental work in psychology is in fact undertaken for this purpose. It may well be vital for linking psychology with neurology, and we should use engineering concepts both to suggest appropriate experiments and to integrate and interpret the available data. For example, studies on tremor take on a new significance within the context of servo-theory, for all error-correcting servos are subject to 'hunting'.

A rather different approach, which we might do well to adopt, is the following: We might look for what we are virtually certain to find and then measure it. Two, rather different, examples must suffice. First, we believe that a system cannot itself gain knowledge without inductive generalisation, and we know that this is impossible without probability estimates. This involves some form of counting, and some form of store for count rates or relative count rates. This at once suggests that the brain should be looked at as in part an inductive machine e.g. (Gregory, 1952). It is probable that no one had actually built an inductive machine until Uttley (1954a and b) built his, specifically as a possible model of brain function, but the man-made induction machine follows standard engineering principles. To go to the next stage and ask whether the brain is the *same sort* of induction machine as Uttley's raises all sorts of difficulties, some of which we have already discussed. The point here is that we believe on *very general grounds* that probabilities must be important to achieve adapted behaviour, and so induction and probability mechanisms really must be found if we look for them.

The second example of this approach is the interesting though more specific problem of 'noise' in the nervous system. It is well known that all communication systems are ultimately limited by random noise, which tends to cause confusion between signals. It seems impossible that the nervous system can be an exception, and so it is hardly a discovery that there is 'noise' in nerve fibres, and in the brain. The assessment of the actual 'noise' level in the various parts of the nervous system (Gregory and Cane, 1955; Gregory, 1956; Barlow, 1956, 1957) and of changes in 'noise' level due to ageing or brain damage (Gregory, 1958) may throw some light on neural function, if only by helping us to apply efficiency criteria to test between rival explanatory models. It is interesting in this connection that Granit (1955) has recently summarised the evidence for random firing of the optic nerve but has not interpreted this as a background 'noise' level against which visual signals must be discriminated, but rather regards it as necessary for keeping the higher centres active. Thus the same observation might be regarded as a necessary evil or a special and useful part of the mechanism. Here the very general properties of communication systems would lead us to the former interpretation, but without these general considerations there would have been no reason to suppose that random firing is not useful to the organism and, so to speak, part of the design. Given the engineering viewpoint, we should ask how the system is designed to *minimise* the effect of the background noise, and this is quite a different sort of question, leading to quite different experiments.

Information rates and noise levels will not in themselves tell us how the ear or the eye gives us useful information—how they work—but such measures are in conformity with the engineer's insistence upon knowing the performance limits, and the reasons for the limits, of his systems. Experimental psychology is currently, and for practical reasons, concerned with the limits of human ability in many directions, e.g. in steering and guiding. These measures may be vital in deciding how the guiding or steering is done. In many cases it is only limits, such as sensory thresholds, which can be used to provide 'engineering' data from complex organisms. Now this idea of looking for properties which are found in all, or at least in most, engineering control systems, and then obtaining quantitative measurements of them under various operating conditions is rather different from the idea of thinking of a physical model as a possible 'analogy' to a behaviour mechanism and then testing this model with observation or experiment. Before we attempt seriously to test specific models of brain function—types of memory store and the like—we might do well to make careful estimates of such things as neural 'noise' levels which we are virtually certain must be there to be found. Having done this, we may be in a stronger position to test specific hypotheses, for we should be able to apply engineering criteria with sufficient rigour to make some hypotheses highly improbable, while others might be shown to be quite possibly true.

These considerations have some relevance to the progress of experimental psychology. If we have no idea of the sort of system we are dealing with, controlled experiment becomes impossible, for we cannot know what to control. On the other hand, a too fixed and particular model tends to blinker the mental eye, making us blind to surprising results and ideas without which advance is impossible.

XIII. THE CODING OF SENSORY MESSAGES

Neurophysiologists' efforts are usually aimed at understanding the workings of some small part of the nervous system in terms of the fundamental physical and chemical processes involved in it, and advances occur most often because of improvements in technique, or as a result of opportunities offered by some new preparation. Such detailed analysis is obviously important, but the trouble is that it may fail to lead one to items of knowledge which are basic to the understanding of the nervous system as a whole. One wants to synthesise a workable nervous system, in one's imagination only of course, from one's knowledge of the working of some small part, but the pure analytic approach shows little prospect of yielding the kind of knowledge which would enable one to do this. For this reason the rôle I have adopted in this chapter is the reverse of the usual one of a physiologist in psychological discussions: instead of dropping hints about the physiological mechanisms underlying psychological phenomena, I have tried to pick up hints about what psychologists expect the relevant physiological mechanisms to do. Unfortunately the gulf between the language used by psychologists to describe the ways in which human beings and animals behave, and the concepts of physics and chemistry which physiologists use, is too wide to be bridged in a single span; the memory of an animal is a very different thing from the 'memory' of a tape recorder, but the attempt to translate the language of psychology into physical terms leads, only too often, to trite analogies of this sort. Information theory may provide the means for removing the triviality from the analogy, and in the hope that this is the case I have first tried to formulate a rather central problem in the physiology and psychology of sensation in cybernetic language, and have then attempted to use the added insight in a re-examination of the problem.

The chapter is divided into four sections. In the first I try to show how existing physiological and anatomical knowledge leads up to, but fails to solve, the problem of how the nervous system learns to make use of combinations of simple properties in order to recognise objects and discriminate between them. In the second, this is translated into cybernetic terms, and the suggestion then arises that compressing the sensory information by reducing the redundancy may be a step towards a solution of the problem; the kind of operations involved in such recoding are briefly examined. In the third section, some simple physiological processes are looked at to see if they may lessen the redundancy of the sensory messages. In the fourth section, problems of recognition and discrimination are re-examined, and it is also suggested that the idea of message compression may help one to understand much more advanced psychological processes than was originally contemplated.

1. *The problem*

One can recognise an object because it possesses a particular combination of simple properties. The number of possible combinations of these simple properties is fabulously large: how does the nervous system manage to pick out one particular combination from amongst this very large number of possible combinations? The problem is less acute with innate, as opposed to learnt, recognition, for this often depends on a single, relatively simple, feature of the stimulus (Tinbergen, 1951) and the range of objects discriminated from each other is small. In contrast, acquired recognition seems to depend on combinations of simple properties, and the range of objects one can learn to recognise is enormous. The solution to the problem involves the mechanisms which sort out and organise the sensory information flowing into the brain and, if the central idea of this chapter is correct, acquired recognition is an inseparable part of the organisation of sensory information.

This problem has been picked on because it is just beyond the range of nervous activities which could be explained fairly

332

easily in terms of known physiological processes. Thus, although there are many details of the mechanisms involved which are not understood, there are no conceptual difficulties in understanding simple reactions evoked by sensory stimulation, such as spinal reflexes. No doubt this is partly because everyone is now familiar with some mechanical device in which pushing a button starts off a sequence of accurately performed actions which bring about some desired result, and this is just what the simpler reflexes do (Sherrington, 1906). An element of sensory discrimination may enter into the reflex without disturbing the analogy. For example, firm pressure applied evenly to the pad of the foot of a cat may produce the 'extensor thrust', in which the limb is extended as if to support the weight of the animal upon it, whereas if the firm pressure is applied with a pin the flexor reflex will be elicited and the cat will withdraw its foot; my wireless can also discriminate, for it produces the sound of an orchestral concert when I press one button, and a commentary on cricket when I press another. The scratch reflex shows a more complicated type of sensory discrimination, for one has to apply a rather elaborate sequence of stimuli lasting some time and spread over a considerable area of the skin before it can be elicited. The analogy of a push button does not help one to understand the new feature here, but one is familiar with mechanical devices which exhibit comparable specificity either accidentally, as in a car which is tricky to start, or by design, as in the dial system at a telephone exchange. Again, details are lacking, but sufficient is known of the properties of a nerve cell and its synapses to see that connections could be made to it in such a way that only one more or less specific 'key' pattern of impulses would 'unlock' it and cause it to discharge. It does not seem necessary to postulate anything different in principle to mediate the sensory discrimination of other hereditarily fixed, unpliable, types of behaviour: it is, for instance, quite plausible to suggest that the specific features of the stimuli which elicit the feeding response of the frog are largely determined by specific features of the neurophysiology of its retina (Barlow, 1953; Marler, **Chapter VI**, this volume).

This train of thought is acceptable so far because the number of 'locks' required is not unmanageably large. A few dozen, perhaps a few hundred, would be required in the spinal cord of a dog to cover the sensory field on the back and flanks from which the scratch reflex can be elicited. Rather more might be needed at higher levels in the C.N.S. for some innate releaser mechanisms, for the 'locks' would have to be extensively duplicated to cover the whole of the sensory field, and where there are several releaser mechanisms several locks would be required for each part of the field, but the total number seems in no danger of rising above the number that could be formed from the nerve cells available in the sensory centres.

It is when one considers the range of acquired, or learnt, sensory discriminations in higher animals that this problem becomes acute. Is one to suppose that there is a 'lock' for each letter of the alphabet in each part of the visual field? And more 'locks' for the numerals, for the capital letters, for the Greek alphabet? Then there are also all the other objects which can be instantly recognised in all parts of the visual field, irrespective of their size and orientation; are there to be separate 'locks' for all of these? It is difficult to produce exact arguments to decide whether this picture is possible or not (see Uttley, 1954a), but one additional requirement seems to make the picture quite impossible. The range of things which one can *actually* discriminate is enormous, but it is small compared to the total range of things one *could* discriminate, given the necessary training. Granted that there might be a 'lock' available for each object that *is* successfully discriminated, but there surely cannot be one available for each object that *could* be discriminated; and if there is not one available, how is one found when the need arises?

The difficulties appear when one comes to consider learned discrimination, and it might be thought that if one knew more about the mechanism of learning itself one would see a way through the difficulties. In its simplest form, learning presumably consists of changes in threshold of an excitable structure in response to excitation of itself or of another structure, but the discovery of a physiological preparation which showed such changes, and in which their cause could be further analysed, might not greatly help our understanding of the present problem. We need to know what task the nerve cells are performing, rather than details of how they are doing it. The problem posed is how the nervous system organises sensory information so that it can be used for present and future decisions, and physiology seems to have little to say on this point at the moment.

The problem of sorting out sensory messages also lies just beyond our range of understanding when one considers its anatomical background. There are about 3 million sensory nerve fibres relaying onto the cerebral cortex. In some short unit of time each fibre can be either active or not active, so that there is a total of $2^{3 \text{ million}}$ possible combinations of activity in these fibres. In contrast, there are less than 10^{10} cells available in the cortex (Sholl, 1956), for sorting out these patterns. We are not greatly helped by supposing that the sorting is done by a structure smaller than a single cortical cell, for the number of possible patterns of activity is not only greater than the number of cells but it is also greater than the number of synapses, or even molecules, in the cortex; in fact it is vastly greater than Eddington's figure (10^{79}) for the numbers of particles in the whole universe! In spite of this enormous discrepancy, one knows that the cortex is able to discriminate between patterns which differ only in some very small detail: a false note in a symphony, a fly on the face of the Mona Lisa, an ashtray displaced on the mantelpiece, are each detectable to the musician, the artist, the housewife. Furthermore, the detection of each of these small differences can lead to a single simple action on the part of the person who has detected it, so that one particular pattern of sensory impulses may result in the activation of a few nerve cells elsewhere in

335

the nervous system, whereas a very closely similar one will fail to do so.

It seems pretty certain that the critical events responsible for these types of discrimination occur in the regions of the brain between the well-defined sensory pathways projecting onto the cortex and the well-defined motor pathways leaving it; anatomists cannot give us any information about these regions which leads to useful ideas about what goes on in them.

Finally the present problem concerns a stage in the working of the brain where it is extremely difficult—at least for a physiologist—to relate the language and theories of psychology to actual or hypothetical mechanisms. The physical properties of sensory stimuli and sense organs set certain limits to performance, and these seem to be the overall limits as long as one confines oneself to simple and familiar stimuli. When complex or unfamiliar stimuli are used it is presumably the central mechanisms which fail, but until one has clearer ideas about the task which is being performed one is unlikely to understand the natural difficulties in performing it, and so one is unlikely to be able to relate the failures to the operation of neural mechanisms.

2. Formulation of the problem for a machine

To show what is meant by using combinations of simple properties to discriminate, let us suppose that there is a box with three push buttons on it labelled a, b and c, and a light at the other side. The box would perform the type of discrimination we are interested in if the light lit up when, say, b and c were pushed, but not for any other combination of buttons. This would not be very difficult to arrange; there are only 8 possible combinations of the buttons, and the device has to be made to respond to one of them. The fraction 1/8 thus gives an idea of the accuracy required in the discriminating mechanism. If the number of buttons is increased to 10, the number of possible combinations is increased to 1024, and the act of discrimination is correspondingly more difficult; an increase to

40 buttons raises the number of possible combinations to a figure greater than the number of cells in the cerebral cortex, and a further increase to 100 buttons raises it above the number of molecules in the brain. This is still much simpler than the brain itself, however, which has some 3×10^6 nerve fibres leading into it.

These considerations show first that the actual task of discrimination becomes very difficult when the number of buttons is increased; but there is also a second point. Suppose that we have boxes with 3 to 100 buttons, and each is capable of discriminating a single combination of buttons; now let somebody start pushing the buttons more or less at random. It will not be long before the correct combination for the 3-button box is pushed, and after about 1000 attempts we may expect the 10-button box to have performed its task of discrimination. But those with more buttons are 'discriminating' such a rare combination of simple events that they will virtually never be used at all. Thus when the number of buttons becomes large, there is not only a difficulty in making it perform its discrimination, there is also the difficulty that, if it can perform the discrimination, it is likely never to be given the opportunity of doing so. This is just another aspect of the fact that the number of possible combinations increases in a geometric series when the number of buttons increases in an arithmetic series.

These difficulties come as a direct result of increasing the number of buttons. The way in which they are used does not affect the difficulty of the problem unless one knows about it in advance, but if one does know, then the problem can be made easier. To take an extreme example: Suppose that half the buttons are not used at all, then the problem could be greatly simplified by omitting all connections to them. One cannot omit these connections until one knows that they are not going to be used, but this does suggest a way in which one could set about simplifying the problem. The difficulties result from the large capacity of the channel leading to the box, rather than from the amount of information passing down the channel, and a possible first step would be to recode the sensory messages in such a way that they could be passed down a smaller channel; the number of buttons required would then be reduced. It has just been pointed out that the number of cells available in the human cortex is not enough to discriminate between all possible combinations of a mere 40 buttons, and it is highly improbable that sensory messages could be compressed to this extent; admittedly, then, this is only a first step, and is not a complete solution to the problem of how to make a machine with the range of discrimination shown by the human brain. Nevertheless, the enormous capacity of the sensory pathways, and the paucity of cells in the analysing mechanism point to the need of the utmost economy in using these cells, and the compression of sensory information would be a step towards achieving such economy.

The suggestion is emerging that sensory messages may be recoded and compressed into a smaller channel before 'recognition' is attempted; it is the redundancy present in the incoming messages which makes this possible, and its reduction is perhaps the main principle underlying the organisation of sensory information. The next three sub-sections are concerned with the means by which this could be done. The first contrasts reversible and irreversible transformations; the second describes some forms of redundancy which make compression possible; and the third considers the possibility of such compression in the nervous system.

(a) *Coding*

A code is a set of relations between input messages and output messages such that, given the output message and the code, the input message can be reconstructed exactly (Woodward, 1953, p. 43); the transformation that the code performs is thus reversible, and the information content of coded and uncoded messages is the same. In contrast, information is lost in irreversible transformations so that the untransformed message cannot be reconstructed exactly from the transformed message. It is pretty certain that sensory messages are transformed irreversibly before they have gone very far into the nervous system (Broadbent, Chapter X). Such transformation probably takes place below the level of consciousness because there seem to be things which our senses receive, but which we cannot reconstruct in our conscious experience; it certainly takes place before sensations lead to any action, because the maximum rate at which sensory messages can influence action is 5 to 25 bits per sec. (Hick, 1952a; Quastler, 1956), which is very much less than the rate at which information is collected by the senses. Nevertheless, the possibility that sensory messages may be compressed by a reversible code, and the advantages of doing so, seem to have been overlooked, and this is what is primarily considered in this chapter. Of course, irreversible transformations must also occur, but this does not diminish the importance of compression by the reduction of redundancy.

(b) *Redundancy of sensory messages*

It may be asked 'How can coding compress a message without at the same time sacrificing information?' This subsection attempts to show how redundancy in the sensory input makes this possible. The concept of redundancy in information theory agrees roughly with its ordinary English usage, but the more exact meaning is illustrated by describing how to detect and measure it. It can be shown that a communication channel is not filled by the messages passing over it, and therefore contains redundancy, if, over a long period of time, the possible messages are not received with equal frequency and in an apparently random order (Shannon and Weaver, 1949, p. 21; Woodward, 1953, p. 23). It is easier to discuss the ways in which this may happen if one has in mind some specific type of message and channel, so the boxes with buttons which are pushed in various combinations and sequences will be considered again. The capacity of the channel is fixed by the number of buttons and the least interval that can elapse between successive pushes of the same button. From a record of the way the buttons have been pushed one can find out whether the full capacity of the channel has been used or not: First divide up the record into intervals of duration equal to the shortest intervals between pushes; in each such interval a button is pushed once or not at all. The type of redundancy which is easiest to show would be present if the total number of times any button was pushed differed greatly from half the number of intervals in the message examined. This follows from the fact that the two messages 'pushed' and 'not pushed' would not have been received in equal numbers from the button under consideration. Secondly, one could estimate the frequency with which a specific button was pushed, given that it was pushed in the preceding interval; if this frequency indicated that the probability of a button being pushed was dependent on what happened in the preceding interval, then one would not be receiving the messages from that button in a random order. Thirdly, one could estimate the probability of a button being pushed given that another specified button was pushed, and if this differed from the overall probability for the first button, it would indicate that the four possible messages from these two buttons were not being received with equal frequency. After this one might go on to find whether the probability of a specified button being pushed depended on whether another button has been pushed at a previous moment. These are the simplest examples of redundancy, and they can be listed as follows:

1. Mean frequency of pushing a button less or greater than half the maximum frequency.
2. Positive or negative serial correlations in probability of pushing a single button.
3. Positive or negative correlations in probability of pushing two different buttons at the same time.
4. Combinations of 2 and 3, for example one button frequently being pushed following another button being pushed.

This list could be extended, and if one asks how long it must be made if it is to include all possible forms of redundancy, one comes across the very large numbers which have already been encountered in this chapter. Taking messages of a certain duration, the number of separate forms the redundancy may assume, or its degree of freedom, is one fewer than the number of possible messages. The classes of redundancy listed above correspond to some simple ways of dividing up these degrees of freedom into a small number of groups, and the number of ways this can be done is even greater than the number of degrees of freedom. A particular redundant feature must, of course, be detected before a code can be devised to eliminate it, and the detection of one particular form amongst the vast number of possible forms appears to require an act of recognition no simpler than those we set out to consider. It might therefore be held that nothing has been gained by disguising the problem in the language of information theory. But this is not a just criticism for the following reason.

If one considers the total *elimination* of a particular form of redundancy in a single stage, then a difficult act of recognition is required and no progress has been made towards understanding the type of mechanism that could achieve it. But it is possible to visualise mechanisms which would *reduce* the redundancy, and this takes one a step towards eliminating it; one can now see a method by which a complex and difficult act can be subdivided into comprehensible stages. The idea that recognition is performed in stages is not, of course, a new one, but it has not been clear that such subdivision makes the task simpler. An example may show how redundancy can be reduced in two stages, and how this is simpler than the same task performed in one stage.

In English the letter *q* is always followed by *u*; the *u* in this position is wholly redundant, and no information is lost by deleting it. Let us suppose that in a first stage of recoding the frequencies of pairs of letters (digrams) are measured; the redundancy of *u*'s following *q*'s is detected, and they are deleted.

Now in a second stage the operations are repeated and the frequencies of the new digrams *qa*, *qb*, *qc*, etc. are measured: it will be found that *q*-vowel digrams occur not infrequently, but *q*-consonant digrams are non-existent. These uneven frequencies constitute a new form of redundancy, the elimination of which would further compress the messages being sent. To detect these forms in a single stage would have required a determination of the trigram structure of English involving 26^3 measurements of frequency. By performing the operation in two stages it has been necessary to determine the digram structure twice, but this requires only 2×26^2 measurements of frequency. This is a particular example chosen to illustrate the merit of subdivision, but it seems to be generally true that any reversible transformation which reduces redundancy makes it easier to detect redundancy remaining in the output.[1] It appears, then, that the reduction of redundancy is a means by which a complex act of recognition can be performed progressively in comprehensible stages. This is not only a conceptual simplification but offers real economies in performing the task which at first seemed so hopelessly difficult.

(c) *Redundancy-reducing codes and noise*

The theoretical possibility of codes which reduce redundancy was established by Shannon for the case of a discrete noiseless channel (Shannon and Weaver, 1949, p. 28). If the physical stimuli are regarded as supplying information at a certain rate, then codes are always possible which will enable the information to be passed down a channel whose capacity is only slightly greater than this rate. If the information is supplied in a redundant form, these codes must eliminate the redundancy, and it is easy to believe that normal sensory messages are

[1] This requires more exact demonstration, but if one takes the number of degrees of freedom or the possible forms of redundancy as a measure of the difficulty of detecting it, then it is easy to see that any code which decreases redundancy decreases the channel capacity required for the output, and hence decreases the number of degrees of freedom available for the residual redundancy in segments of the output corresponding to input messages of a certain length.

highly redundant. It might, however, be questioned whether the discrete noiseless case is the appropriate one in the present context. There is not much doubt that the discrete rather than the continuous case is appropriate to the nervous system, both because of the all-or-none nature of nerve impulses, and because the nervous system is made up of a large number of almost independent, parallel channels. Noise, however, requires further consideration.

The working of the nervous system may be perturbed by noise entering at two levels: it may be present in the signals reaching the sense organs, or it may be introduced as a result of indeterminancy in the nervous system itself. The latter type may well be important, and there may be aspects of the functioning of the nervous system which can only be understood as noise-combatting measures—the retention or reintroduction of redundancy each provides possible examples. It seems premature, however, to consider these aspects until the primary steps in the analysis of sensory information are better understood. Noise in the signals reaching the sense organs is a different matter, and since it is always present it might be suggested that the case of interest is that of coding information on to a discrete channel in the presence of noise. But this suggestion would be based on a misunderstanding. When Shannon and others consider coding for a noisy channel they assume that knowledge of the statistical structure both of the signals to be sent, and of the noise to which they will be exposed, is available before the code is devised. This knowledge is not available to the nervous system, and it can only determine the statistical structure of the messages received—the signals after they have been degraded by the noise. The rules for using noisy channels at their full capacity are largely irrelevant in this context.

If no independent knowledge is available to show what is signal and what is noise, how can the nervous system ever distinguish between them? This must be done partly by genetically determined features (pain, for instance, is very rarely treated as irrelevant 'noise'), and also by associations built up

343

on basic reflexes as in the familiar conditioning experiments. These mechanisms really distinguish between 'valuable' and 'useless' messages, and are beyond the scope of this chapter but there is one way of distinguishing signals from noise which should be mentioned. Noise, except in the less exact sense of 'useless signals', is something which occurs at random and is unpredictable: it follows that the redundancy in sensory messages cannot result from noise, and a mechanism which detected redundancy would thus help to pick out the signals. The importance of redundancy in this context is distinct from its importance in enabling sensory messages to be represented in a more compact form.

To summarise: it has been suggested in this section that acts of recognition can be broken down into successive coding operations in which the redundancy of the sensory messages is reduced. During the discussion it has incidentally emerged that the progressive, stage by stage, reduction of redundancy decreases the complexity of the task, and also that the detection of redundancy is important in discriminating signals from noise. The reduction of redundancy is a sufficiently simple, well-defined operation to look for in the action of physiological preparations, and in the next section some evidence that it occurs will be presented.

3. *Physiological mechanisms for compressing messages*

The first step in looking for examples of redundancy-reducing codes is to transfer the analogy of the previous section to the nervous system. If the messages are patterns of buttons being pushed, then compression consists of recoding so that they can be transferred to a smaller number of buttons. Nerve messages are carried by impulses in fibres, and if a set of messages contains redundancy, recoding should enable the same messages to be sent *either* down fewer nerve fibres, *or* by means of fewer impulses in the same number of fibres. The former possibility would correspond more closely to the push-button analogy, but, in the examples which follow, it is the latter which the nervous system seems to do.

There are two alternative ways of looking at recoding in which the reduction of redundancy is used to decrease the number of impulses but not the number of fibres. It could be held that the average number of impulses passing down a nerve defines its capacity: in this case the recoded messages would be passing down a channel of reduced capacity, and it would be correct to say that the redundancy of the messages was reduced. On the other hand, it might be said that the capacity is a fixed attribute of the nerve which is not affected by the way it is used: in this case reducing the number of impulses would not reduce the redundancy, for the same amount of information would be passing down a channel of the same capacity as the input. It would, however, be converting it into a standard form corresponding to class 1 in the list on p. 340, and since redundancy is so easily identifiable in that form, this operation has many of the advantages of actually decreasing it. In the rest of this section physiological transformations of patterns of nerve impulses are brought forward which will reduce redundancy corresponding to classes 2, 3 and 4 on p. 340: alternatively, they could be described as converting these into the readily identifiable class 1.

The adaptation of sensory discharges first suggests itself as an example of a transformation which economises the number of impulses used to transmit information. When a physical stimulus is applied, the nerve coming from a sense organ responds with a burst of impulses, but the frequency of the impulses declines even if the strength of the stimulus is maintained at a constant value (Adrian, 1928). This is a widespread, though not universal, property of such discharges, and Adrian relates it to Sherrington's (1906) concepts of reflex fatigue and 'phasic' as opposed to 'postural' reflex mechanisms. This implies that adaptation makes the sense organ more useful to the animal by adjusting the characteristics of the discharge to suit some central mechanism, and he goes on to point out that

'It might well be inconvenient if our central nervous system were to be continually flooded with messages from every point of the skin surface ...' Instead of relating it to the largely unknown requirements of more central parts of the nervous system, the present hypothesis regards it as a means of improving the matching of the physical stimulus to the nerve fibre so that information about the stimulus can be transmitted with fewer impulses. This explanation of the part played by adaptation can be illustrated by defining what a sequence of impulses should look like if the matching were perfect.

If there are no restrictions on the time of occurrence of impulses except that the mean frequency is known, then the entropy of the set of impulse intervals will be maximum when (a) impulse intervals are distributed exponentially, and (b) there is zero correlation between the values of successive intervals (see, for instance, Woodward, 1953, p. 24). For the information carried by a fibre discharging at a certain average rate to be maximum, these same conditions should be fulfilled: hence if the matching were perfect, a sequence of impulses recorded from a nerve in the body in everyday use should also have these properties. It is not, unfortunately, easy to extrapolate from the conditions of laboratory experiments in order to assess how accurately such a natural discharge would fulfil the requirements, but it certainly seems likely that adaptation brings the train of impulses into closer agreement with the two conditions than would otherwise be the case. It must frequently happen that a physical stimulus is held at a rather constant value for considerable periods of time, as, for example, the pressure applied to the skin when sitting down. Without adaptation this would cause a large number of impulses separated by intervals whose values would be highly correlated. Adaptation changes this to a brief vigorous burst when the pressure is applied, followed by smaller bursts when the pressure changes, but the redundant impulses signalling the maintenance of the stimulus at a steady value are cut out.

Adaptations of sense organs, then, may be regarded as a

mechanism for compressing sensory messages into fewer impulses by decreasing the tendency for serial correlations in the values of impulse intervals (redundancy of class 2 in the list on p. 340). There must be many occasions when neighbouring sense organs are subjected to the same stimulus, and this will lead to correlations in the values of impulses in neighbouring fibres (redundancy of class 3 in the list on p. 340). This offers scope for further economy of impulses, and one might look for a mechanism to perform the appropriate recoding. It would be the spatial analogue of adaptation whose existence Adrian (1947) forecast: it would diminish the correlation between impulse intervals spatially separated, whereas normal adaptation decreases the correlation between impulse intervals which are temporally separated.

A mechanism which seems designed to do just this has been found in the compound eye of *Limulus* (Hartline, 1949; Hartline, Wagner and Ratliff, 1956). Light falling on an ommatidium causes a discharge of impulses in the nerve fibre connected to that ommatidium, but the number of impulses is reduced if light falls on a neighbouring ommatidium at the same time. A mechanism which must have similar functional results is found in the frog's and cat's retina (Barlow, 1953; Kuffler, 1953), but there is no possibility that this mechanism is related phylogenetically to that in *Limulus*, for the structures involved are totally different. Evidence is also accumulating that 'lateral inhibition' analogous to that found in eyes occurs elsewhere in the nervous system. Galambos (1944) found that if a single unit in the cochlear nucleus (Davis, 1951) was responding to stimulation by a tone of the frequency to which it was most sensitive, the rate of response could be reduced by adding a tone of slightly different frequency. Mountcastle (1957) and Towe and Amassian (1958), recording from units in the 'touch' area of the cerebral cortex, found that the discharges characteristically evoked by stimulation of one part of the skin could be diminished by simultaneously stimulating a neighbouring region. The common occurrence of lateral inhibition suggests that the mechanism has rather a general significance, as would be the case if its function were to eliminate redundant impulses.

The outcome of lateral inhibition in the eye will be that units at the edge of an evenly illuminated area will be more active than those at the centre, which will lead to the borders of objects being accentuated in the nerve message. This is clearly related to the phenomena of simultaneous contrast, and to the fact that things seem to be largely recognised by their outlines. As in the case of adaptation, it would be possible to explain these mechanisms by referring to higher nervous functions which utilise them, but this argument strongly tends to become circular: the extraction of outlines is useful to higher centres, therefore a peripheral mechanism for extracting outlines has survival value and has been evolved—but this begs the question why the extraction of outlines is useful. It would be extraordinary if a mechanism found in a sense organ was *not* used by the central nervous system, and the question one should ask is 'Why does the extraction of outlines have survival value?' The answer to this question may be that it allows the sensory messages to be compressed into fewer impulses and it is the advantages of compression which are of survival value. According to this suggestion, the extraction of outlines, and simultaneous contrast, are by-products of compression, and are not ends in themselves.

Adaptation of the sense organ diminishes the correlation in values of successive intervals in the same fibre; lateral inhibition reduces the correlation in intervals in neighbouring fibres. Is there a mechanism for compressing the message by taking advantage of the correlations between impulse intervals in one fibre and those in another fibre at a later instant (redundancy of class 4 in the list on p. 340)? Correlations of this sort would results from a moving stimulus, and the fact that the retina responds to changes of illumination rather than to its actual intensity (Hartline, 1938, 1940; Granit, 1947) makes it well adapted to signal such events, but the mechanism postulated now must do rather more than simply detect movement; the idea is that a sequence of impulses in two or more fibres can be represented by fewer impulses if there are mechanisms specifically allotted the task of signalling such sequences. There is no direct physiological evidence for such a mechanism,[1] but its existence is suggested by several lines of indirect evidence.

The simplest stimulus which would cause a group of impulse intervals in one fibre to be followed by a similar group in another fibre would consist of a light flashed first on one region of the retina, then on another region; this is, of course, a type of stimulus which gives rise to the *phi* phenomenon of apparent movement (Graham, 1951), and it suggests that there is a mechanism 'looking for' successive correlations of this kind, excitation of which gives rise to a sense of directed movement. Now movement of an image over the retina will cause a vigorous and sustained discharge of impulses in the optic nerve, but once the direction and velocity have been determined these impulses are redundant; the hypothesis leads one to expect that they will be eliminated, for instance by arranging that units which detect directed movement inhibit units which would otherwise be active. This may sound a rash speculation, but it is only a little more complicated than the mechanism of lateral inhibition, in which, if the 'predicted' uniformity of illumination occurs, the impulses which would have been elicited by stimulation of the centre of the uniform area are inhibited, leaving the units at the edge to carry the information.

It very often happens that large sections of the retinal image move in a co-ordinated way over the retina; for example, the movement of the image of the ground when walking with the eyes directed ahead, or the movement of the image of the land-scape when looking out of a train window. Such co-ordinated movements, which can be of various types, would cause higher

---

[1] Since this was written, Hubel (1959) has recorded from single units in the cat's striate cortex, and shown that some of them respond, not just to any movement, but only to movement in a specific direction. Hubel and Wiesel (1959) have subsequently found units whose spatial pattern or connection to the visual field are of very great interest from the point of view of coding to reduce redundancy.

order correlations in the discharges from different parts of the physiological system responsible for the sense of movement, and these will give still further opportunities for reducing the number of impulses. Mechanisms for spotting such co-ordinated movements have been suggested by Gibson (1950), and it is, incidentally, hard to account for the following movements of the eyes without postulating something of the sort. When the eye looks at a scene which is moving in relation to the head, it tends to move in such a way as to minimise the movements of the image over the retina, so that a mechanism for detecting co-ordinated movements of the image is precisely what the centres controlling eye-movements require.

The suggestion that the analysis of movement depends on mechanisms whose prime function is to compress the nervous messages is clearly speculative. Nevertheless, it fits well with the hard fact that, on the one hand, the existence of co-ordinated movements of the retinal image offers scope for message compression, and on the other hand, such movements are detected and acted upon by the nervous system.

4. *Automatic compression*

Adaptation of sense organs, lateral inhibition in the retina, and the systems for movement perception, all look like mechanisms for compressing the sensory message by eliminating redundant impulses, but a new principle must be invoked before contemplating the possibility that recognition, or learned discrimination, also depends on the operation of similar mechanisms at higher levels in the nervous system. The problem originally brought forward for physiological explanation was 'recognition by the use of combinations of simple properties', but although the detection of co-ordinated movements is certainly a task of the right order of complexity, it is not a good example of the type of recognition that was contemplated. The universal occurrence, and importance, of movement make it quite likely that mechanisms for detecting and analysing it would be developed in the course of evolution, but no one would seriously suggest that the recognition of the letter *a*, for example, depends on the action of a mechanism specifically evolved for recognising *a*'s, and it is this type of recognition which is most difficult to understand. The point of this distinction is brought out by asking the question 'How do the mechanisms for message compression come into existence?' One answer is that each specific mechanism has been evolved because the specific type of redundancy it eliminates occurs in the sensory messages of each individual, so the specific mechanism for dealing with it has survival value. The alternative and more interesting answer is that the mechanism for removing redundancy starts by being unspecific and becomes specific in response to the particular redundant features of the messages received; the code is formed anew by each individual in response to its own experience.

Bearing in mind the possibility that a particular form of compression may arise in response to a particular form of redundancy in the input, it is worth looking again at the phenomenon of lateral inhibition. It was naturally assumed at first that this was an innate, genetically determined, property of the retina, but a new fact has recently turned up which suggests a different interpretation. Although lateral inhibition is constantly present in light-adapted retinae in which stimulus lights are superimposed upon a uniform, steady, background illumination, it does not occur in retinae which have been completely dark-adapted where there is no background illumination (Barlow, Fitzhugh and Kuffler, 1957). This disappearance of lateral inhibition could be explained teleologically; after dark-adaptation the utmost sensitivity is desirable, and dispensing with all types of inhibition might help to achieve this. But it also seems possible that the disappearance results from the adaptability of the coder; uniform background illumination will cause many receptors to become active together, and this correlated activity might itself cause the development of mutual inhibition between neighbouring retinal regions, which would decrease the correlation between the activities of neighbouring output fibres. Thus the system of lateral inhibition might be the result of automatic message compression in the retina.

This speculation can be tested experimentally, but meanwhile one can look for evidence of similar mechanisms working at higher levels in the nervous system. First notice that the development of lateral inhibition could be put in the following terms: In darkness all retinal units behave independently, but when the general level of illumination is raised, the receptors all tend to become active together. This would cause a correlated discharge in the nerve fibres coming from the retina, but to avoid this the retina develops a system whereby only spatial differences of illumination are signalled; the uniform illumination is, as it were, taken for granted, and the retina economises by signalling only differences of illumination between one retinal region and the neighbouring regions. Now transfer this verbal description to the mechanism for detection of movements. In the absence of movement, all parts behave independently, and movement is detected and signalled where and when it occurs. Suppose that uniform movement is imposed on a part of the visual field; at first this will be detected and signalled through to higher centres, but just as uniform illumination comes to be taken for granted and the retina economises by signalling only lack of uniformity, so one might expect that uniform movement would be taken for granted, and economies would be effected by signalling only change of uniform movement. The after-image of seen movement, or the waterfall phenomenon (Wohlgemuth, 1911) might depend on the action of a system of this sort.

Many examples of adaptation of the senses to constant features of the stimulus are known (Köhler and Wallach, 1944), and it may be possible to devise experimental tests for explanations along the lines indicated above. The essential idea is that the alteration in perception is a specific act of the nervous system aimed at removing a specific form of redundancy, rather than just a passive non-specific adaptation. The constant feature of the stimulus causes a specific type of redundancy in the sensory messages, and the nervous system modifies its code in an equally specific way in order to eliminate it. It should be pointed out, however, that the explanation becomes very involved as soon as one comes to consider the rate of adjustment of the code at different levels in the nervous system. To return to the waterfall phenomenon, the illusion of reverse movement would not occur if the higher centres had up-to-date knowledge of the code used by the lower centres. The experimental facts certainly show that the rate of adaptation is very different for different types of constant feature. In the case of the waterfall phenomenon, it takes only 20–30 seconds to establish the state of adaptation which leads to the illusion of reverse movement when the real movement is stopped, whereas it takes 10 days or longer to adapt to inverting spectacles up to the point at which the visual world appears the right way up while the spectacles are in place, but appears inverted when they are removed.

Errors of perception of the type shown up by adaptation experiments may be the penalty one pays for having a versatile coder, but the overall effect of compressing sensory information is not at all easy to see if one tries to analyse the mechanism by looking at the errors it makes. It is perhaps easier to see where such mechanisms might lead by comparing the properties of compressed and uncompressed messages. In the remaining part of this chapter three rather general consequences of automatic message compression are compared with the results of perceptual processes. The comparison is necessarily very sketchy and is only intended to show that the idea of automatic, reversible, reduction of redundancy penetrates deeply into this field.

First consider the question of the extent to which the code is reversible. If the encoding of sensory information is achieved without loss of reversibility, it should be possible to reconstruct the exact physical stimulus which gives rise to any sensory message. This certainly cannot be done; for example, there is a physical difference between a light of wave-length 600 mμ and a mixture of wave-lengths 650 and 530 mμ, yet they can be made to look exactly alike. From this it follows that information is lost after the light enters the eye and before the judgements about the sensory messages are made. Psycho-physical experiments can be regarded as explorations of the extent and nature of such losses, so that there is a large body of evidence available showing *what* information is lost; information showing *where* it is lost is very scanty, and this aspect is the subject of hypothesis rather than proof. The physiologist naïvely attempts to relate the loss of information to some physical property of the sense organs—for example the spectral sensitivity curves of the colour-sensitive elements. Now if irreversible coding took place on a large scale between the sense organs and the level of the judgements in psycho-physical experiments, such naïve attempts would be doomed to failure, yet on the whole they are remarkably successful. In psycho-physical experiments simple and familiar stimuli are observed by trained subjects, and the fact that the physiologist can with impunity neglect loss of information central to the sense organs suggests that these losses are actually small: under these conditions, central coding appears to be almost reversible.

On the other hand, it is easy to find tasks which are well within the physical limits of the senses but which cannot be performed, so the coding is not always reversible. One simple example is the inability to count more than about seven objects presented to the eye for a short time (Hunter and Sigler, 1940); another would be the inability of a novice to discriminate from each other the letters of an unfamiliar alphabet. On examination these two examples are found to differ from each other in an important way. The first seems to be a fundamental limit of the perceptual processes which cannot be overcome by training, whereas nobody but a complete novice has any difficulty in discriminating letters of the alphabet. In other words, it must be admitted that the counting limit is the result of an irreversible coding of the sensory information, whereas the second is only irreversible for as long as the messages are un-familiar: so there are some things which *cannot* be reversibly coded, and others which *are* not. Irreversible coding will be discussed briefly at the end of this chapter, but the conclusion can be drawn at this point that coding is sometimes, but not always, reversible.

The second point of comparison between input messages and that part of the output from the supposed recoder which reaches our consciousness follows very directly. If coding is not always reversible, and if the object of the coding is to compress the messages, what predictions can be made about the features of the sensory messages for which the code will be reversible? The answer to this is quite precise: they will be features which carry a lot of information. The argument leading to this con-clusion is as follows: The output from the coder can be split up into simple parts (for example, an impulse in one nerve fibre), and one such part corresponds to some feature of the input message. This feature need not be at all simple, but since only those codes in which the input completely determines the out-put are being considered, the presence or absence of a particular symbol in the output is necessarily determined by some definable property of the input, and this definable property is what is meant by the feature. As long as the message in the output is compressed, each of the simple parts which compose it will carry a lot of information, and hence the features of the input to which each part corresponds must also carry a lot of information.

In general terms this does seem to be true of the elements into which we break down our sensations. The objects, events, sensations, concepts which we use to describe our subjective life are all terms which, we like to think, carry much information. Perhaps more significant is the fact that one can correctly predict what will happen when the probability of occurrence of a feature in the sensory input is altered. If this probability is very low, then the average information associated with it is also very low (because $-p \log p \to o$ as $p \to o$), and consequently one might expect that the sensory system would not be reversible for that feature of the sensory stimulus; this might account for the failures of reversibility due to unfamiliarity, for example the inability to discriminate between the letters of an unknown alphabet. At the other extreme, if the probability of occurrence is very high, the information associated with the feature also becomes very low ($-p \log p \to 0$ as $p \to 1$), and consequently one might expect failures of reversibility here also; adaptation to constant features of the stimulus, and habituation, seem to be examples of this happening.

There is one more general point of comparison between the output of the hypothetical compressing coder and our own conscious mental processes. If the message has been fully compressed, it will contain no redundancies, and it is a property of such messages that any arrangement of the symbols is a possible one, and has meaning. In contrast, with highly redundant messages, relatively few of the possible arrangements of the symbols have meaning; for example, some arrangements of the letters of the alphabet are completely meaningless, but almost all arrangements of shorthand symbols have meaning. It is hard to be sure about the symbols of our sensory experience, but it does not seem easy to arrange them in completely meaningless sequences—at least surrealist artists charge high fees for attempting to do so. In contrast, if a screen were filled with dots put down in a completely random manner, there is a high probability that the result would be completely meaningless.

The idea of compressing the sensory messages by adaptable, reversible, recoding was originally put forward because it seemed the logical way to set about simplifying the more complicated tasks performed by the central nervous system, but it now seems possible that the mere repetition of the same process might actually accomplish the tasks that were to be simplified. The features of the sensory input which are recognised, or discriminated from each other, appear to be the very features which enable the sensory message to be compressed.

It is interesting that the principle of compression seems to go even beyond this, and may be relevant to 'thinking' at an advanced level. Many problems of the type found in intelligence tests require the subject to spot a general feature of some data presented to him; this feature allows him to solve the problem, but it would also allow him to compress the data into a smaller message. Other problems seem aimed at testing the reversibility of the codes used by the subject, for example those which require for their solution an exact knowledge of the meaning of words. So, if one may judge by the nature of the problems used to test for it, 'intelligence' is simply the ability to compress by reversible recoding. At an even higher level, scientific laws and hypotheses are attempts at compressing many empirical facts into a single statement or equation, and Occam's Razor is nothing but the same principle applied to hypotheses themselves.

Memory may be another subject to which these same ideas are relevant, though it is a long way from our original problem. The state of the code at any time is dependent on the nature of the message received in the past; to what extent can 'memory' be explained as the state of the code? At first sight it seems that it would be only a sort of average memory of everything that had happened, since it is the average properties of the message which are needed to devise the code, and which are used to achieve compression. It would therefore be lacking any record of the sequence of past events, or any details of any one event. It has been shown, however, that the code is modified as time goes along. These modifications start in a few seconds in the case of the perception of movement, yet it takes a fortnight or longer to adapt fully to the distortions such as those produced by prismatic spectacles, and many of the more subtle features of the code must take much longer to acquire. The sensory messages are therefore not simply averaged all together in compiling the code; just as local redundancies can be eliminated before tackling ones which occupy a large fraction of the message, so it is possible to tackle redundancies which become apparent in a short time before looking for those which are spread over long periods. If the messages are averaged with short as well as long time constants, then a comparison of the short with the long time constant average shows which events took place recently, which long ago. The code would, in fact, contain an indirect record of the temporal sequence of events. In addition, the occurrence of ordered temporal sequences of impulses, such as those resulting from movement, was one of the types of correlation which, it was suggested, might be made use of in compressing the message: consequently there must be a direct record, in some form, of the occurrence of such temporal sequences. An attractive feature of this notion of the nature of memory is that it spreads it out over a large part of the nervous system, which is very much what neuro-anatomical investigations independently suggest is really the case (Lashley, 1950; Zangwill, Chapter III this volume).

Finally it must be pointed out that compression by reversible coding is not in any way incompatible with 'cutting' the sensory messages by irreversible coding or 'filtering' (Broadbent, Chapter X; Marler, Chapter VI this volume). At various points in this chapter compression has been contrasted with filtering, but this was because explanations of sensory phenomena in terms of filtering seem to have gained acceptance, whereas alternative explanations of the same phenomena in terms of compression have not been considered. I have tried to look at the working of the nervous system from the sensory end because this is my field of interest. When approached from this side compression without loss of information seems to be the first step required, for it is clearly more efficient to compress before filtering, rather than vice versa. Someone approaching from, say, the field of animal behaviour might prefer to look upon compression as one function of the filtering mechanism which lies between the sense organs and the parts of the nervous system which decide upon the animal's actions. What has been called 'compression' could then be looked upon as the case where the filter changes the things it passes in response to characteristics of the messages it receives, rather than in response to the requirements of the animal, or to the results of passing a certain type of information. But as soon as one departs from the idea of compression by a reversible code one loses sight of the clue it offers about the way a highly complicated act of discrimination could be subdivided into simple acts of an order of complexity which might not be beyond the powers of a single nerve cell. On the other hand, it must be admitted that completely reversible compression may only be an ideal towards which the more highly developed nervous systems tend; the serious attempt to construct an accurate description of the causes of sensation is, after all, an activity confined to that small academic minority of mankind which is popularly supposed to have the most highly developed nervous system in the whole animal kingdom.

5. *Conclusion*

Recognition by using combinations of simple properties was picked from the psychological hat as the trick that most needed a physiological explanation. The idea of removing redundancy and so compressing the sensory message into the smallest possible channel was brought forward because it seemed that the subsequent step of learning to discriminate different sensory stimuli would be easier if the sensory information was presented in this compressed form. The possibility that adaptation and lateral inhibition are devices for compressing messages throws some new light on their function, and similar devices at higher levels might also fit in with some of the elementary psychological facts about failures of discrimination, adaptation to complex features of the stimulus, and the illusions that may result from such adaptation. Message compression leads one on to consider the much more elaborate psychological functions underlying intelligence, and surprisingly it leads to the idea that the code which compresses the messages also constitutes the memory.

There are two encouraging results of considering the operation of the sensory side of the nervous system as a reversible
coder for removing redundancy. One is that this is a subdivisible
operation and so could be brought about by the large number
of semi-independent units which appear to make up the
central nervous system; the other is that mental acts, when
broken down in this way, may be reduced to a level of simplicity
which can be understood and investigated by the physiologist.

Sources and references

The references quoted in the text would be incomplete without some
indication of the origin of the ideas about the organisation of sensory information which I have put forward. In trying to apply the ideas of information
theory to the nervous system I have been helped enormously by discussions
with many people, but particularly by talks with Uttley (1954a, b, 1955,
1958) about his papers, which deal with matters closely related to the
subject of this chapter. But the central idea really seems to go back to
Craik's (1943) monograph in which he puts forward the view that it is the
function of the brain to set up a working model of reality. A model is a
reduced copy which preserves the essential features of the original; one
might say that a code which compresses sensory messages by reducing
their redundancy must include such a working model, and the correspondence would be still closer if one added to Craik's idea and said that it was
the function of the brain to construct an *economical* working model of the
causes of sensations.

When I wrote this essay (1956) I was not aware of the paper by Attneave
(1954) in which he presents a similar argument. It is interesting that,
approaching from different directions, we have converged on the same
solution.

XIV. SOME WAYS OF DESCRIBING BEHAVIOUR

1. *Introduction*

It quite often happens, at any rate in psychology, that a result which seems firmly established by one experimenter is contradicted by the work of another performing an apparently similar experiment. It might of course be argued that behaviour is necessarily so variable that firmly established results are not to be expected; but correct statistical treatment should take account of irrelevant variation and, unless we are to believe that behaviour is nothing more than a sequence of random actions, there must be some part of it that is systematic and capable of exact description. How then are such discrepancies to be explained? The reason is very often seen to be, and must surely always be in fact, that the description of the experiment was inadequate. As a simple example, consider the statement that a child's score on a given intelligence test tends to increase with age; this is true on the whole of children with a European type of background and schooling, it is not true of East African children. An important determiner of the behaviour observed, the environment and previous experience of the experimental group, has been omitted. Similarly, wild birds and birds hand-reared in a laboratory differ greatly in their method of learning a new response (Vince, 1960), and any account of how birds learn would have to include a specification—however rough—of their past history. These two examples show that where too few variables are used to describe some piece of behaviour, apparent contradictions may occur. A more detailed example is discussed later in this chapter.

The choice of variables for describing a given sort of behaviour is clearly difficult, and is usually decided by the scientific intuition of the experimenter and his theoretical background (cf. Lehrman, 1953). The latter itself may lead him into error; for example Bendig (1951), in attempting to check a statement of Hull on reinforcement, reported the results of an experiment on rewarded and non-rewarded guessing in terms of alternation, on the ground that Hull's theory would predict that a rewarded response is more likely to be repeated; inconsistencies within the experiment itself suggest that the results should be interpreted in terms of guessing habits, that is, that the total previous experience of the subjects is again a relevant condition. Difficulties in fact arise not only in doing an experiment exactly but also in reporting and interpreting it. In experiments on behaviour it is not easy to draw a line between these last two activities, for since only a few aspects of behaviour can be observed and reported, the making of a choice between the various possibilities is itself a sort of interpretation.

Another sort of discrepancy may arise through the use of experimental methods or descriptions of results that are thought to be equivalent but are not so in fact; for instance, it can be shown that different methods of obtaining thresholds would give different results even if the subject's behaviour remained the same. Thus such differences, which are customarily interpreted as the effect on the subject of the change in the psychological situation, must be in part at least simply differences in statistical technique. Similarly, Hall and Kobrick (1952) have shown that the various measures of response strength are not correlated to any extent, and an experiment involving that concept would give results depending largely on the measure used.

There is an obverse of the preceding case, often quite difficult to discover but always of great importance if experimental results are used to formulate or support theories of behaviour, which may be briefly described as too much correlation. If two measures are found to be always, or almost always, correlated it may mean that they describe two sorts of behaviour that are linked, or it may mean that the measures are not statistically independent. For example, Himmelweit (1947) found a high correlation between goal discrepancy and attainment discrepancy scores in a tracking task and concluded that 'they measure virtually the same factors'; an examination of the way in which these two concepts are defined shows that they are measured by almost the same combination of observed scores. It is often useful to investigate how the choice of measures may influence the interpretation of an experiment by setting up an urn model, or a random number model, which would imitate the sort of behaviour one thinks is involved, and trying out the various possible measures on results obtained from that. An example of such a procedure is given later.

It remains to ask what sort of properties we should require of the variables used to describe behaviour. First, they should clearly be adequate in number, so that anyone replicating the experiment in the prescribed form, and using the prescribed measure, should obtain substantially the same results; this of course does not preclude a more sensitive experiment, with more complete description, from giving better results. Secondly, these variables should be formally independent (that is, not two ways of measuring the same thing) for otherwise we should get an apparently very consistent linking of two sorts of behaviour which in reality means nothing. Thirdly, they should summarise as much information from the data as possible.

As an example of how methods of description may be analysed, I shall consider first the detailed recording of a sequence of behaviour in an unchanging situation. This is the basis of certain ethological studies, e.g. the nest-building behaviour of canaries (Hinde, 1958*b*), parental fanning in the male Three-spined Stickleback (van Iersel, 1953) and courtship in *Drosophila melanogaster* (Bastock and Manning, 1955).

2. *Records of behaviour*

These are obtained from continuous observation of the animal concerned over a fairly long period, say half an hour, during which the external situation is unchanged and the internal state of the animal is assumed to be relatively constant. Records made from different individuals, in different experimental conditions, or at times when the internal state of the animals may be supposed different are then compared. Although the observation is continuous, the recording of course is not. Generally, the experimenter defines a certain number of types of behaviour that he is interested in, say—in the case of nest-building—gathering, carrying and sitting, labels them G, C and S and then records at small constant intervals of time what the animal is actually doing. For observations every second throughout half an hour there would be a record of 1,800 observations like this:

$$GGGCSSGCSCC \ldots$$

These are the basic data of the experiment, and a corresponding model has only to produce such records. Some differences between observers may occur even at this level, in that their definitions of types of behaviour may be slightly different. I shall assume that this is relatively unimportant.

3. Summarising records

It is clearly impracticable, not to say confusing, to work with so much data, and a method of summarising such records is necessary. The three methods of description that are commonly used are mean length of bout (a bout is a sequence of one type of behaviour only), the number of bouts, and the total number of observations (total time) for a given sort of activity.

Suppose, for simplicity, that only one sort of activity, labelled S_1, is of interest (e.g. fanning in van Iersel's experiments) and that any other sort of activity is labelled S_2. Then for the sequence

$$S_1 \, S_2 \, S_1 \, S_1 \, S_1 \, S_2 \, S_2 \, S_1 \, S_2 \, S_2 \, S_1 \, S_1$$

the number of bouts for S_1 is 4, the mean length of bout is 7/4 and the total number of entries S_1 is 7. These three measures are clearly not independent since the last is the product of the first two. Thus two of them at most can be used to give independent information, and the best two to choose will depend generally on what will give the clearest description of the experiment. An example of this dependence is in Tables I and II of Bastock and Manning's paper. They find that both the mean length of courtship bout and the total time courting are longer in the response of male *Drosophila* to virgin female flies than to fertilised females. If the number of bouts is calculated for the two cases it will be seen that they are almost equal for the two females assigned to each male, which suggests that the male may alternate from one to the other. The length of courtship bout towards the virgin female is markedly longer; the increase in total courtship time is simply a consequence of this. There are, then, at most two independent variables in this set of three; if, for example, number of bouts is discarded, there remains mean length of bout and total time. In the literature quoted above, however, it is almost always found that these increase or decrease together. It is therefore possible that they may be linked statistically.

4. *Connection with Markoff chains*

It is worth looking at the description of records in terms of bouts in a slightly different way, which brings it into line with the idea of transition probabilities in Markoff chains. (A Markoff chain is a sequence of trials for which the probability of a particular result at any given trial depends only on what occurred at the preceding trial: it can be completely described in terms of the probabilities of the various outcomes of the first trial together with transition probabilities of the form AB, where this is to be read as: the chance that event B will follow when event A has occurred. For further information, see e.g. Feller, 1950.) A bout of S_1 ends whenever S_1 is followed by S_2 or if S_1 is the last entry in the record. Therefore the number of bouts is the number of occurrences of the pair S_1S_2, possibly plus one; for a long record, the odd 'one' will be relatively unimportant unless there are very few S_1 terms and I shall neglect it. Also a bout begins whenever the pair S_2S_1 occurs or if S_1 is the first entry in the record. Therefore, apart from an odd one somewhere, the number of S_1S_2 equals the number of S_2S_1. Also, since S_1 must be followed either by S_1 or S_2, the number of S_1 is equal to the sum of the number of S_1S_1 and the number of S_1S_2. If the length of the record is fixed, the information about the number of bouts, mean length of bout, and frequency for both S_1 and S_2 is completely given by:

(i) number of S_1S_1
(ii) number of S_1S_2

Thus the description of the data is essentially in terms of the first order transitional probabilities defined by:

$$P_{11} = \frac{n(S_1S_1)}{n(S_1)} \qquad P_{12} = \frac{n(S_1S_2)}{n(S_1)}$$

$$P_{21} = \frac{n(S_2S_1)}{n(S_2)} \qquad P_{22} = \frac{n(S_2S_2)}{n(S_2)}$$

where $n(S_1)$ is written for the number of S_1 and so on, and where $P_{11} + P_{12} = P_{21} + P_{22} = 1$.

5. Urn model for random behaviour

To see whether the mean length of bout and the total time are statistically dependent we should consider the sort of results that would be obtained if the behaviour were purely random. One way of doing this is by making up an urn model. Suppose, for instance, that $1/5$ of the total number of entries is to be S_1 and $4/5$ is to be S_2. We should take some container—usually described as an urn—with one red and four white balls in it, take out one at random, record S_1 if it is red and S_2 if it is white, return it, and repeat the process 1,800 times; this could be done for various values of p_1, the proportion of red balls. This would be rather tedious for a long record (though it might be very useful for short ones) and it is more comfortable to rely on algebra. In a long record we expect very nearly 1,800/5, i.e.

360 of S_1, and of these we expect $360 \times 4/5$ to be followed by S_2 and this gives the number of bouts. Therefore the number of bouts should be 288, the number of S_1 should be 360, and the mean length of bout for S_1 would then be 360/288, i.e. $1\frac{1}{4}$. Similarly, for any p_1, the number of bouts should be 1,800 $p_1(1 - p_1)$, the number of S_1 should be 1,800 p_1, and the mean length of bout would then be $1/1 - p_1$. The interesting thing about this result is that if any change in the condition causes S_1 to occur more frequently (i.e. causes an increase in p_1) then the mean length of bout will increase automatically with it. This is really obvious: the more S_1 there are in the record, the more likely they are to occur as neighbours, so that the mean length of bout will increase. The number of bouts, 1,800 $p_1(1 - p_1)$, will be small if *either* the number of S_1 is small *or* the number of S_1 is large—that is, if there are very few S_1 in the record or if there are so many that there is little room for breaks between them. It is clear that the number of bouts is a measure to be handled with care since a given number can arise in very different circumstances.

Thus the observed result noted earlier, that mean length of bout increases with the total frequency of a given type of behaviour, would be obtained if the behaviour within a given recording period were purely random, and the only difference between records lay in the frequency.

6. *More complicated model*

The previous model is clearly unrealistic in that it assumes that the behaviour of the organism is made up of elements each taking no longer than the interval of observation to perform, and that at each observation some new element is 'triggered off' by the internal or external environment. It would be more convincing to think that, once begun, a piece of behaviour would be completed, and this might take several of the observation intervals. Suppose, for example, that S_1 took at least two seconds to complete (assuming the record to be made at intervals of one second) and might take as long as 6 seconds, and that the average time was 3 seconds; and suppose the average time for S_2 was 4 seconds. Then whenever a red ball was pulled out we should enter 3 S_1 in the record to allow for the average effect, and for a white ball we should enter 4 S_2. Suppose that at the completion of each piece of activity the animal begins at random on a new piece, so that the model of taking balls from an urn still applies. I shall describe the end of one piece of behaviour and the beginning of another as a trigger-point. In a long record there will be, say, N trigger-points; $N/5$ will initiate S_1 behaviour and will give $(N/5) \times 3$ entries of S_1; similarly there will be $(4N/5) \times 4$ entries of S_2, and N can be found from the fact that $3N/5 + 16N/5 = 1,800$ (hence $N = 5 \times 1,800/19$) and so S_1 will be approximately 284, S_2 will be 1,516. The number of bouts will again be the number of times S_1 is followed by S_2, i.e. $(N/5) \times (4/5) = 76$, and the mean length of bout will be

$$\frac{3N/5}{4N/25},$$

i.e. $3 \times 5/4 = 15/4$ for S_1, and 20 for S_2.

More generally, if p_1 is the probability that a piece of behaviour S_1 will be initiated, so that $p_2 = 1 - p_1$ is the probability for S_2, and if m_1, m_2 are the average lengths of a piece of behaviour of type S_1 or S_2, then the number of entries of S_1 will be Np_1m_1, the number of entries of S_2 will be Np_2m_2, and $N(p_1m_1 + p_2m_2) = 1,800$; the number of bouts will be Np_1p_2 and the observed mean length of bout will be

$$m_1/(1 - p_1), \; m_2/(1 - p_2)$$

for S_1, S_2 respectively. On substituting for N we get:

number of $S_1 = 1,800 \, m_1p_1/(m_1p_1 + m_2p_2)$
number of $S_2 = 1,800 \, m_2p_2/(m_1p_1 + m_2p_2)$
number of bouts $= 1,800 \, p_1p_2/(m_1p_1 + m_2p_2)$
mean length of bout for $S_1 = m_1/(1 - p_1)$
mean length of bout for $S_2 = m_2/(1 - p_2)$

The sort of records obtained from such a model will probably not look much like observed records since no allowance has been made for the variation in length of a piece of behaviour; such variation could of course be introduced but in a long record it would make no difference to the above results.

I pointed out earlier that the measures used implied at most two pieces of information and it may seem rather contrary to express these in terms of the three variables m_1, m_2, p_1, since to a certain extent one of these could be given an arbitrary value. There are two reasons for this: first, if more than two types of behaviour are considered, fewer variables are used in this model than would be in a general Markoff chain, and second, if it can be shown (in the case of two types of behaviour) that the changing of only one of these variables produces results similar to those observed, then it is quite possible that these measures give only one piece of information. Thus they should not be treated as though each supplied independent information without further examination of the data.

Figure 20 shows the three possible cases that arise: Figure 20a is plotted by allowing p_1 to vary and keeping $m_1 = 3$, $m_2 = 4$; the mean length of bout for S_1 increases with the proportion of time spent on S_1 and the number of bouts first increases and then decreases. In Figure 20b, p_1 is kept constant at the value $1/5$, $m_2 = 4$. As m_1 varies, the mean length of bout for S_1 increases with the frequency of S_1, the mean length of bout for S_2 remains constant, and the number of bouts steadily declines. Figure 20c, with m_2 varying, is simply the reverse of b, so that in this case the mean length of bout for S_1 remains constant, and the number of bouts increases with the frequency of S_1.

We should presumably expect some sort of mixture of the effects as shown in Figure 20a, b and c; but by considering graphs of this sort it should be possible to decide whether the important thing is a change in the basic time occupied by a piece of behaviour, or whether it is a change in the probability of its occurrence. For example, in van Iersel's work on the behaviour of the male Three-spined Stickleback when hatching eggs, there is an experiment designed to investigate whether the amount of oxygen in the water affects the way in which the stickleback fans water across the eggs with its fins. The mean length of fanning bout is found to be almost constant in different conditions although the proportion of time spent fanning changes. This suggests that the case shown in Figure 20c applies,

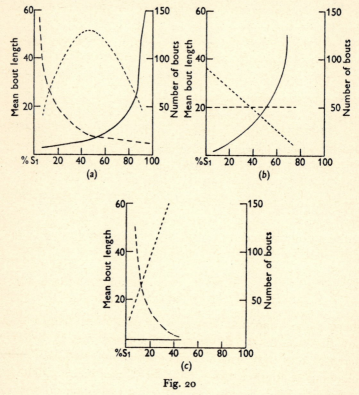

Fig. 20

i.e. that there is some change in the alternative activity as the quality of the water changes, and this is borne out by van Iersel's investigation (1953, Table 3).

It is clear when behaviour is considered even in this elementary way that the concept of 'tendency' (i.e. probability of occurrence) suggested by Russell, Mead and Hayes (1954) is too simple; three, at least, of the measures they propose for it could be altered in two ways:

(i) by a change in the probability that a certain sort of behaviour will be triggered off, and

(ii) by a change in the persistence with which it is continued, once begun.

It is by no means certain that these are two aspects of the same thing.

In our three cases shown in Figure 20, the mean length of bout increases or remains constant with increasing frequency of the action. Hinde (1958), however, finds that under certain conditions the mean length of bout decreases with greater activity. This can only occur if a decrease in one or both of m_1, p_1 (leading to a decrease in length of bout) is accompanied by a decrease in m_2 which more than compensates for the effect of the other decreases on the frequency of S_1. Since this would generally imply some sort of interaction between the two activities, it is perhaps better considered in the way described below. There is in any case a limit to the amount of such compensation in that m_2 can never be less than one, and might have a natural limit greater than one.

7. Switching and organised behaviour

There is one important way in which the mean length of bout may decrease as the frequency of a given sort of behaviour increases, and that is if the behaviour of one sort itself induces behaviour of another sort, i.e. the fact that a piece of behaviour S_1 has just been performed is itself an element in the total situation that triggers off the next piece of behaviour, its effect being to encourage a switch to that behaviour. A suitable model for this requires the use of two urns, one labelled R, the other W, each containing red and white balls. Some arbitrary rule is adopted for deciding which urn to choose from first; at all later turns, if the previous ball was white a ball is taken from the urn labelled W, if it was red one is taken from the urn labelled R. In addition, the proportion of red balls in the urn R is to be less than in the urn W, so that the design of the model, is this:

Proportion of

	Red balls	White balls
Urn R	$p_1 - \theta$	$p_2 + \theta$
Urn W	$p_1 + \varphi$	$p_2 - \varphi$

where p_1 is to be thought of as the environmental effect, θ and ϕ as the triggering effects of S_1 and S_2 respectively. To return to the case $p_1 = 1/5$ (or $2/10$), $p_2 = 4/5$ (or $8/10$), the urn R corresponding to S_1, might contain one red and 9 white balls and the urn W, corresponding to S_2, might contain 4 red and 6 white balls.

Suppose that during the whole urn experiment N_1 balls are taken from the urn R (and, since the colour of the preceding ball decides the urn, this means that N_1 red balls have been chosen), then 9 times out of 10 these will be white; therefore the number of S_1S_2 will be $9/10 \, N_1$. Similarly, if N_2 balls are picked from the urn W, 4 times out of 10 these will be red, therefore the number of S_2S_1 will be $4/10 \, N_2$, and as before these two numbers must be equal, so that $9/10 \, N_1 = 4/10 \, N_2$. Also a red ball has been chosen N_1 times and a white one N_2 times. If the average length of a piece of S_1 behaviour is 3 seconds and of a piece of S_2 behaviour is 4 seconds, then the record will contain $3 \, N_1$ entries of S_1 and $4 \, N_2$ entries of S_2, so that $3 \, N_1 + 4 \, N_2 = 1,800$. From these two equations,

$$N_1 = 1,800/12 = 150.$$

Thus the number of S_1 entries is $3 \times 150 = 450$, and the number of S_2 entries is $1,350$. The number of bouts is

$$9/10 \, N_1 = 135,$$

and the mean length of bout for S_1 and S_2 respectively is $3\frac{1}{3}$ and 10. In the previous case, the number of S_1 was 284 and the mean

372

length of bout $3\frac{3}{4}$, so this process has given an increase in the number of S_1 and a decrease in the mean length of bout.

In the general case, with arbitrary probabilities, the results are:

$$\text{number of } S_1 = \quad 1,800m_1(p_1 + \phi)/$$
$$(m_1(p_1 + \phi) + m_2(p_2 + \theta))$$
$$\text{number of bouts} = \quad 1,800(p_1 + \phi)(p_2 + \theta)/$$
$$(m_1(p_1 + \phi) + m_2(p_2 + \theta))$$
$$\text{mean length of bout for } S_1 = m_1/(1 - p_1 + \theta)$$
$$\text{mean length of bout for } S_2 = m_2/(1 - p_2 + \phi)$$

As an example of a more complicated case, suppose the observer defines two sorts of behaviour he is interested in, S_1 and S_2, and that S_3 is any residual behaviour. When actions are triggered off at random, let p_1, p_2, p_3, $(p_1 + p_2 + p_3 = 1)$ be the probabilities that these types will respectively occur. The means lengths of bout will be $m_1/(1 - p_1)$, $m_2/(1 - p_2)$, $m_3/(1 - p_3)$. Now suppose that performance of S_1 tends to lead to performance of S_2. The transition probabilities might become:

	Following		
	S_1	S_2	S_3
Preceding S_1	kp_1	$p_2 + (1 - k)p_1$	p_3
S_2	p_1	p_2	p_3
S_3	p_1	p_2	p_3

where k is less than one. The mean lengths of bout would become $m_1/(1 - kp_1)$, $m_2/(1 - p_2)$, $m_3/(1 - p_3)$: that is, the mean length of bout of S_1 would decrease, the other two remaining constant.

It should be emphasised that the suggestions given above are not unique interpretations of the data; their relevance could be established if more statistics (for instance, the variance of length of bout) were calculated. They do, however, lead to the following conclusions:

(i) the mean length of bout will certainly increase with the frequency of the activity if the only variable factor is the probability of the occurrence of that activity;

(ii) if the mean length of bout decreases markedly as the frequency of the activity increases, the most likely inference is that the behaviour is becoming more organised.

8. *Some examples*

In Figure 7 of van Iersel's book, the mean length of fanning bout is shown to increase with the total time spent fanning. This suggests that either there is an increased probability of fanning (case (a)), or the basic length of a piece of fanning is increasing (case (b)). However, his Table I shows that the number of bouts also increases with the frequency of fanning, so it is case (a) that seems to be relevant here. This can be roughly checked by plotting the mean length of bout against the ratio

$$\frac{\text{total time fanning}}{\text{total time not fanning,}}$$

because from the earlier equations can be derived:

$$\text{mean length of bout} = m_1 + m_2 \times \frac{\text{number of } S_1.}{\text{number of } S_2}$$

All the points on the graph relating to the fanning of one clutch of eggs, and all but the last three days for the fanning of 3 clutches, lie approximately on the same straight line. This line has an intercept on the length-of-bout axis of about 12 and a slope of about 6, which gives the values of m_1 and m_2. The points for the last 3 days of the fanning of 3 clutches appear to lie on a parallel straight line with intercept 6. It would appear that all the changes in observed behaviour could be interpreted in terms of the change in probability of fanning except those occurring towards the end of the more strenuous experiment (for the stickleback), when there is an actual change in the length of time devoted to a piece of fanning.

374

A similar graph can be plotted from the data of his Table 14, which relates various measures of fanning with the date of the hatching of the eggs. There are two interesting points about these results. First, there is a decline in the number of bouts at the highest level of fanning activity, which would be expected in case (a). Second, although the mean length of bout and the frequency can be regarded as dependent—that is, related by the probability that fanning will occur—one of these measures (the mean length of bout) correlates much more closely with the date of hatching; greater fanning activity, it would seem, is more effective in hatching the eggs *because* it produces longer bouts. This is an example of how one measure may be better, for a particular purpose, than another even though they are statistically dependent.

Bastock and Manning (1955) use the same sort of measures in investigating the courtship of *Drosophila melanogaster*. Their results might be expected to look rather different from those of van Iersel since on the whole courtship seems to occupy between half and all the time of observation (see their Figure 4), whereas fanning usually occupies less than half the time. Moreover, many of their graphs relate to the differences between individual flies, rather than to the comparison of records obtained in different conditions. They use their data to support a theory involving successively higher thresholds, so it is worth considering how far this support is real.

They describe the course of courtship in terms of two activities, orientation and vibration plus licking, the second activity being thought to have a higher threshold than the first. In their Figure 2, they plot the mean length of bout for orientation against the mean length of bout for vibration plus licking and obtain a negative correlation (-0.61) between these. They suggest that this means that 'when the level is for most of the time above the threshold for the "middle" element (orientation) it rises more often above that necessary for the "highest" one (vibration plus licking).' This statement could be properly investigated by considering not the mean length of bout of these activities, but the actual time spent on them. The statement would be justified if the percentage of courtship time spent in vibration plus licking increased as the total courtship time increased. The correlation they give does not support it for the following reason.

If the observed mean length of bout is controlled only by the probability of the occurrence of orientation, p_0, or of vibration plus licking, p_v, then the respective mean lengths of bout will be $m_0/(1 - p_0)$ and $m_v/(1 - p_v)$. If p_0 is near 1, p_v must be very small, and conversely, so that in any case these lengths could not be completely independent. We can investigate how such a dependence might induce a correlation between the mean lengths of bout for the two activities by setting up a suitable urn model; or, more simply, a random number model as follows (assuming that m_1 and m_2 are equal): from a random number table take two successive pairs of numbers to represent the percentage of observations allocated to the two activities. If these are, say, 37 and 25, this would mean a total courtship time occupying 62% of the observation period and mean lengths of bout proportional to $1/(1 - 0.37)$ and $1/(1 - 0.25)$ respectively. If this is repeated, say, 50 times, it will be equivalent to taking records from 50 insects. Any pair whose sum is greater than 100 must be discarded; so also must any pair of which one member is 00 since if an activity does not occur it is difficult to say what is meant by the mean length of bout (in fact I discarded any pair which contained one member less than 05 on the assumption that each activity would occur at least a small proportion of the time). In view of their Figure 4, which shows that courtship occupies at least half the total time of observation, it is reasonable to reject also any pair whose sum is less than 50. From such a sample of 50 pairs I found a correlation between mean lengths of bout of -0.45. The theoretical correlation is -0.65.

Thus the correlation found by Bastock and Manning may simply reflect the fact that the total time spent on courtship must be divided between these two activities. The triangular distribution of the points on the graph suggests the same conclusion. From this graph it would seem that, for a given length of vibration bout, almost any length of orientation bout is equally likely up to a certain cut-off value. If any probability of orienting is equally likely to appear with any probability of vibrating, subject to the limitations of the total courtship time, then it cannot be said that there is any relationship between them, or any successive levels of excitation.

In the same way, the average bout length of vibration plus licking should increase with the total courtship time, for the first is proportional to $1/(1 - p_v)$ and the second to $(p_v + p_0)$ and each of these will increase with p_v. Using the preceding random number data I found the following:

Total courtship time (%)	50–60	60–70	70–80	80–90	90–100
Mean bout length	1·43	1·62	1·80	2·03	2·77

which is rather like Bastock and Manning's results when the lower line is multiplied by a suitable constant representing m_v. These two measures cannot, at least on the evidence presented, be taken as independent measures of sexual excitation.

Bastock and Manning find a correlation of 0·65 between mean bout length of vibration plus licking and percentage licking in courtship; the correlation, for the random model, between the same mean bout length and the percentage of total courtship spent in vibration plus licking is 0·7. It is thus reasonable to regard the observed result as indicating no more than that a fixed proportion of the time spent vibrating includes licking.

In their Figure 7 they show that a difference exists between males courting *melanogaster* females and those courting *simulans* females; for a given bout length of vibration plus licking the former show a shorter bout length of orientation than the latter do, and the authors conclude that 'there is something about *simulans* females which causes males to . . . stand oriented for longer before they vibrate'. This sort of diagram can be very misleading when the variable treated as independent—in this case, vibration bout length—is not controlled but is itself subject to random variation. Since the two scatter diagrams from which these results are abstracted are presumably like those in their Figure 2, it is likely that for a given amount of orientation they would find longer bouts of vibration towards *simulans* females, and this would completely reverse their conclusions. This would appear to be an example of a theoretical background determining the selection and interpretation of data.

It is the use of the mean length of a bout as independent variable that causes trouble. For suppose that in general the proportion of total time that the males spend courting is less for *melanogaster* females than for *simulans* females. When sequences with the same amount of vibration behaviour are compared (this corresponds roughly to those with the same mean length of bout) those which are obtained from males with *melanogaster* females will have less orientation behaviour and so shorter bouts of orientation; moreover, the proportion of the courtship time spent on vibration, and hence presumably on licking, will be higher for these males. Thus all the results of Bastock and Manning's Figure 7 could be the consequence of *less* courtship by the males of *melanogaster* females. What the authors should consider is the proportion of total courtship time spent vibrating and the proportion of vibrating time spent licking.

Hinde's (1958b) paper is particularly interesting in that it describes the change of nest-building behaviour over several days and shows how it is altered towards the date of laying the first egg. The mean length of bout for gathering and carrying (of nest-building material) tends to decrease towards this date (see his Figure 1b), although the time spent on these activities increases, and this would seem to indicate that the behaviour is becoming more organised. This interpretation is probably borne out by the data relating to complete building sequences, i.e. those in which gathering, carrying and placing of nest-building material occur in that order; he shows, in his Table 10, that the shorter the mean length of bout, the fewer the incomplete building sequences. Since the shorter bouts are, on the whole, observed in the day or two before the first egg is laid, this would mean that complete sequences occur more often on these days, and this should surely be regarded as more organised behaviour.

Hinde (Table 9) finds a positive correlation between the number of times material is placed in the nest—which he uses as a measure of effectiveness of the behaviour—and the total time spent building; this table, like that of Bastock and Manning relating the mean bout lengths of two activities, seems to show the effect of a cut-off—that is, a given total building time appears to impose a definite upper limit on the number of placings possible. Hinde himself suggests that the relation may in part be due to the fact that, when the total time building is longer, there will be more gathering and carrying and thus a greater chance that one will succeed the other, a necessary condition for placing to occur. In fact a very large part of the observed effect may be due to this. If the differences in basic length of bout for the various kinds of behaviour are neglected, then the total building time is $1,800(p_g + p_c + p_s)$, where p_g, p_c and p_s represent the probability of gathering, carrying and sitting respectively, and the total number of gathering bouts which will be immediately followed by carrying is $1,800\,p_g p_c$; for a given total building time, B, this number will have a maximum value of $B^2/4 \times 1,800$, and the bounding line of the observed results seems to be of this shape. It might in fact be possible to decide whether the number of placings is simply related to the total time spent on gathering and on carrying, or whether a longer time induces placings over and above the expected number, by fitting a suitable regression line. If placing occurs with a constant probability, given the sequence 'gathering and carrying', we expect

log (number of placings)
$$= \text{log (time spent gathering)}$$
$$+ \text{log (time spent carrying)}$$
$$+ \text{constant.}$$

It would also be useful to sort the data by days, to see whether a long time building on a day well before the laying of the first egg is accompanied by as many placings as a similar time just before the laying of the first egg.

The rather curious curves relating the mean bout length to the number of bouts (Hinde, Figure 10) are again the sort of thing that would be expected if the probability of the response were the chief, or perhaps the only variable. The number of bouts should be small if the mean bout length is either very long or very short. The longest of all, theoretically, which correspond to cases where almost the whole of the observation time is spent on one activity, probably do not occur in actuality. Thus the figure obtained is very likely the result of averaging for a curve such as that in Figure 21. If high or low frequencies were equally likely, the averaged curve would be the dotted line, but if some of the very high frequencies were missing, the averaged curve would drop considerably at the left-hand end. It is rather confusing to plot the curve in the way chosen simply because the number of bouts does have this property of first increasing and then decreasing as the mean length of bout increases.

In general it might be said that many of the differences found, in this sort of record, between subjects or between conditions may be explained simply in terms of a different propensity to perform a given action. (It should perhaps be noted that most of the methods of analysis suggested still work even when there are no real differences in propensity, and the experimenter has sorted into classes those organisms which merely by chance have shown a greater or less responsiveness. For long records, however, random variations in response would not be great enough to give the differences in total response time that are observed.) Until it can be shown that this is not the only factor, complicated explanations in terms of successive thresholds, instinctive centres and so on are purely imaginative. There are cases, however, for instance that of nest-building behaviour, in which changes in response cannot be related merely to overall probability of response; these could be most simply described in terms of an increase in organisation of behaviour, without of course necessarily implying that this organisation is conscious. Such a change might, as Hinde suggests, be the result of practice.

These examples are all taken from ethology since, as far as I know, it is only in this field that such records are obtained. They might, however, be of value in other investigations. Increasing organisation or disorganisation of behaviour may occur, for example, in mental defectives who are being trained

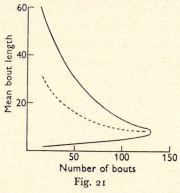

Fig. 21

on relatively difficult tasks, in patients suffering progressive deterioration through brain injury, or in trainees acquiring an industrial skill. For instance, someone learning a new skill might very well pass through two stages: an appreciation of the correct sequence of responses, and a development of manual control that would enable him to spend less time on each response; and which of these came first might depend on the method of instruction used as well as on the sort of skill. Successive records analysed in the way described above would make it possible to differentiate between these stages.

9. *The ordering of experimental conditions*

This idea, that there may be two or more stages in acquiring a skill, requiring two or more variables to describe the process, leads to another important question, that of how far results obtained in different conditions or from different subjects may be ordered. Haldane (1946) has considered this in the case where the performance differs in different environments, but where the performance is at least stable within one environment. The introduction of a change in behaviour through learning increases the complexity of the problem. As an example, it is impossible to say that one variety of wheat is better than another if the first gives a greater yield in a cold climate and the second gives a greater yield in a hot climate; if, in addition, each variety changes its yield over successive years the question of which is to be preferred may become almost unanswerable. The behaviour of animals or human beings is frequently modified by learning or adaptation, and yet the ordering of subjects or of conditions is often attempted and even sometimes required. There are two practical problems where ordering seems of importance—in selection procedures and in transfer of training experiments.

(a) Selection

When from a group of people we select a sub-group who will, we hope, be good at a particular job or benefit from a given sort of education, we are trying to order them into two sets, the more suitable and the less suitable. Selection is usually based on tests which are thought to relate to the performance required. Suppose that in fact the test is performance on the actual job for which the selection is to be made, so that we have to decide what the final ordering of a group of subjects would be in terms of their behaviour on a few trials. If the trials are scored in some way, we might have as successive scores from two subjects:

Subject X:	20	30	40	50	60	70	80	90
Subject Y:	5	20	35	50	65	80	95	100

In such a case there are clearly two factors to be discriminated: original skill, which might depend on familiarity with the test situation, and rate of improvement. Ordering, however, must be in terms of one variable so the tester has to select one score or combination of scores to compress the results of the test into a single number. If only one trial were given, X would be classified as better than Y; if more than four practice trials were allowed before the test trial, Y would be better; if the sum of the scores for the first few tests were used, X would be better on up to six trials, Y would be better on eight or more. Thus, where two variables are operating, different measures will be different compounds of them and will lead to different results, and any single measure may give the same score to two quite different performances.

If it is possible for improvement to continue indefinitely, then the rate of improvement will become the dominant factor and initial performance will ultimately be irrelevant. De Jong (1957) has shown that industrial skills, as measured by productivity, may continue to improve over many years, and that the rate of increase is roughly linear if performance is plotted against the logarithm of time. Again, Cross (1958) finds that the trend of a pupil's marks in successive examinations is often a better predictor of later performance than his mark on any one occasion. Therefore, if the ultimate degree of skill is the criterion, there is much to be said for using the rate of improvement as a measure. In general, however, it is preferable to sort out the relevant variables and consider them separately. In theoretical investigations one tries to relate observed results to experimental or natural conditions, and if the conditions influence different variables in different ways the effect will be obscured whenever an arbitrary combination of the variables is used as a measure. For example, it might be possible to show for a particular task that older subjects were better to begin with but learned less quickly; each effect might be ordered perfectly with respect to age, but a combined score would not be.

Even in practical work, where the measure to be used is generally defined by the problem posed, it may be better to work with more variables. If, say, we wished to choose those people who would most quickly reach a given level of competence, it might be preferable to predict the time taken to reach this level from a knowledge of initial performance and rate of improvement rather than to find this time by experiment, or attempt to estimate it by other tests.

(b) *Transfer of training*

The question of whether training on one task improves performance on a related task is of practical importance because it is often cheaper or safer to use a training device than to set a novice to work with the machine he will finally be required to use. It is then clearly desirable to have some general ideas on what is a suitable training device for a given job, what is the best length of time for training, and so on. The results of transfer of training experiments in the laboratory should provide such ideas; the impression to be gained from them is, however, mainly one of confusion. Gagné *et al.* (1948) have surveyed a 'voluminous literature' on the subject and found that results are often not comparable because of the variety of measures used of 'amount of transfer'. Similar difficulties over a suitable measure are found with most experiments involving learning or extinction.

In a transfer of training experiment a group of subjects is first trained on task A to a given time and then is required to perform task B. Another group simply performs task B. Yet another group might be trained on C before performing on B. The sort of statements that are made from the results of such experiments are, for instance, that there is a transfer from A to B (if the first group's performance on B is better than the second group's), or that there is more transfer from A than from C (if the first group does better on B than the third). More general statements are sometimes made—such as that there is more transfer from a difficult task to an easy one than vice versa. All these statements imply that performances can be ordered. As before, the ordering implies the use of a single variable, either one observation or a combination of a number of observations; the examples below are designed to show that the measure chosen may have a very great influence on the interpretation of results. In these examples it is assumed that a perfect performance (score 100) can be reached in a fairly short number of trials.

Consider first the simplest question—whether there is any transfer from task A to task B. Two groups are compared in respect of performance on B, one having first received training on A. (The question of the amount of training to be given is a further complication, introducing another variable.) If the results are expressed as percentage of correct responses at each trial, successive trials might give:

B alone	40	50	60	70	80	90	100
B following A	70	80	90	100	100	100	100

On any criterion, the second group is more successful; but suppose the results were:

B alone	40	50	60	70	80	90	100	100	100	100	100
B following A	50	55	60	65	70	75	80	85	90	95	100

If the number of trials to reach a perfect performance is used as the criterion, then the first group is better than the second; if the total number of correct responses is used, then the result will depend on the number of trials considered. If the experiment included only the first four trials, the sum of the correct responses would be higher for the second group than for the first; if it included six or more, the first group would give more correct responses. It is clear that, in this example, practice on A makes B easier to deal with at first (this sort of thing might happen if familiarity with the apparatus were important) but makes subsequent learning more difficult.

In asking the more complicated question whether there is more transfer from A to B than from B to A, we might have to consider the following results:

Group 1

A first	20	30	40	50	60	70	80	90	100			(9 trials to learn)
B second	0	6	12	18	24	30	36	42	48	54	60	(18 trials to learn)

Group 2

B first	0	5	10	15	20	........................(21 trials to learn)
A second	0	20	40	60	80	100 ....(6 trials to learn)

These results show that training on *A* slightly increases the rate of improvement on *B*, although it does not improve the initial performance, whereas training on *B* reduces the initial score on *A* but enables it to be learnt much more rapidly. One can imagine such results arising if tasks *A* and *B* involved the use of similar apparatus but task *B* required some unusual movement which would have to be omitted in performing *A*. There are various ways used to compare such results, of which three are given here:

(i) Total number of trials to learn both tasks; this is the same for both groups.

(ii) Percentage gain in time to learn; this is $(3/21) \times 100$ for *B* following *A* as against *B* first, and $(3/9) \times 100$ for *A* following *B* as against *A* first. This suggests more transfer from *B* to *A* than from *A* to *B*.

(iii) $\dfrac{A \text{ second} - A \text{ first}}{100 - A \text{ first}}$ at a given trial number, and a similar measure for *B*. Using the scores at successive trials we get:

Trial number	1	2	3	4	5	6
Transfer, *B* to *A*	−25	−14	0	20	50	100%
Transfer, *A* to *B*	0	1	2	4	5	7%

so that there is at first negative transfer from *B* to *A* and after that a high positive transfer. There will appear to be more transfer from *A* to *B* if up to three trials are given, more from *B* to *A* otherwise.

It will be seen that the answer given depends both on the measure used and the extent of the experiment; in fact, the question is meaningless except in reference to a particular practical situation. Szafran and Welford (1950), for instance, find that when an easy task follows a difficult one, transfer will tend to be positive, and when a difficult task follows an easy one, transfer will tend to be negative. It is difficult to know how

386

far the results obtained depend on the fact that all the data were collected from fifty trials in each condition.

In regard to statements about the relative value of training on 'difficult' and on 'easy' tasks, there is one further point that needs discussion. 'Easy' is usually defined either by what is customary (e.g. a control and a dial moving in the same sense, as in an experiment of Gibbs (1951)) or in terms of high initial scores. By either definition it is likely that the term will apply to tasks already well practised in everyday life, so that any further small amount of practice will make very little difference to the degree of skill. In the same way, 'difficult' almost by definition means 'new'; it must demand something further, for instance a higher degree of manual control, and this is likely to be used when the easy task is subsequently presented. Thus, to say that there is more transfer from a difficult task to an easy one than vice versa may mean no more than that training on a difficult task may make for more accurate performance on a simpler version of it. This is not very helpful if the difficult task is the one that is to be performed, which may take years to learn. Belbin, Belbin and Hills (1957) have shown that the mending of finely woven woollen fabrics can be taught in a few weeks by training operatives on very coarsely woven stuffs, whereas the learning of the difficult task may take two years.

The examples above were made up by choosing values for two variables, the initial score and the rate of improvement. If the rate of learning is not constant, yet another variable is required. It may, however, be possible to find a transformation of the time scale which will make the rate of improvement roughly linear for all conditions or subjects, and this would reduce the problem to the simpler case considered here. In certain kinds of practical work such complications may not be necessary; but a general statement, such as that it is better to learn the more difficult of two tasks first, necessarily involves a consideration of the exact meaning of the terms involved in it, and when results cannot always be ordered, the words 'better' and 'worse' do not necessarily apply.

10. *Conclusion*

In work on behaviour, where the measures available for describing results are virtually unlimited in number, it is particularly necessary that the ones chosen should be suitable for their purpose. This purpose is taken to be the clearest possible description of an experiment in a few terms, and either its interpretation in terms of an existing theory or its use as a basis for a new theory. It is suggested in this chapter that two of the criteria for deciding which measures to choose should be (1) that there are enough of them, (2) that they should be independent, or if dependent, at least known to be so. Whether measures satisfy this criteria can usually be discovered by an urn or random number experiment, or by making up imaginary examples of a kind similar to the experiments envisaged. The measures used in transfer of training experiments often fail to meet the first criterion, and those used in summarising ethological records often fail to meet the second. In both cases, complications arise in the theoretical description of results.

REFERENCES

ADAMS, J. A. (1955). A source of decrement in psychomotor performance. *J. exp. Psychol.* **49**, 390–4. 256.

ADRIAN, E. D. (1928). *The Basis of Sensation.* London. 345.

ADRIAN, E. D. (1947). *The Physical Background of Perception.* Oxford. 347.

AHRENS, R. (1954). Beiträge zur Entwicklung des Physiognomie—und Mimik-erkennes. *Z. Exp. Angew. Psychol.* **2**, 412–54 and 599–633. 190, 191.

ALDERSTEIN, A. & FEHRER, E. (1955). The effect of food deprivation on exploratory behaviour in a complex maze. *J. comp. physiol. Psychol.* **48**, 250–3. 282.

ALLEE, W. C. (1931). *Animal Aggregations.* Chicago. 129.

ANDREW, R. J. (1956a). Some remarks on conflict situations, with special reference to *Emberiza* spp. *Brit. J. Anim. Behav.* **4**, 41–45. 114.

ANDREW, R. J. (1956b). Normal and irrelevant toilet behaviour in *Emberiza* spp. *Brit. J. Anim. Behav.* **4**, 85–91. 131, 135.

ANDREW, R. J. (1956c). Intention movements of flight in certain passerines, and their use in systematics. *Behaviour,* **10**, 179–204. 102, 109.

ANDREW, R. J. (1957). The aggressive and courtship behaviour of certain Emberizinae. *Behaviour,* **10**, 255–308. 113, 118, 127.

ARMSTRONG, E. A. (1950). The nature and function of displacement activities. *Symp. Soc. exp. Biol.* **4**, 361–87.

ARMSTRONG, E. A. (1951). The nature and function of animal mimesis. *Bull. Anim. Behav.* No. 9, 46–58. 135.

ARMSTRONG, E. A. (1955). *The Wren.* London. 147.

ATTNEAVE, F. (1954). Informational aspects of visual perception. *Psychol. Rev.* **61**, 183–93. 360.

AUTRUM, H. (1950). Die Belichtigungspotentiale und das Sehen der Insekten. *Z. vergl. Physiol.* **32**, 176–277. 159.

AUTRUM, H. & GALLWITZ, U. (1951). Zur Analyse der Belichtungspotentiale des Insektenauges. *Z. vergl. Physiol.* **33**, 407–435. 12.

AUTRUM, H. & STOECKER, M. (1950). Die Verschmelzensfrequenzen des Bienenauges. *Z. Naturforschg.* **5**b, 38–43. 12.

BAERENDS, G. P. (1950). Specializations in organs and movements with a releasing function. *Symp. Soc. exp. Biol.* **4**, 337–60. 95.

BAERENDS, G. P. (1955). Egg recognition in the herring gull. *Proc. 14th Int. Congr. Psychol.* Montreal, 93. 163.

BAERENDS, G. P. (1956). Aufbau des tierischen Verhaltens. *Handbuch der Zoologie,* **10** (3), 1–32. 115.

BAERENDS, G. P. (1959). The ethological analysis of incubation behaviour. *Ibis,* **101**, 357–68. 58.

BAERENDS, G. P., BROUWER, R. & WATERBOLK, H. TJ. (1955). Ethological studies on *Lebistes reticulatus* (Peters). I. An analysis of the male courtship pattern. *Behaviour,* **8**, 249–334. 115.

BALDUS, K. (1926). Experimentelle Untersuchungen über die Entfernungs-localisation der Libellen (*Aeschna cyanea*). *Z. vergl. Physiol.* **3**, 475–505. 153.

BARD, P. (ed.). (1956). *Medical Physiology.* London. 31.

BARLOW, H. B. (1953). Summation and inhibition in the frog's retina. *J. Physiol.* **119**, 69–88. 49, 54, 160, 334, 347.

BARLOW, H. B. (1956). Retinal noise and absolute threshold. *J. opt. Soc. Amer.* **46**, 634–9. 328.

BARLOW, H. B. (1957) Increment thresholds at low intensities considered as signal/noise discrimination. *J. Physiol.* **136**, 469–88. 328.

BARLOW, H. B., FITZHUGH, R. & KUFFLER, S. W. (1957). Change of organization in the receptive fields of the cat's retina during dark adaptation. *J. Physiol.* **137**, 338–54. 351.

BARTLETT, F. C. (1932). *Remembering*. Cambridge. 198.

BASTOCK, MARGARET & BLEST, A. D. (1958). An analysis of behaviour sequences in *Automeris aurantiaca* (Weym.) (Lepidoptera). *Behaviour*, **12**, 243–84. 113, 116, 118, 121.

BASTOCK, M. & MANNING, A. (1955). The courtship of *Drosophila melanogaster*. *Behaviour*, **8**, 86–111. 27, 363, 375.

BEACH, F. A. (1937). The neural basis of innate behaviour. I. Effects of cortical lesions upon the maternal behaviour pattern in the rat. *J. comp. Psychol.* **24**, 393–439. 76, 151.

BEACH, F. A. (1938). The neural basis of innate behaviour. II. Relative effects of partial decortication in adulthood and infancy upon the maternal behaviour of the primiparous rat. *J. genet. Psychol.* **53**, 108–48. 76.

BEACH, F. A. (1940). Effects of cortical lesions upon the copulatory behaviour of male rats. *J. comp. Psychol.* **29**, 193–245. 66, 76.

BEACH, F. A. (1948). *Hormones and Behaviour*. New York. 178.

BEACH, F. A. & JAYNES, J. (1954). Effects of early experience upon the behaviour of animals. *Psychol. Bull.* **51**, 239–63. 231, 237.

BEACH, F. A. & JAYNES, J. (1956). Studies of maternal retrieving in rats. III. Sensory cues involved in the lactating female's response to her young. *Behaviour*, **10**, 104–25. 151.

BEER, G. R. DE (1944). *Vertebrate Zoology* (5th imp.). London. 21.

BELBIN, E., BELBIN, R. M. & HILLS, F. (1957). A comparison between the results of three different methods of operator training. *Ergonomics*, **1**, 39–50. 387.

BENDER, M. B. & KRIEGER, H. P. (1951). Visual function in perimetrically blind fields. *Arch. Neurol. Psychiat.*, Chicago, **65**, 72–79. 37, 51.

BENDER, M. B. & TEUBER, H. L. (1947). Spatial organization of visual perception following injury to the brain. *Arch. Neurol. Psychiat.*, Chicago, **58**, 721–39. 52.

BENDER, M. B. & TEUBER, H. L. (1948). Spatial organization of visual perception following injury to the brain. *Arch. Neurol. Psychiat.*, Chicago, **59**, 39–62. 52, 57.

BENDER, M. B. & TEUBER, H. L. (1949). Psychopathology of vision. In *Progress in Neurology and Psychiatry*. New York. 52, 54, 57.

BENDIG, A. W. (1951). The effect of reinforcement on the alternation of guesses. *J. exp. Psychol.* **41**, 105–7. 362,

BERKUN, M. M., KESSEN, M. L. & MILLER, N. E. (1952). Hunger reducing effects of food by stomach fistula versus food by mouth measured by consummatory response. *J. comp. physiol. Psychol.* **45**, 550–4. 291.

BERLYNE, D. E. (1951). Attention, perception and behaviour theory. *Psychol. Rev.* **58**, 137–46. 255.

BERLYNE, D. E. (1955). The arousal and satiation of perceptual curiosity in the rat. *J. comp. physiol. Psychol.* **48**, 238–46. 282.

BERLYNE, D. E. & SLATER, J. (1957). Perceptual curiosity, exploratory behaviour and maze learning. *J. comp. physiol. Psychol.* **50**, 228–32. 284.

BIEL, W. C. (1940). Early age differences in maze performance in the albino rat. *J. genet. Psychol.* **56**, 439–53. 235.

BINGHAM, W. E. & GRIFFITHS, W. J. (1952). The effects of different environments during infancy on adult behaviour in the rat. *J. comp. physiol. Psychol.* **45**, 307–12. 238.

BIRUKOW, G. (1953). Photogeomenotaxische Transpositionen bei *Geotrupes sylvaticus*. *Rev. Suisse Zool.* **60**, 534–40. 121.

BISHOP, P. O. & DAVIS, R. (1953). Bilateral interaction in the lateral geniculate body. *Science*, **118**, 241–3. 54.

BLEST, A. D. (1956). Protective coloration and animal behaviour. *Nature*, **178**, 1190–1. 102.

BLEST, A. D. (1957a). The function of eyespot patterns in the Lepidoptera. *Behaviour*, **11**, 209–55. 118.

BLEST, A. D. (1957b). The evolution of protective displays in the Saturnioidea and Sphingidae (Lepidoptera). *Behaviour*, **11**, 257–310. 102, 104, 105, 110, 111, 113, 116, 117, 118, 119.

BLEST, A. D. (1958a). Interaction between consecutive responses in a Hemileucid moth, and the evolution of insect communication. *Nature*, **181**, 1077–8. 121.

BLEST, A. D. (1958b). Some interactions between flight, protective display, and oviposition behaviour in *Callosamia* and *Rothschildia* spp. *Behaviour*, **13**, 297–318. 111, 113, 116, 117.

BLEST, A. D. (1959). Central control of interactions between behaviour patterns in a Hemileucine moth. *Nature*, **184**, 1164–5. 121.

BLUM, R. A. & BLUM, J. S. (1949). Factual issues in the continuity controversy. *Psychol. Rev.* **56**, 33–50. 261.

BLUM, J. S., CHOW, K. L. & PRIBRAM, K. H. (1950). A behavioural analysis of the organization of the parieto-temporo-preoccipital cortex. *J. comp. Neurol.* **93**, 53–100. 79.

BONIN, G. VON (1942). The striate area of primates. *J. comp. Neurol.* **77**, 405–29. 39.

BORING, E. G. (1929). *A History of Experimental Psychology*. New York. 59.

BOWLBY, J. (1952a). Maternal Care and Mental Health. *World Health Organisation Monograph*, Series 2. 192.

BOWLBY, J. (1952b). Critical phases in the development of social responses in man and other animals. In *Prospects in Psychiatric Research*, ed. J. M. Tanner. Oxford. 199.

BOWLBY, J. (1955). A note on the selection of love-objects in man. *Brit. J. Anim. Behav.* **3**, 122. 191.

BOWLBY, J. (1958). The nature of the child's tie to his mother. *Int. J. Psychoanalysis*, **39**, 1–24. 191.

BOYCOTT, B. B. (1953). The chromatophore system in Cephalopoda. *Proc. Linn. Soc. Lond.* **164**, 235–40. 111.

BOYD, H. (1956). Unpublished—*see also* Fabricius and Boyd (1954). 203.

BRIDGES, K. M. B. (1932). Emotional development in early infancy. *Child Development*, **3**, 324–34. 191.

BRIDGMAN, C. S. & SMITH, K. U. (1942). The absolute threshold of vision in cat and man with observations on its relation to the optic cortex. *Amer. J. Physiol.* **136**, 463–6. 35.

BRISTOWE, W. S. (1941). *The Comity of Spiders.* VOL. II. London. 153.

BROADBENT, D. E. (1953*a*). Neglect of the surroundings in relation to fatigue decrements in output. In *Ergonomics Symposium on Fatigue*. London. 257.

BROADBENT, D. E. (1953*b*). Noise, paced performance, and vigilance tasks. *Brit. J. Psychol.* **44**, 295–303. 257.

BROADBENT, D. E. (1954). The role of auditory localisation in attention and memory span. *J. exp. Psychol.* **47**, 191–6. 265.

BROADBENT, D. E. (1956*a*). Listening between and during practised auditory distractions. *Brit. J. Psychol.* **47**, 51–60. 257, 264.

BROADBENT, D. E. (1956*b*) Growing points in multi-channel Communication. *J. Acoust. Soc. Amer.* **28**, 533–5. 257, 264.

BROADBENT, D. E. (1956*c*). Successive responses to simultaneous stimuli. *Quart. J. exp. Psychol.* **8**, 145–52. 265.

BROADBENT, D. E. (1956*d*). The bass cutting of frequency transposed speech. *M.R.C. Applied Psychology Unit Report* No. 223. 268.

BRODAL, A. (1957). *The Reticular Formation of the Brain Stem*. London. 26.

BROUWER, E. (1928). Harmonische analyse van temperatuurcurven. *Ned. Tijdschr. Geneesk.* **72**, 5319–42. 213.

BROWN, F. A., JNR. (1954*a*). Biological clocks and the Fiddler Crab. *Sci. Amer.* April, 34–37.

BROWN, F. A., JNR. (1954*b*). Persistent activity rhythms in the Oyster. *Amer. J. Physiol.* **178**, 510–4.

BROWN, F. A., JNR., BENNETT, M. F. & RALPH, C. L. (1955). Apparent reversible influence of cosmic ray induced showers upon a biological system. *Proc. Soc. exp. Biol.* **89**, 332–7. 217.

BROWN, F. A., JNR., BENNETT, M. F., WEBB, H. M. & RALPH, C. L. (1956). Persistent daily, monthly and 27-day cycles of activity in the Oyster and the Quahog. *J. exp. Zool.* **131**, 235–62. 218.

BROWN, F. A., JNR., FINGERMAN, M. & HINES, M. H. (1954). A study of the mechanism involved in shifting of the phases of the endogenous daily rhythm of light stimuli. *Biol. Bull.* **106**, 308–17. 215.

BROWN, F. A., JNR., FREELAND, R. C. & RALPH, C. L. (1955). Persistent rhythms of O_2 consumption in potatoes, carrots and the seaweed *Fucus*. *Plant Physiol.* **30**, 280–92. 219.

BROWN, F. A., JNR. & SANDEEN, M. I. (1948). Responses of the chromatophores of the fiddler crab *Uca* to light and temperature. *Physiol. Zool.* **21**, 361–71.

BROWN, F. A., JNR. & WEBB, H. M. (1948). Temperature relations of an endogenous daily rhythmicity in the fiddler crab, *Uca*. *Physiol. Zool.* **21**, 271–81. 214.

BROWN, F. A., JNR. & WEBB, H. M. (1949). Studies of the daily rhythmicity of the fiddler crab, *Uca*. Modifications by light. *Physiol. Zool.* **22**, 136–48. 214.

REFERENCES

Brown, F. A., Jnr., Webb, H. M., Bennett, M. F. & Sandeen, M. (1954). Temperature-independence of the endogenous tidal rhythms of *Uca*. *Physiol. Zool.* **27**, 345–9. 216.

Brückner, G. H. (1933). Untersuchungen zur Tierpsychologie insbesondere zur Auflösung der Familie. *Z. Psychol.* **128**, 1–110. 177.

Bruesch, S. R. & Arey, L. B. (1942). The number of myelinated and unmyelinated fibers in the optic nerve of vertebrates. *J. comp. Neurol.* **77**, 631–65. 48.

Buchholtz, C. (1951). Untersuchungen an der Libellen-Gattung *Calopteryx* Leach unter besonderen Berüchsichtigung ethologischen Fragen. *Z. Tierpsychol.* **8**, 273–93. 163.

Buchholtz, C. (1955). Eine vergleichende Ethologie der orientalische Calopterigiden (Odonata) als Beitrag zu ihren systematischen Deutung. *Z. Tierpsychol.* **12**, 364–86. 163.

Buck, J. B. (1937). Studies on the firefly. II. The signal system and color vision in *Photinus pyralis*. *Physiol. Zool.* **10**, 412–9. 163.

Bühler, C. (1933). The social behaviour of children. In *Handbook of Child Psychology*, M. Murchison. 2nd ed. Worcester: Clark University Press. 191.

Bullock, T. H. (1953). Predator recognition and escape responses of some intertidal gastropods in the presence of starfish. *Behaviour*, **5**, 130–40. 153.

Burckhardt, D. (1944). Mowenbeobachtungen in Basle. *Orn. Beob.* **41**, 50–76. 139.

Burns, B. Delisle (1957). Electrophysiologic basis of normal and psychotic function. In Garattini, S. and Ghetti, V. (eds.) *Psychotropic Drugs*. Amsterdam. 56.

Burtt, E. T. & Catton, W. T. (1954). Visual perception of movement in the Locust. *J. Physiol.* **125**, 566–80. 12.

Butenandt, A. (1955). Über Wirkstoffe des Insektenreiches. II. Zur Kenntnis der Sexual-Lockstoffe. *Naturw. Rdsch.* **12**, 457–64. 153, 158.

Butler, R. A. (1957). The effect of deprivation of visual incentives on visual exploration motivation in monkeys. *J. comp. physiol. Psychol.* **50**, 177–9. 287.

Cajal, R. y & Sanchez, D. (1915). Contribucion al conocimiento de los centros nervosos de los insectos. *Trab. Lab. Invest. biol. Univ. Madr.* **13**, 1–164. 7.

Cane, V. R. (1956). Some statistical problems in experimental psychology. *J. R. Statist. Soc.* B., **18**, 177–201. 264.

Chambers, R. M. (1956a). The effects of intravenous glucose injections upon learning, general activity and hunger drive. *J. comp. physiol. Psychol.* **49**, 558–64. 291.

Chambers, R. M. (1956b). Some physiological bases for reinforcing properties of reward injections. *J. comp. physiol. Psychol.* **49**, 565–68. 291.

Chapman, L. F., Thetford, W. N., Berlin, L., Guthrie, T. C. & Wolff, H. G. (1958). Highest integrative functions in man during stress. *Proc. Ass. Res. nerv. Dis.* **36**, 491–534. 82.

Chapman, R. M. & Levy, N. (1957). Hunger drive and reinforcing effect of novel stimuli. *J. comp. physiol. Psychol.* **50**, 233–8. 282.

Cherry, E. C. (1953). Some experiments on the recognition of speech with one and with two ears. *J. Acoust. Soc. Amer.* **25**, 975–9. 263.

CHOW, K. L. (1952a). Further studies on selective ablation of associative cortex in relation to visually mediated behaviour. *J. comp. physiol. Psychol.* **45**, 109–18. 31.

CHOW, K. L. (1952b). Conditions influencing the recovery of visual discriminative habits in monkeys following temporal neo-cortical ablations. *J. comp. physiol. Psychol.* **45**, 430–7. 79, 80.

CHOW, K. L., BLUM, J. S. & BLUM, R. A. (1950). Cell ratios in the thalamo-cortical visual system of *Macaca mulatta*. *J. comp. Neurol.* **92**, 227–39. 48.

CHOW, K. L. & HUTT, P. J. (1953). The 'association cortex' of *Macaca mulatta*: a review of recent contributions to its anatomy and functions. *Brain*, **76**, 625–77. 79, 81.

CLARK, G. (1948). The mode of representation in the motor cortex. *Brain*, **71**, 320–31. 53.

CLARK, W. E. LE GROS (1942). The visual centres of the brain and their connexions. *Physiol. Rev.* **22**, 205–32. 48.

CLARK, W. E. LE GROS & PENMAN, G. G. (1934). The projection of the retina in the lateral geniculate body. *Proc. Roy. Soc.* B, **114**, 291–313. 31, 39.

COLE, J. & GLEES, P. (1954). Effects of small lesions in sensory cortex in trained monkeys. *J. Neurophysiol.* **17**, 1–13. 53.

COLLIAS, N. E. (1952). The development of social behaviour in birds. *Auk*, **69**, 127–59. 186, 206.

COLLIAS, N. E. & COLLIAS, E. C. (1956). Some mechanisms of family integration in ducks. *Auk*, **73**, 378–400. 177, 179, 185.

COLLIAS, N. E. & JOOS, M. (1953). The spectrographic analysis of sound signals of the domestic fowl. *Behaviour*, **5**, 175–88. 175.

CONDER, P. (1949). Individual distance. *Ibis*, **91**, 649–56. 139.

COPPOCK, H. W. & CHAMBERS, R. M. (1954). Reinforcement of position preference by automatic intravenous injections of glucose. *J. comp. physiol. Psychol.* **47**, 355–7. 291.

CORBET, P. S. (1955). A critical response to changing length of day in an insect. *Nature*, **175**, 338. 218.

COWEY, A. (1958). Unpublished research report, Psychological Laboratory, University of Cambridge. 47.

CRAGG, J. B. (1956). The olfactory behaviour of *Lucilia* spp. (Diptera) under natural conditions. *Ann. appl. Biol.* **44**, 467–77. 105.

CRAGG, J. B. & COLE, P. (1956). Laboratory studies in the chemosensory reactions of blowflies. *Ann. appl. Biol.* **44**, 478–91. 105.

CRAIG, W. (1918). Appetites and aversions as constituents of instincts. *Biol. Bull.* **34**, 91–107. 130.

CRAIK, K. J. W. (1943). *The Nature of Explanation.* Cambridge. 360.

CRAIK, K. J. W. (1947). Theory of the human operator in control systems. I. The operator as an engineering system. *Brit. J. Psychol.* **38**, 56–61. 303.

CRAIK, K. J. W. (1948). *Ibid.* II. Man as an element in a control system. *Brit. J. Psychol.* **39**, 142–8. 303.

CRANE, JOCELYN (1941). Crabs of the genus *Uca* from the West Coast of Central America. *Zoologica*, **26**, 145–208. 102.

CRANE, JOCELYN (1949). The comparative biology of Salticid spiders at Rancho Grande, Venezuela. IV. An analysis of display. *Zoologica*, **34**, 159–214. 102, 107, 117.

CRANE, JOCELYN (1952). A comparative study of innate defensive behaviour in Trinidad Mantids (Orthoptera, Mantoidea). *Zoologica*, **37**, 259–93. 102.

CRANE, JOCELYN (1955). Imaginal behaviour of a Trinidad butterfly, *Heliconius crato hydara* Hewitson, with special reference to the social use of color. *Zoologica*, **40**, 167–96. 164.

CRANE, JOCELYN (1957). Basic patterns of display in fiddler crabs (Ocypodidae, genus *Uca*). *Zoologica*, **42**, 69–82. 102, 163.

CRAWFORD, M. P. (1939). The social psychology of vertebrates. *Psych. Bull.* **36**, 407–46. 135.

CREED, R. S., DENNY-BROWN, F., ECCLES, J. C., LIDDELL, E. G. T. & SHERRINGTON, C. S. (1932). *Reflex Activity of the Spinal Cord.* London. 15. 117.

CRESPI, L. P. (1942). Quantitative variation of incentive and performance in the white rat. *Amer. J. Psychol.* **55**, 467–517. 294, 298.

CRITCHLEY, M. (1953). *The Parietal Lobes.* London. 84.

CROOK, J. H. (1953). An observational study of the Gulls of Southampton Water. *Brit. Birds*, **46**, 386–97. 139.

CROOK, J. H. (1958). Studies on the comparative ethology and social organisation of the Weaver Birds (Ploceinae). Ph.D. thesis, Cambridge Univ. Library. 127, 131.

CROOK, J. H. (1959). Behaviour study and the classification of West African Weaver Birds. *Proc. Linn. Soc. Lond.* Session 170, 1957–8, Pt. 2, 147–153. 130.

CROOK, J. H. (1959). Studies on the social behaviour of *Quelea q. quelea* (Linn.) in French West Africa. (*In press.*) 130.

CROSS, G. R. (1958). A study of the test performance of secondary school pupils at the age of the transfer examinations and their correlation with ability and attainment at later stages of development. *Unpublished dissertation.* 383.

CROSSMAN, E. R. F. W. (1953). Entropy and choice time: the effect of frequency unbalance on choice response. *Quart. J. exp. Psychol.* **5**, 41–51. 314.

CROSSMAN, E. R. F. W. (1955). The measurement of discriminability. *Quart. J. exp. Psychol.* **7**, 176–95. 268.

CROSSMAN, E. R. F. W. & SZAFRAN, J. (1956). Changes with age in the speed of information intake and discrimination. *Proc. Internat. Gerontol. Asscn. Rsch. Cttee.* (European Section.) *Symposium on Experimental Research on Ageing* (in press, quoted Welford). 201.

CRUZE, W. W. (1938). Maturity and learning ability. *Psychol. Monog.* **50**, 49–65. 235.

CULLEN, E. (1957). Adaptations in the kittiwake to cliff-nesting. *Ibis*, **99**, 275—302. 108.

CULLER, E. & METTLER, F. A. (1934). Conditioned behaviour in a decorticate dog. *J. comp. Psychol.* **18**, 291–303. 42.

DAANJE, A. (1950). On the locomotory movements in birds and the intention movements derived from them. *Behaviour*, **3**, 48–98. 102, 103, 104, 109.

D'AMATO, M. R. (1955). Secondary reinforcement and magnitude of primary reinforcement. *J. comp. physiol. Psychol.* **48**, 378–80. 298.

DARWIN, CHARLES (1872). *The Expression of the Emotions in Man and Animals.* London. 103.

DAVIS, D. R. (1957). *Introduction to Psychopathology.* London. 68.

DAVIS, H. (1951). Psychophysiology of Hearing and Deafness. In *Handbook of Experimental Psychology*, ed. S. S. Stevens. New York. 347.

DAVITZ, J. R. (1955). Reinforcement of fear at the beginning and at the end of shock. *J. comp. physiol. Psychol.* **48**, 152–5. 280.

DEESE, J. (1951). The extinction of a discrimination without performance of the choice response. *J. comp. physiol. Psychol.* **44**, 362–6. 253.

DELACOUR, J. & MAYR, E. (1945). The family Anatidae. *Wilson Bull.* **57**, 3–55. 108.

DEMBER, W. N. (1956). Response by the rat to environmental change. *J. comp. physiol. Psychol.* **49**, 93–95. 284.

DENNY-BROWN, D. & CHAMBERS, R. A. (1958). The parietal lobe and behaviour. *Proc. Ass. Res. nerv. Dis.* **36**, 35–117. 83.

DERNOWA-YARMOLENKO, A. A. (1933). The fundamentals of a method of investigating the function of the nervous system as revealed in overt behaviour. *J. genet. Psychol.* **42**, 319–38. 197.

DETHIER, V. G. (1947a). Chemoreceptors in the ovipositor of *Nemeritis.* *J. Exp. Zool.* **105**, 199–208. 17.

DETHIER, V. G. (1947b). *Chemical Insect Attractants and Repellents.* London. 152.

DETHIER, V. G. (1955). The physiology and histology of the contact chemoreceptors of the blowfly. *Quart. Rev. Biol.* **30**, 348–71. 10, 158.

DETHIER, V. G. (1957). Communication by insects: physiology of dancing. *Science*, **125**, 331–6. 120, 121.

DEUTSCH, J. A. (1953). A new type of behaviour theory. *Brit. J. Psychol.* **44**, 304–17. 293, 300.

DEUTSCH, J. A. (1956a). The inadequacy of the Hullian derivation of reasoning and latent learning. *Psychol. Rev.* **63**, 389–9. 296.

DEUTSCH, J. A. (1956b). A theory of insight reasoning and latent learning. *Brit. J. Psychol.* **47**, 115–25. 300.

DE VALLOIS, R. L., SMITH, C. J., KITAI, S. T. & KAROLY, A. J. (1957). Responses from different layers of the monkey lateral geniculate nucleus to monochromatic light stimuli. *Amer. Psychologist*, **12**, 468. 55.

DITCHBURN, R. W. & GINSBORG, B. L. (1953). Involuntary eye movements during fixation. *J. Physiol.* **119**, 1–17. 10.

DREES, O. (1952). Untersuchungen über die angeborenen Verhaltensweisen bei Springspinnen (Salticidae). *Z. Tierpsychol.* **9**, 169–207. 27, 151, 163.

DUKE-ELDER, W. S. (1949). *Textbook of Ophthalmology.* Vol. IV. London. 54.

DUSSER DE BARENNE, J. G. (1933). Corticalization of function and functional localization in the cerebral cortex. *Arch. Neurol. Psychiat.* **30**, 884–901. 38.

EIBL-EIBESFELDT, I. (1952). Nahrungswerb und Beuteschema der Erdkröte (*Bufo bufo* L.). *Behaviour*, **4**, 1–35. 153.

EIDMANN, H. (1955). Ueber rhythmische Erscheinungen bei Stabheuschrecke *Carausius morosus* Br. *Z. vergl. Physiol.* **38**, 370. 215.

EMLEN, J. T. (1952). Flocking behaviour in birds. *Auk*, **69**, 160–70. 127, 130, 141.

EWER, R. F. (1957). Ethological concepts. *Science*, **126**, 599–603. 164.

FABER, A. (1953). Zur Laut und Gebärdensprache bei Insekten. Stuttgart. 111.

REFERENCES

FABRICIUS, E. (1951). Zur Ethologie junger Anatiden. *Acta zool. Fenn.* **68**, 1–175. 176, 177, 179, 185.

FABRICIUS, E. & BOYD, H. (1954). Experiments on the following-reactions of ducklings. *Wildfowl Trust Annual Report* 1952–3, 84–89. 176, 177, 183, 184.

FAUST, C. (1955). *Die zerebralen Herdstörungen bei Hinterhauptsverletzungen und ihre Beurteilung.* Stuttgart. 83.

FEHRER, E. (1956). The effects of hunger and familiarity of locale on exploration. *J. comp. physiol. Psychol.* **49**, 549–52. 282.

FELLER, W. (1950). *An introduction to probability theory and its applications.* London. 365.

FERRIER, D. (1886). *The Functions of the Brain.* London. 45.

FINLAYSON, L. H. & LOWENSTEINO (1955). A proprioceptor in the body musculature of Lepidoptera. *Nature,* **176**, 1031. 16.

FINLAYSON, L. H. & LOWENSTEIN, O. (1958). The structure and function of abdominal stretch receptors in insects. *Proc. Roy. Soc. (Lond.)* B. **148**, 433–449. 16.

FINLEY, C. B. (1941). Equivalent losses in accuracy of response after central and after peripheral sense deprivation. *J. comp. Neurol.* **74**, 203–37. 70, 71.

FISHER, A. E. (1956). Maternal sexual behaviour induced by intracranial chemical stimulation. *Science,* **124**, 228–9. 117.

FLOURENS, P. (1824). *Recherches expérimentales sur les propriétés et les fonctions du système nerveux dans les animaux vertébrés.* Paris. 59.

FORGAYS, D. G. & FORGAYS, J. W. (1952). The nature of the effect of free environmental experience in the rat. *J. comp. physiol. Psychol.* **45**, 322–8. 238.

FORGUS, R. H. (1955). Early visual and motor experience as determiners of complex maze learning ability under rich and reduced stimulation. *J. comp. physiol. Psychol.* **48**, 215–20. 239.

FORGUS, R. H. (1956). Advantage of early over late perceptual experiences in improving form discrimination. *Canad. J. Psychol.* **10**, 147–55. 238.

FREUD, S. (1910). *Drei Abhandlungen zur Sexualtheorie.* 2nd ed. Leipzig and Vienna. 178.

FRISCH, K. VON (1950). *Bees, their vision, chemical senses and language.* Ithaca, New York. 120.

FRY, D. B. & WHETNALL, E. (1954). The auditory approach in the training of deaf children. *Lancet,* **266**, 583–7. 198.

FULLER, J. L., EASLER, C. A. & BANKS, E. M. (1950). Formation of conditioned avoidance in young puppies. *Amer. J. Physiol.* **160**, 462–6. 208.

FULTON, J. F. (1949). *Physiology of the Nervous System.* New York. 46.

GAGNÉ, R. M. (1953). In *Ergonomics Symposium on Fatigue,* W. F. Floyd and A. T. Welford, Eds. London. 256.

GAGNÉ, R. M., FOSTER, H. & CRAWLEY, M. E. (1948). The measurement of transfer of training. *Psychol. Bull.* **45**, 97–130. 384.

GALAMBOS, R. (1944). Inhibition of activity in single auditory nerve fibres by acoustic stimulation. *J. Neurophysiol.* **7**, 287–303. 347.

GALANTER, E. H. (1955). Place and response learning: Learning to alternate. *J. comp. physiol. Psychol.* **48**, 17–18. 281.

GESELL, A. (1954). The ontogenesis of infant behaviour. In *Manual of Child Psychology,* 2nd ed. Ed. L. Carmichael. New York. 191.

397

GIBBS, C. B. (1951). Transfer of training and skill assumptions in tracking tasks. *Quart. J. exp. Psychol.* **3**, 99–110. 387.

GIBSON, J. J. (1950). *The Perception of the Visual World.* New York. 350.

GILBERT, J. C. (1941). Memory loss in senescence. *J. abnorm. and soc. Psychol.* **36**, 73–86. 200.

GLANZER, M. (1953*a*). The role of stimulus satiation in spontaneous alternation. *J. exp. Psychol.* **45**, 387–93. 254.

GLANZER, M. (1953*b*). Stimulus satiation: an explanation of spontaneous alternation and related phenomena. *Psychol. Rev.* **60**, 257–68. 254.

GLEES, P. (1941). The termination of optic fibres in the lateral geniculate body of the cat. *J. Anat.* **75**, 434–40. 48.

GLEES, P. (1942). The termination of optic fibres in the lateral geniculate body of the rabbit. *J. Anat.* **76**, 313–18. 48.

GLEES, P. & CLARK, W. LE GROS (1941). The termination of optic fibres in the lateral geniculate body of the monkey. *J. Anat.* **75**, 295–308. 48.

GLEES, P. & COLE, J. (1950). Recovery of skilled motor function after small repeated lesions of motor cortex in Macaque. *J. Neurophysiol.* **13**, 137–48. 53.

GLEES, P., COLE, J., LIDDELL, E. G. T. & PHILLIPS, C. G. (1950). Beobachtungen über die motorische Rinde des Affen. *Arch. f. Psychiatr. u. Z. Neur.* **185**, 675–89. 53.

GOETHE, F. (1954). Vergleichende Beobachtungen über das Verhalten der Silbermowe (*Larus a. argentatus*) und der Heringsmowe (*Larus f. fuscus*). *Proc. XI Int. Orn. Congr.* 557–82. 105.

GOLTZ, F. (1881). Über die Verrichtungen des Grosshirns. *Pflüg. Arch. ges. Physiol.* **26**, 1–49. 60, 74.

GOODWIN, J., LONG, L. & WELCH, L. (1945). Generalisation in memory. *J. Exp. Psychol.* **35**, 71–75. 196.

GRABOWSKI, C. T. & DETHIER, V. G. (1954). The structure of the tarsal chemoreceptors of the blowfly *Phormia regina*. *J. Morph.* **94**, 1–20. 10.

GRAHAM, C. H. (1951). Visual Perception. In *Handbook of Experimental Psychology.* Ed. S. S. Stevens. New York. 349.

GRANIT, R. (1942). Spectral properties of the visual receptor elements of the guinea pig. *Acta physiol. scand.* **3**, 318–28.

GRANIT, R. (1947). *The Sensory Mechanisms of the Retina.* Oxford. 348.

GRANIT, R. (1955). *Receptors and Sensory Perception.* New Haven. 16, 328.

GRANIT, R. & TANSLEY, KATHERINE (1948). Rods, cones and the localization of pre-excitatory inhibition in the mammalian retina. *J. Physiol.* **107**, 54–66. 55.

GREGORY, R. L. (1952). A speculative account of the brain in terms of probability and induction. *Medical Research Council Applied Psychology Unit Report* No. 183. 328.

GREGORY, R. L. (1953). Physical model explanations in psychology. *Brit. J. Phil. Sci.* **4**, 192–7. 308, 327.

GREGORY, R. L. (1956). An experimental treatment of vision as an information source and noisy channel. *Third Lond. Information Symposium.* Ed. Cherry. London. 201, 328.

GREGORY, R. L. (1958). Increase in 'Neurological Noise' as a factor in ageing. In *Proceedings of Fourth International Gerontological Congress.* 328.

GREGORY, R. & CANE, V. (1955). Noise and the visual threshold. *Nature,* **176**, 1272. 328.

GRIFFIN, D. R. & NOVICK, A. (1955). Acoustic orientation of neotropical bats. *J. exp. Zool.* **130**, 251–300. 153.

GROSS, C. G. & WEISKRANTZ, L. (1959). A note on the perception of total luminous flux in monkeys and man. *Quart. J. exp. Psychol.* **11**, 49–53. 41.

GUHL, A. M. (1958). The development of social organisation in the domestic chick. *Anim. Behav.* **6**, 92–111. 148.

HALDANE, J. B. S. (1946). Interaction of nature and nurture. *Ann. Eug.* **13**, 197–205. 382.

HALDANE, J. B. S. & SPURWAY, H. (1954). A statistical analysis of communication in *Apis mellifera,* and a comparison with communication in other animals. *Insectes Sociaux,* **1**, 247–83. 103.

HALE, E. B. (1956a). Social facilitation and forebrain function in the maze performance of Green Sunfish (*Lepomis cyanellus*). *Physiol. Zool.* **29**, 93–107. 106.

HALE, E. B. (1956b). Effects of forebrain lesions on the aggressive behaviour of Green Sunfish (*Lepomis cyanellus*). *Physiol. Zool.* **29**, 107–27. 106.

HALL, J. F. & KOBRICK, J. L. (1952). The relation of three measures of response strength. *J. comp. physiol. Psychol.* **45**, 280–2. 362.

HARKER, J. E. (1953). The diurnal rhythm of activity of mayfly nymphs. *J. exp. Biol.* **30**, 525–33. 215.

HARKER, J. E. (1954). Diurnal rhythms in *Periplaneta americana.* L. *Nature,* **173**, 689. 215.

HARKER, J. E. (1956). Factors controlling the diurnal rhythm of activity of *Periplanta americana. J. exp. Biol.* **33**, 224–234. 215.

HARKER, J. E. (1958). Diurnal rhythms in the animal kingdom. *Biol. Rev.* **33**, 1–52. 216, 218, 220.

HARLOW, H. F. (1939). Recovery of pattern discrimination in monkeys following unilateral occipital lobectomy. *J. comp. Psychol.* **27**, 467–89. 44.

HARLOW, H. F. (1953). Higher functions of the nervous system. *Ann. Rev. Physiol.* **15**, 493–514. 80.

HARLOW, H. F. (1959). Affectional responses in the infant monkey. *Science,* **130**, 421–432. 193.

HARLOW, H. F., DAVIS, R. T. SETTLAGE, P. H. & MEYER, D. R. (1952). Analysis of frontal and posterior association syndromes in brain-damaged monkeys. *J. comp. physiol. Psychol.* **45**, 419–29. 79.

HARTLINE, H. K. (1938). The response of single optic nerve fibres of the vertebrate eye to illumination of the retina. *Amer. J. Physiol.* **121**, 400–15. 55, 348.

HARTLINE, H. K. (1940). The receptive fields of optic nerve fibres. *Amer. J. Physiol.* **130**, 690–9. 348.

HARTLINE, H. K. (1949). Inhibition of activity of visual receptors by illuminating nearby retinal areas in *Limulus. Fed. Proc.* **8**, 69. 347.

HARTLINE, H. K., WAGNER, H. G. & RATLIFF, F. (1956). Inhibition in the eye of *Limulus. J. gen. Physiol.* **40**, 357–76. 347.

REFERENCES

HARTRIDGE, R. (1947). The visual perception of fine detail. *Philos. Trans.* B, **232**, 519–671. 10.

HASKELL, P. T. (1956). Hearing in certain Orthoptera. II. The nature of the response of certain receptors to natural and imitation stridulation. *J. Exp. Biol.* **33**, 767–76. 12, 157, 163.

HASKELL, P. T. (1957). Stridulation and associated behaviour in certain Orthoptera. I. Analysis of the stridulation of, and behaviour between, males. *Brit. J. Anim. Behav.* **5**, 139–48. 111.

HASSENSTEIN, B. (1951). Ommatidienraster und afferente Bewegungsintegration. *Z. vergl. Physiol.* **33**, 301–26. 12.

HEAD, H. (1926). *Aphasia and Kindred Disorders of Speech.* 2 vols. London. 67.

HEBB, D. O. (1942). The effect of early and later brain injury upon test scores and the nature of normal adult intelligence. *Proc. Amer. Philos. Soc.* **85**, 275–92. 198.

HEBB, D. O. (1949). *The Organisation of Behaviour.* New York. XI, 9, 60.

HEBB, D. O. & PENFIELD, W. (1940). Human behaviour after extensive bilateral removal from the frontal lobes. *Arch. Neurol. Psychiat. Chicago*, **44**, 421–38. 84.

HEBB, D. O. & WILLIAMS, K. (1946). A method of rating animal intelligence. *J. gen. Psychol.* **34**, 59–65. 77.

HEDIGER, N. (1950). *Wild Animals in Captivity* (English edition of *Wildtiere in Gefangenschaft*, 1942). London. 138.

HEILBRUN, A. B. (1956). Psychological test performance as a function of lateral localization of cerebral lesion. *J. comp. physiol. Psychol.* **49**, 10–14. 82.

HEINROTH, O. (1910). Beiträge zur Biologie, namentlich Ethologie und Physiologie der Anatiden. *Verhl. 5. Int. Orn. Kongr.* 589—702. 102, 167, 175.

HEINROTH, O. (1924–33). *Die Vogel Mitteleuropas.* Vol. 1. Berlin. 147.

HEINZ, H. (1949). Vergleichende Beobachtungen über die Putzhandlung bei Dipteren im allgemeinen und bei *Sarcophaga*. *Z. Tierpsychol.* **6**, 330–371. 19.

HELLWIG, H. & LUDWIG, W. (1951). Versuche zum Frage der Arterkennung bei Insekten. *Z. Tierpsychol.* **9**, 456–62. 153.

HERB, F. H. (1940). Latent learning. Non-reward followed by food in blinds. *J. comp. Psychol.* **29**, 247–55. 295.

HERNÁNDEZ-PÉON, R., SCHERRER, H. & JOUVET, M. (1956). Modification of electrical activity in cochlear nucleus during 'attention' in unanesthetised cats. *Science*, **123**, 331–2. 269.

HERRICK, C. J. (1931). *Introduction to Neurology* (5th ed.). New York. 22.

HERTZ, M. (1929). Die Organisation des optischen Feldes der Biene. I. *Z. vergl. Physiol.* **8**, 693–748. 12.

HERTZ, M. (1930). ditto. II. *Z. vergl. Physiol.* **11**, 107–145. 12.

HERTZ, M. (1931). ditto. III. *Z. vergl. Physiol.* **14**, 629–674. 12.

HERTZ, M. (1933). Über figurale Intensitäten und Qualitäten in der optischen Wahrnehmung der Biene. *Biol. Zbl.* **53**, 10–40. 12.

HERTZ, M. (1934). Zur Physiologie des Formen und Bewegungssehens. III. Figurale Unterscheidung und reziproke Dressuren bei der Biene. *Z. vergl. Physiol.* **21**, 604–615. 12.

HERTZ, M. (1935). Die Untersuchungen über die Formensinn der Honigbiene. *Naturwissenschaften*, **23**, 619—24. 12.

HERTZ, M. (1937). Beitrag zum Farbensinn und Formensinn der Biene. *Z. vergl. Physiol.* **24**, 413–21. 12.

HESS, E. H. (1955). An experimental analysis of imprinting—a form of learning. *Prog. Rep. U.S. Pub. Health Service Grant Ab.M.* 776. 207.

HESS, E. H. (1959). Imprinting. *Science,* **130**, 133–41. 193.

HICK, W. E. (1952a). On the rate of gain of information. *Quart. J. exp. Psychol.* **4**, 11–26. 268, 314, 339.

HICK, W. E. (1952b). Why the human operator? *Trans. Soc. Instrum. Tech.* **4**, 67–77. 339.

HIMMELWEIT, H. T. (1947). A comparative study of the level of aspiration of normal and neurotic persons. *Brit. J. Psychol.* **37**, 41–59. 362.

HINDE, R. A. (1952). The behaviour of the Great Tit (*Parus major*) and some other related species. *Behaviour Supplement* No. 2, 1–201. 128, 129, 179, 187.

HINDE, R. A. (1953). The term "Mimesis". *Brit. J. Anim. Behav.* **1**, 7–12. 128, 135.

HINDE, R. A. (1954). Factors governing the changes in strength of a partially inborn response, as shown by the mobbing behaviour of the Chaffinch (*Fringilla coelebs*). I and II. *Proc. Roy. Soc.* B, **142**, 306–31, 331–58. 113, 117, 164, 258.

HINDE, R. A. (1955). The modifiability of instinctive behaviour. *Advanc. Sci.* **12**, 19–24. 126, 127, 176, 184, 189.

HINDE, R. A. (1955–56). A comparative study of the courtship of certain finches (Fringillidae). *Ibis,* **98**, 701–45. 113, 119, 129.

HINDE, R. A. (1956). Ethological models and the concept of 'drive'. *Brit. J. Philos. Sci.* **6**, 321–31. 115, 127.

HINDE, R. A. (1958a). Alternative motor patterns in Chaffinch song. *Anim. Behav.,* in press. 210.

HINDE, R. A. (1958b). The nest-building behaviour of domesticated canaries. *Proc. zool. Soc. Lond.* **131**, 1–48. 117, 363, 378.

HINDE, R. A., THORPE, W. H. & VINCE, M. A. (1956). The following response of young coots and moorhens. *Behaviour,* **9**, 214–42. 177, 179, 182, 183, 184, 203.

HINSCHE, G. (1928). Kampfreaktionen bei einheimischen Anuren. *Biol. Zbl.* **48**, 575–617. 102.

HODGSON, E. S. (1957a). A comparative electrophysiological analysis of chemo-receptors in arthropods. *Anat. Rec.* **128**, 565–6. 158.

HODGSON, E. S. (1957b). Electrophysiological studies of arthropod chemorecep-tion. II. Responses of labellar chemoreceptors of the blowfly to stimulation by carbohydrates. *J. ins. Physiol.* **1**, 240–7. 158.

HODGSON, E. S., LETTVIN, J. Y. & ROEDER, K. D. (1955). Physiology of a primary chemoreceptor unit. *Science,* **122**, 417–18. 10.

HODGSON, E. S. & ROEDER, K. D. (1956). Electrophysiological studies of arthro-pod chemoreception. I. General properties of the labellar chemoreceptors of Diptera. *J. cell. comp. Physiol.* **48**, 51–76. 158.

HOLMES, G. (1919). The cortical localization of vision. *Brit. med. J.* **2**, 193–9. 51, 57.

HOLMES, G. (1956). *Selected Papers,* ed. F. M. R. Walshe. London. 83.

HOLMES, G. & LISTER, W. T. (1916). Disturbances of vision from cerebral lesions, with special reference to the cortical representation of macula. *Brain*, **39**, 34–73. 51.

HOLST, E. VON (1936). Versuche zur Theorie der relativen Koordination. *Pflüg. Archiv*, **235**, 345–59. 110.

HOLZAPFEL, M. (1939). Analyse des Sperrens und Pickens in der Entwicklung des Stares. *J. f. Orn.* **87**, 525–53. 177.

HOLZAPFEL, M. MEYER (1956). Das Spiel bei Saugetieren. In Kukenthal's *Handbuch der Zoologie*, 8. 191.

HOPKINS, C. O. (1955). Effectiveness of secondary reinforcing stimuli as a function of the quantity and quality of food reinforcement. *J. exp. Psychol.* **50**, 339–42. 298.

HOYLE, G. (1955a). Function of the insect ocellar nerve. *J. Exp. Biol.* **32**, 397–407. 11.

HOYLE, G. (1955b). Neuromuscular mechanisms of a Locust skeletal muscle. *Proc. Roy. Soc. (Lond.)* B, **143**, 343–67. 16.

HOYLE, G. (1957). *Comparative physiology of the nervous control of muscular contraction.* Cambridge. 15.

HUBBERT, H. B. (1915). The effect of age on habit formation in the albino rat. *Behav. Monog.* 2, No. **6**, 1–55. 234.

HUBEL, D. H. (1959). Single unit activity in striate cortex of unrestrained cats. *J. Physiol.* **147**, 226–238. 349.

HUBEL, D. H. & WIESEL, T. N. (1959). Receptive fields of single neurones in the cat's striate cortex. *J. Physiol.* **148**, 574–591. 349.

HUBER, F. (1955a). Sitz und Bedeutung nervoser Zentren für Instinkthandlungen bei Mannchens der *Gryllus campestris*. *Z. Tierpsychol.* **12**, 12–48. 18, 27.

HUBER, F. (1955b). Über die Funktion der Pilzkörper beim Gesang des Mannchens der Keulenheuschrecke *Gomphocerus rufus. Naturwissenschaften*, **42**, 566–7. 27.

HUDSON, W. H. (1892). *The Naturalist in La Plata.* London. 190.

HUGHES, B. (1954). *The Visual Fields.* Oxford. 51.

HUGHES, G. M. (1957). The coordination of insect movements. II. *J. exp. Biol.* **34**, 306–33. 18.

HULL, C. L. (1943). *The Principles of Behaviour.* New York. 251, 274.

HULL, C. L. (1952). *A Behaviour System.* New Haven. 274.

HUMPHREY, M. E. & ZANGWILL, O. L. (1952). Effects of a right-sided occipito-parietal brain injury in a left-handed man. *Brain*, **75**, 312–24. 83.

HUNTER, W. S. (1930). A consideration of Lashley's theory of the equipotentiality of cerebral action. *J. gen. Psychol.* **3**, 455–68. 63.

HUNTER, W. S. & SIGLER, M. (1940). The span of visual discrimination as function of time and intensity of stimulation. *J. exp. Psychol.* **26**, 160–79. 354.

HURWITZ, H. M. B. (1955). Response elimination without performance. *Quart. J. exp. Psychol.* **7**, 1–7. 253.

HUXLEY, J. S. (1923). Courtship activities in the red-throated diver (*Colymbus stellatus* Pontopp.), together with a discussion of the evolution of courtship in birds. *J. Linn. Soc. Lond.* **25**, 253–92. 112.

HYMOVITCH, B. (1952). The effects of experimental variations on problem solving in the rat. *J. comp. physiol. Psychol.* **45**, 313–21. 238.

IERSEL, J. J. A. VAN (1953). An analysis of the parental behaviour of the male Three-spined Stickleback. *Behaviour Supplement* No. 3, 1–159. 185, 163.

ILSE, D. (1941). The colour vision of insects. *Proc. Roy. Phil. Glasgow,* **65**, 68–82. 164.

ISSERLIN, M. (1923). Über Störungen des Gedächtnisses bei Hirngeschädigten. *Z. ges. Neurol. Psychiat.* **85**, 84–97. 83.

JACOBS, W. (1953). *Verhaltensbiologische Studien an Feldheuschrecken.* Berlin and Hamburg. 12.

JACOBSON, C. F. (1936). Studies of cerebral function in primates. I. The functions of the frontal association areas in monkeys. *Comp. Psychol. Monog.* 13, No. 63, 3–60. 78, 79.

JAMES, WILLIAM (1890). *Principles of Psychology.* 2 vols. New York. 60, 204.

JAMES, WILLIAM (1892). *Textbook of Psychology.* London. 167, 178, 184.

JASPER, H., GLOOR, P. & MILNER, B. (1956). Higher functions of the nervous system. *Ann. Rev. Physiol.* **18**, 359–86. 81.

JAYNES, J. (1956). Imprinting: the interaction of learned and innate behaviour. I. Development and generalisation. *J. comp. physiol. Psychol.* **49**, 201–6. 177, 182, 183, 205.

JAYNES, J. (1957). Imprinting: the interaction of learned and innate behaviour. II. *J. comp. physiol. Psychol.* **50**, 6–10. 184, 185.

DE JONG, J. R. (1957). The effects of increasing skill on cycle time and its consequences for time standards. *Ergonomics,* **1**, 51–60. 383.

JUNG, R. & BAUMGARTNER, G. (1955). Hemmungsmechanismen und bremsende Stabilisierung an einzelnen Neuronen des optischen Cortex. *Pflüg. Arch.* **261**, 434–56. 55.

KANTROW, R. W. (1937). An investigation of conditioned feeding responses and concomitant adaptive behaviour in young infants. *Univ. Iowa Stud. Child Welfare* 13, No. 3. 196.

KARLI, P. (1956). The Norway rat's killing response to the white mouse: an experimental analysis. *Behaviour,* **10**, 81–103. 106.

KAY, H. (1953). Experimental Studies of Adult Learning. *Unpublished Ph.D. Thesis.* Cambridge. 200.

KEENLEYSIDE, M. H. A. (1955). Some aspects of the schooling behaviour of fish. *Behaviour,* **8**, 183–249. 126.

KENNARD, MARGARET A. (1938). Reorganisation of motor function in the cerebral cortex of monkeys deprived of motor and premotor areas in infancy. *J. Neurophysiol.* **1**, 477–96. 53.

KENNARD, MARGARET A. & McCULLOCH, W. S. (1943). Motor response to stimulation of cerebral cortex in absence of areas 4 and 6 (*Macaca mulatta*). *J. Neurophysiol.* **6**, 146–59. 53.

KENNEDY, J. L. (1939). The effects of complete and partial occipital lobectomy upon thresholds of visual real movement. *J. genet. Psychol.* **54**, 119–149. 34.

KIESEL, A. (1894). Untersuchungen zur Physiologie des facettierten Auges. *S.B. Akad. Wiss. Wien,* **103**, 97–139. 213.

KIMBLE, G. A. & KENDALL, J. W. (1953). A comparison of two methods of producing experimental extinction. *J. exp. Psychol.* **45**, 87–90. 253.

KING, J. A. (1958). Parameters relevant to determining the effect of early experience upon the adult behaviour of animals. *Psychol. Bull.* **55**, 46–58. 231, 237.

KISH, G. B. (1955). Learning when the onset of illumination is used as reinforcing stimulus. *J. comp. physiol. Psychol.* **48**, 261–4. 288.

KIVY, P. N., EARL, R. W. & WALKER, E. L. (1956). Stimulus content and satiation. *J. comp. physiol. Psychol.* **49**, 90–92. 284.

KLEIN, M. (1932). *Psychoanalysis of Children.* London. 191.

KLEIN, M & OTHERS (1952). *Developments in Psychoanalysis.* London. 191.

KLOPFER, P. H. (1959). Development of sound-signal preferences in ducks. *Wilson Bull.* **71**, 262–266. 177, 179.

KLUVER, H. (1927). Visual disturbances after cerebral lesions. *Psychol. Bull.* **24**, 316–58. 51, 57.

KLUVER, H. (1936). An analysis of the effects of removal of the occipital lobes in monkeys. *J. Psychol.* **2**, 49–61. 36.

KLUVER, H. (1937). Certain effects of lesions of the occipital lobes in macaques. *J. Psychol.* **4**, 383–401. 36, 45.

KLUVER, H. (1941). Visual functions after removal of the occipital lobes. *J. Psychol.* **11**, 23–45. 36.

KLUVER, H. (1942). Functional significance of the geniculostriate system. *Biol. Symposia* **7**, 253–99. 36.

KÖHLER, W. (1920). Physical Gestalten, in *Source Book of Gestalt psychology*, ed. W. D. Ellis, 1938. London. (Translation of *Die physischen Gestalten in Ruhe und im stationären Zustand, Eine naturphilosophische Untersuchung*, Erlangen.) 308.

KÖHLER, W. (1925). *The Mentality of Apes.* London. 248.

KÖHLER, W. & WALLACH, H. (1944). Figural after-effects: an investigation of visual processes. *Proc. Amer. phil. Soc.* **88**, 269–357. 352.

KOHN, M. (1951). Satiation of hunger from food injected directly into the stomach versus food ingested by mouth. *J. comp. physiol. Psychol.* **44**, 412–22. 291.

KOLLROSS, J. (1942). Localised maturation of lid-closure reflex mechanism by thyroid implants into the tadpole hind-brain. *Proc. Soc. exp. Biol. Med.* **49**, 204–6. 117.

KOLLROSS, J. (1943). Experimental studies on the development of the corneal reflex in Amphibia. II. Localised maturation of the reflex mechanisms effected by thyroxin-agar implants into the hind-brain. *Physiol. Zool.* **16**, 269–79. 117.

KRAMER, G. & ST. PAUL, U. VON (1951). Über angeborenes und erworbenes Feinderkennen beim Gimpel (*Pyrrhula pyrrhula*). *Behaviour*, **4**, 243–55. 178.

KRECHEVSKY, I. (1935). Brain mechanisms and 'hypotheses'. *J. comp. Psychol.* **19**, 425–68. 77.

KRYNAUW, R. A. (1950). Infantile hemiplegia treated by removing one cerebral hemisphere. *J. Neurol. Neurosurg. Psychiat.* **13**, 243–67. 82.

KUFFLER, S. W. (1953). Discharge patterns and functional organization of mammalian retina. *J. Neurophysiol.* **16**, 37–68. 347.

REFERENCES

LACK, D. (1954). *The Natural Regulation of Animal Numbers*. Oxford. 129.

LAMBERCIER, M. & REY, A. (1935). Contribution à l'étude de l'intelligence pratique chez l'enfant. *Arch. Psychol.* **25**, 1–59. 242.

LANSDELL, H. C. (1953). Effect of brain damage on ntelligence in rats. *J. comp. physiol. Psychol.* **46**, 461–4. 77.

LASHLEY, K. S. (1920). Studies of cerebral function in learning. *Psychobiology*, **2**, 55–135. 60, 62.

LASHLEY, K. S. (1922). Studies of cerebral function in learning. IV. Vicarious function after destruction of the visual areas. *Amer. J. Physiol.* **59**, 44–71. 41, 62.

LASHLEY, K. S. (1926). Studies of cerebral function in learning. VII. The relation between cerebral mass, learning and retention. *J. comp. Neurol.* **41**, 1–48. 61, 62, 68, 78.

LASHLEY, K. S. (1929). *Brain Mechanisms and Intelligence: a quantitative study of injuries to the brain.* Chicago. 62, 64, 67, 68, 69, 70, 74, 76, 81, 82.

LASHLEY, K. S. (1931a). The mechanism of vision. IV. The cerebral areas necessary for pattern vision in the rat. *J. comp. Neurol.* **53**, 419–78. 39, 42, 44, 56.

LASHLEY, K. S. (1931b). Mass action in cerebral function. *Science,* **73**, 245–54. 65, 67, 69, 76.

LASHLEY, K. S. (1931c). Cerebral control versus reflexology: a reply to Professor Hunter. *J. gen. Psychol.* **5**, 3–20. 63, 74.

LASHLEY, K. S. (1932). Studies of cerebral function in learning. VIII. A re-analysis of the data on mass action of the visual cortex. *J. comp. Neurol.* **54**, 77–84. 69.

LASHLEY, K. S. (1934). The mechanism of vision. VIII. The projection of the retina upon the cerebral cortex of the rat. *J. comp. Neurol.* **60**, 57–79. 32, 41, 42.

LASHLEY, K. S. (1935). The mechanism of vision. XII. Nervous structures concerned in the acquisition and retention of habits based on reactions to light. *Comp. Psychol. Mon.* 11, No. 52, 43–79. 39, 43, 44, 69.

LASHLEY, K. S. (1938a). Factors limiting recovery after central nervous lesions. *J. nerv. ment. Dis.* **88**, 733–55. 81.

LASHLEY, K. S. (1938b). Experimental analysis of instinctive behaviour. *Psychol. Rev.* **45**, 445–71. 81, 150.

LASHLEY, K. S. (1939). The mechanism of vision. XVI. The functioning of small remnants of the visual cortex. *J. comp. Neurol.* **70**, 45–67. 41, 44, 50, 69.

LASHLEY, K. S. (1942). The mechanism of vision. XVII. Autonomy of the visual cortex. *J. genet. Psychol.* **60**, 197–221. 41, 75.

LASHLEY, K. S. (1943). Studies of cerebral function in learning. XII. Loss of the maze habit after occipital lesions in blind rats. *J. comp. Neurol.* **79**, 431–62. 70, 73, 75.

LASHLEY, K. S. (1944). Studies of cerebral function in learning. XIII. Apparent absence of trans-cortical association in maze learning. *J. comp. Neurol.* **80**, 257–81. 66, 68, 78.

LASHLEY, K. S. (1948). The mechanism of vision. XVIII. Effects of destroying the visual 'associative areas' of the monkey. *Genet. Psychol. Mon.* **37**, 107–66. 78, 81.

LASHLEY, K. S. (1950). In search of the engram. *S.E.B. Symposia*, **4**, 454–82. 66, 67, 68, 78, 80, 358.

LASHLEY, K. S. (1952). Functional interpretation of anatomic patterns. *Res. Publ. Ass. nerv. ment. Dis.* **30**, 529–47. 80.

LASHLEY, K. S. & FRANK, M. (1934). The mechanism of vision. X. Post-operative disturbances of habits based on detail vision in the rat after lesions in the cerebral visual areas. *J. comp. Psychol.* **17**, 355–93. 41, 69.

LASHLEY, K. S. & FRANZ, S. I. (1917). The effects of cerebral destruction upon habit-formation and retention in the albino rat. *Psychobiology,* **1**, 71–139. 62, 72.

LASHLEY, K. S. & WILEY, J. (1933). Studies of cerebral function in learning. IX. Mass action in relation to the number of elements in the problem to be learned. *J. comp. Neurol.* **57**, 3–56. 69, 71.

LAWRENCE, D. H. & MASON, W. A. (1955). Systematic behaviour during discrimination reversal and change of dimensions. *J. comp. physiol. Psychol.* **48**, 1–7. 265.

LAWSON, R. (1953). Amount of primary reward and strength of secondary reward. *J. exp. Psychol.* **46**, 183–7. 298.

LAWSON, R. (1957). Brightness discrimination performance and secondary reward strength as a function of primary reward amount. *J. comp. physiol. Psychol.* **50**, 35–39. 298.

LEHRMAN, D. S. (1953). A critique of Konrad Lorenz's theory of instinctive behaviour. *Quart. Rev. Biol.* **28**, 337–63. 155, 362.

LEHRMAN, D. S. (1955). The physiological basis of parental feeding responses in the ring-dove (*Streptopelia risoria*). *Behaviour,* **7**, 241–86. 117.

LEHRMAN, D. S. (1956). On the organisation of maternal behaviour and the problem of instinct. In *L'instinct dans le comportement des animaux et de l'homme,* 475–520. Fondation Singer-Polignac, Paris. 117.

LEVINE, R., CHEIN, L. & MURPHY, G. (1942). The relation of the intensity of a need to the amount of perceptual distortion: a preliminary report. *J. Psychol.* **13**, 283–93. 269.

LEWIS, P. R. & LOBBAN, M. C. (1956). Patterns of electrolyte excretion in human subjects during a prolonged period of life on a 22-hour day. *J. Physiol.* **133**, 670–80. 212.

LEWIS, P. R., LOBBAN, M. C. & SHAW, T. I. (1956). Patterns of urine flow in human subjects during a prolonged period of life on a 22-hour day. *J. Physiol.* **133**, 659–69. 212, 213.

LINDSLEY, D. B. (1957). Psychophysiology and perception. In *Current Trends in Psychology*. Pittsburgh. 49.

LISSMANN, H. W. (1932). Die Umwelt des Kampfisches (*Betta splendens* Regan). *Z. vergl. Physiol.* **18**, 65–111. 153.

LIU, S. Y. (1928). The relation of age to the learning ability of the white rat. *J. comp. Psychol.* **8**, 75–85. 235.

LOEB, J. (1901). *Comparative Physiology of the Brain and Comparative Psychology*. London. 1, 60.

LOHER, W. (1957). Untersuchungen über den Aufbau und die Entstehung der Gesänge einiger Feldheuschreckenarten und den Einfluss von Lautzeichen auf das akustische Verhalten. *Z. verg. Physiol.* **39**, 313–56. 163.

LOOMIS, W. F. (1955). Glutathione control of the specific feeding reaction of *Hydra*. *Ann. N.Y. Acad. Sci.* **62**, 209–28. 154.

LORENZ, K. (1935). Der Kumpan in der Umwelt des Vögels. *J. f. Orn.* **83**, 137–214, 289–413. 128, 135, 150, 155, 175, 177, 184, 188.

LORENZ, K. (1937). The companion in the bird's world. *Auk*, **54**, 245–73. 128. 135, 175, 186, 188.

LORENZ, K. (1939). Vergleichende Verhaltensforschung. *Zool. Ann. Supp.* **12**, 69–102. 153.

LORENZ, K. (1941). Vergleichende Bewegungstudien an Anatinen. *J. Orn.* **89**, 194–294. 94, 102, 108.

LORENZ, K. (1950). The comparative method in studying innate behaviour patterns. *Symp. Soc. exp. Biol.* **4**, 221–68. 102, 113.

LOUCKS, R. B. (1931). Efficacy of the rat's motor cortex in delayed alternation. *J. comp. Neurol.* **53**, 511–67. 72.

LURIA, A. R. (1932). *The Nature of Human Conflicts.* New York. 196, 240.

LYNIP, A. W. (1951). The use of magnetic devices in the collection and analysis of the preverbal utterances of an infant. *Genet. Psychol. Monogr.* **44**, 221–62. 192.

MCCULLOCH, T. L. & HASLERUD, G. M. (1939). Affective response of an infant chimpanzee reared in isolation from its kind. *J. comp. Psychol.* **28**, 437–45. 191.

MCCULLOCH, T. L. & PRATT, J. G. (1934). A study of the presolution period in weight discrimination by white rats. *J. comp. Psychol.* **18**, 271–90. 267.

MCDERMOTT, F. A. (1917). Observations on the light emission of American Lampyridae. *Canad. Ent.* **49**, 53–61. 163.

MCDOUGALL, W. (1905). *Physiological Psychology.* London. 1.

MCDOUGALL, W. (1923). *An Outline of Psychology.* London. 91.

MCFIE, J. & PIERCY, M. F. (1952). Intellectual impairment with localized cerebral lesions. *Brain*, **75**, 292–311. 82, 83.

MACKAY, D. M. & MCCULLOCH, W. S. (1952). The limiting capacity of a neuronal link. *Bull. Math. Biophysics.* **14**, 127–35. 314.

MACKWORTH, N. H. (1950). Researches on the measurement of human performance. *M.R.C. Special Report Series* No. 286. 257.

MACKWORTH, N. H. (1956). Work design and training for future industrial skills. *J. Inst. Prod. Eng.* **35**, 214–40.

MADDEN, E. H. (1957). A logical analysis of 'Psychological Isomorphism'. *Brit. J. Philos. Sci.* **8**, 177–91. 308.

MAGNUS, D. (1954). Zum Problem der 'überoptimalen' Schlüsselreize. *Verh. dtsch. zool. Ges.* (Tubingen, 1954), 317–25. 159, 162.

MAGNUS, D. (1958). Experimentelle Untersuchungen zur Bionomie und Ethologie des Kaisermantels *Argynnis paphia*. *Z. f. Tierpsych.*, **15**, 397–426. 159.

MAIER, N. R. F. (1931). Reasoning in humans: the solution of a problem and its appearance in consciousness. *J. comp. Psychol.* **12**, 181–94. 249.

MAIER, N. R. F. (1932a). Cortical destruction of the posterior part of the brain and its effect on reasoning in rats. *J. comp. Neurol.* **56**, 179—208. 72.

MAIER, N. R. F. (1932b). Age and intelligence in rats. *J. comp. Psychol.* **13**, 1–6. 235.

MAIER, N. R. F. & SCHNEIRLA, T. C. (1935). *Principles of Animal Psychology.* New York. 26.

MANTON, S. M. (1950). The evolution of arthropodan locomotory mechanisms. I. The locomotion of *Peripatus*. *J. linn. Soc. Lond.* **41**, 529–70. 123.

MANTON, S. M. (1952*a*). II. General introduction to the locomotory mechanisms of the Arthropoda. *J. linn. Soc. Lond.* **42**, 93–117. 123.

MANTON, S. M. (1952*b*). III. The locomotion of the Chilopoda and Pauropoda. *J. linn. Soc. Lond.* **42**, 118–67. 123.

MANTON, S. M. (1954). IV. The structure, habits and evolution of the Diplopoda. *J. linn. Soc. Lond.* **42**, 299–368. 123.

MARINESCO, G. & KREINDLER, A. (1933). Des réflexes conditionnels: I. L'organisation des réflexes conditionnels chez l'enfant. *J. Psychol., Paris*, **30**, 855–996. 196.

MARLER, P. (1956*a*). Behaviour of the Chaffinch (*Fringilla coelebs*). *Behaviour Supplement No.* 5, 1–184. 102, 103, 126, 127, 147, 187.

MARLER, P. (1956*b*). Studies of fighting in Chaffinches. (3) Proximity as a cause of aggression. *Brit. J. Anim. Behav.* **5**, 23–30. 102, 140.

MARLER, P. (1957*a*). Specific distinctiveness in the communication signals of birds. *Behaviour*, **11**, 13–39. 119.

MARLER, P. (1957*b*). Studies of fighting in chaffinches. (4) Appetitive and consummatory behaviour. *Brit. J. Anim. Behav.* **5**, 29–37. 148.

MARQUIS, D. G. (1934). Effects of removal of the visual cortex in mammals, with observations on the retention of light discrimination in dogs. *Res. Pub. Ass. nerv. ment. Dis.* **13**, 558–92. 37.

MARQUIS, D. G. (1935). Phylogenetic interpretation of the functions of the visual cortex. *Arch. Neurol. Psychiat.* **33**, 807–15. 33.

MARQUIS, D. G. (1942). The neurology of learning. In Moss (ed.), *Comparative Psychology*. New York. 33.

MARQUIS, D. G. & HILGARD, E. R. (1936). Conditioned lid responses to light in dogs after removal of the visual cortex. *J. comp. Psychol.* **22**, 157–78. 35, 37, 42.

MARQUIS, D. G. & HILGARD, E. R. (1937). Conditioned responses to light in monkeys after removal of the occipital lobes. *Brain*, **60**, 1–12. 35, 37.

MARSHALL, A. J. (1954). *Bower Birds*. London. 120.

MARSHALL, W. H. & TALBOT, S. A. (1940). Recovery cycle of the lateral geniculate of the nembutalized cat. *Amer. J. Physiol.* **129**, 417–8. 54.

MARSHALL, W. H. & TALBOT, S. A. (1942). Recent evidence for neural mechanisms in vision leading to a general theory of sensory acuity. *Biol. Symposia*, **7**, 117–64. 46.

MARX, M. H., HENDERSON, R. L. & ROBERTS, C. L. (1955). Positive reinforcement of bar-pressing response by a light stimulus following dark operant pretests with no after effect. *J. comp. physiol. Psychol.* **48**, 73–76. 288.

MASON, W. A., BLAZEK, N. C. & HARLOW, H. F. (1956). Learning capacities of the infant rhesus monkey. *J. comp. physiol. Psychol.* **49**, 449–52. 202.

MATEER, F. (1918). *Child Behaviour*. Boston. (*Not seen.*) 197.

MAYNARD-SMITH, J. (1952). The importance of the nervous system in the evolution of animal flight. *Evolution*, **6**, 127–9. 123.

MAYNARD-SMITH, J. & SAVAGE, R. J. G. (1956). Some locomotory adaptations in animals. *J. linn. Soc. Lond.* **42**, 604–22. 123.

MILLER, G. A. (1951). Speech and language. In *Handbook of Experimental Psychology*, ed. S. S. Stevens. New York. 263.

MILLER, N. E. & KESSEN, M. L. (1952). Reward effects of food via stomach fistula compared with those of food via mouth. *J. comp. physiol. Psychol.* **45**, 555–64. 291.

MILLER, N. E., SAMPLINER, R. I. & WOODROW, P. (1957). Thirst reducing effects of water by stomach fistula versus water by mouth measured by both a consummatory and instrumental response. *J. comp. physiol. Psychol.* **50**, 1–5. 291.

MILLS, J. N. & STANBURY, S. W. (1952). Persistent 24-hour renal excretory rhythm on a 12-hour cycle of activity. *J. Physiol.* **117**, 22–37. 212.

MILLS, J. N., THOMAS, S. & YATES, P. A. (1954). Reappearance of renal excretory rhythm after forced disruption. *J. Physiol.* **125**, 466–74. 212.

MILNER, B. (1954). Intellectual function of the temporal lobes. *Psychol. Bull.* **51**, 42–62. 79, 80.

MISHKIN, M. (1954). Visual discrimination performance following partial ablation of the temporal lobe. II. Ventral surface vs. hippocampus. *J. comp. physiol. Psychol.* **47**, 187–93. 31.

MISHKIN, M. & PRIBRAM, K. H. (1954). Visual discrimination performance following partial ablations of the temporal lobe. I. Ventral vs. lateral. *J. comp. physiol. Psychol.* **47**, 14–20. 31.

MISHKIN, M. & WEISKRANTZ, L. (1959). The effects of cortical lesions in monkeys on critical fusion frequency. *J. comp. physiol. Psychol.* **52**, 660–6. 45.

MITTELSTAEDT, H. (1957). Prey capture in Mantids. Recent advances in invertebrate physiology. Univ. Oregon Publications. 20.

MONAKOW, C. VON (1914). *Die Localisation im Grosshirn.* Wiesbaden.

MONTGOMERY, K. C. (1951). The relation between exploratory behaviour and spontaneous alternation in the white rat. *J. comp. physiol. Psychol.* **44**, 582–9. 281.

MONTGOMERY, K. C. (1952a). Exploratory behaviour and its relation to spontaneous alternation in a series of maze exposures. *J. comp. physiol. Psychol.* **45**, 50–57. 281.

MONTGOMERY, K. C. (1952b). A test of two explanations of spontaneous alternation. *J. comp. physiol. Psychol.* **45**, 287–93. 254, 281.

MONTGOMERY, K. C. (1953a). The effect of activity deprivation on exploratory behaviour. *J. comp. physiol. Psychol.* **46**, 438–41. 285.

MONTGOMERY, K. C. (1953b). The effect of hunger and thirst drives on exploratory behaviour. *J. comp. physiol. Psychol.* **46**, 315–19. 282.

MONTGOMERY, K. C. (1954). The role of the exploratory drive in learning. *J. comp. physiol. Psychol.* **47**, 60–64. 283, 286.

MONTGOMERY, K. C. (1955). The relation between fear induced by novel stimulation and exploratory behaviour. *J. comp. physiol. Psychol.* **48**, 254–60. 282, 283.

MONTGOMERY, K. C. & SEGALL, M. (1955). Discrimination learning based upon the exploratory drive. *J. comp. physiol. Psychol.* **48**, 225–8. 283.

MORGAN, C. LLOYD (1894). *Introduction to Comparative Psychology.* London.

MORGAN, C. LLOYD (1896). *Habit and Instinct.* London. 1.

MORGAN, C. T. & STELLAR, E. (1950). *Physiological Psychology*. New York. 35, 71, 75.

MORRIS, D. (1954). The reproductive behaviour of the Zebra Finch (*Poephila guttata*) with special reference to pseudo-female behaviour and displacement activities. *Behaviour*, **6**, 271–322. 102, 110, 114, 127.

MORRIS, D. (1956a). The feather postures of birds and the problem of the origin of social signals. *Behaviour*, **9**, 75–113. 102.

MORRIS, D. (1956b). The function and causation of courtship ceremonies. In *L'instinct dans le comportement des animaux et de l'homme*, 261–86. Fondation Singer-Polignac. Paris. 102, 113, 114, 125, 128, 136.

MORRIS, D. (1957a). 'Typical intensity' and its relationship to the problem of ritualisation. *Behaviour*, **11**, 1–13. 93, 102, 110, 113.

MORRIS, D. (1957b). The reproductive behaviour of the Bronze Mannikin (*Lonchura cucullata*). *Behaviour*, **11**, 156–201. 102, 103, 104, 109, 127.

MORRIS, D. The comparative ethology of grassfinches (Erythrurae) and Mannikins (Amadinae): a preliminary statement. (*In preparation*, 1960.) 102, 110.

MOUNTCASTLE, V. B. (1957). Modality and topographic properties of single neurons in the cat's somatic sensory cortex. *J. Neurophysiol.* **20**, 408–34. 347.

MOWRER, O. H. (1956). Two factor learning theory reconsidered, with special reference to secondary reinforcement and the concept of habit. *Psychol. Rev.* **63**, 114–28. 280.

MOWRER, O. H. & AIKEN, E. G. (1954). Contiguity vs. drive reduction in fear conditioning: Temporal variations in conditioned and unconditioned stimulus. *Amer. J. Psychol.* **67**, 26–38. 280.

MOWRER, O. H. & LAMOUREAUX, R. R. (1942). Avoidance conditioning and signal duration: a study of secondary motivation and reward. *Psychol. Monog.* 54, No. 5. 279.

MOWRER, O. H. & LAMOUREAUX, R. R. (1946). Fear as an intervening variable in avoidance conditioning. *J. comp. Psychol.* **39**, 29–49. 279.

MOYNIHAN, M. (1955a). Some aspects of reproductive behaviour in the black-headed gull (*Larus r. ridibundus* L.) and related species. *Behaviour Supplement* No. 4, 1–201. 117.

MOYNIHAN, M. (1955b). Remarks on the original sources of display. *Auk.* **72**, 240–46. 102.

MOYNIHAN, M. (1955c). Types of hostile display. *Auk*, **72**, 247–59. 102.

MOYNIHAN, M. (1958). Notes on the behaviour of some N. American gulls. II. Non-aerial hostile behaviour of adults. *Behaviour*, **12**, 95–182. 117.

MOYNIHAN, M. & HALL, F. (1953). Hostile sexual and other social behaviour patterns of the Spice Finch (*Lonchura punctulata*) in captivity. *Behaviour*, **7**, 33–77. 125, 127, 130, 135, 137.

MUNN, N. L. (1950). *Handbook of Psychological Research on the Rat*. Boston. 202, 235.

MUNN, N. L. (1954). Learning in Children, being Ch. 7 of *Manual of Child Psychology*, ed. Leonard Carmichael, 2nd ed. New York and London. 196, 199.

NEIRMARK, E. & SALTZMAN, I. J. (1953). Intentional and incidental learning with different rates of stimulus presentation. *Amer. J. Psychol.* **66**, 618–21. 265.

NICE, M. M. (1943). Studies in the life history of the Song Sparrow. II. *Trans. Linn. Soc. N.Y.* **6**, 1–328. 128, 135, 147, 178.

NICE, M. M. (1953). Some experiences in imprinting ducklings. *Condor*, **55**, 33–37. 177.

NICE, M. M. & TER PELKWYK, J. J. (1941). Enemy recognition by the Song Sparrow. *Auk.* **58**, 195–214. 164.

NOBLE, G. K. (1941). The effect of forebrain lesions on the sexual and fighting behaviour of *Betta splendens* and other fishes. *Anat. Rec., Suppl.* **79**, 49. 106.

OBERHOLZER, R. J. H. & HUBER, F. (1957) Methodik der elektrischen Reizung und Ausschaltung im Oberschlundganglion nicht narkotisiert Grillen. *Helv. Physiol. Acta*, **15**, 185–92. 27.

OGLE, K. N. (1950). *Binocular Vision.* Philadelphia. 54.

O'LEARY, J. L. (1940). A structural analysis of the lateral geniculate nucleus of the cat. *J. comp. Neurol.* **73**, 405–30. 31, 49.

OSIPOVA, V. N. (1926). (*See* Razran, 1933.) 197.

PANCHENKOVA, E. F. (1956). The ontogenetic development of conditioned reflexes in the white rat. (*In Russian.*) *Zhurnal Visshei Nerv'noi Deyatelnosti*, **6**, 312–8. 241.

PANTIN, C. F. A. (1952). The elementary nervous system. *Proc. Roy. Soc. Lond.* B, **140**, 147–68. 28.

PARAMONOVA, M. P. (1955). On the question of the development of the physiological mechanism of movement. (*In Russian.*) *Voprosy psichologii*, **1**, No. **3**, 51–62. 241.

PARRY, D. A. (1947). The function of the insect ocellus. *J. exp. Biol.* **24**, 211–9. 11.

PAVLOV, I. P. (1927). *Conditioned Reflexes.* Oxford. 226, 246.

PELKWIJK, J. J. TER & TINBERGEN, N. (1937). Eine reizbiologische Analyse einiger Verhaltensweisen von *Gasterosteus aculeatus* L. *Z. Tierpsychol.* **1**, 193–204. 151.

PERDECK, A. C. (1957). The isolating value of specific song patterns in two sibling species of grasshoppers (*Chorthippus brunneus* Thunb. and *C. biguttulus* L.). *Behaviour*, **12**, 1–72. 163.

PESETSKY, I. & KOLLROSS, I. J. (1956). A comparison of the influence of locally applied thyroxine upon Mauthner's cell and adjacent neurons. *Exp. Cell. Res.* **11**, 477–82. 117.

PETERSON, B., TÖRNBLUM, O. & BODIN, N. O. (1952). Verhaltenstudien am Rapsweissling und Bergweissling (*Pieris napi* L. and *P. bryoniae* Ochs.). *Behaviour*, **4**, 67–84. 162.

PIAGET, J. (1929). *The Child's Conception of the World.* London. 191.

PICKETT, J. M. (1952). Non-equipotential cortical function in maze learning. *Amer. J. Psychol.* **65**, 177–95. 71, 74, 77.

PITTENDRIGH, C. S. (1954). On temperature independence in the clock system controlling emergence time in *Drosophila. Proc. nat. Acad. Sci.* **40**, 1018–29. 213, 216.

POLIAK, S. (1927). An experimental study of the associational, callosal and projectional fibers of the cerebral cortex of the cat. *J. comp. Neurol.* **44**, 197–257. 32.

POLIAK, S. (1932). *Afferent Fiber Systems in Primate Cerebral Cortex*. Berkeley, California. 32.

POLIAK, S. (1933). A contribution to the cerebral representation of the retina. *J. comp. Neurol.* **57**, 541–617. 32, 45.

POLIAK, S. (1941). *The Retina*. Chicago. 46.

POLIAK, S. & HAYASHI, R. (1936). The cerebral representation of the retina in the chimpanzee. *Brain*, **59**, 51–60. 32, 52.

POWELL, D. R. & PERKINS, C. C. (1957). Strength of secondary reinforcement as a determiner of the effects of duration of goal response on learning. *J. exp. Psychol.* **53**, 106–12. 298.

PRATT, K. (1954). The neonate, in *Manual of Child Psychology*, ed. L. Carmichael. 2nd ed. New York. 191.

PRECHT, H. (1952). Über das angeborene Verhalten im Tieren. Versuch an Springspinnen (Salticidae). *Z. Tierpsychol.* **9**, 207–30. 117.

PRIBRAM, K. H. & BAGSHAW, M. (1953). Further analysis of the temporal lobe syndrome utilizing fronto-temporal ablations. *J. comp. Neurol.* **99**, 347–76. 79.

PRINGLE, J. W. S. (1938). Proprioception in insects. II. The action of the campaniform sensillae. *J. exp. Biol.* **15**, 114—31. 11.

PRINGLE, J. W. S. (1954). A physiological analysis of Cicada song. *J. exp. Biol.* **31**, 525–60. 12, 111.

PRINGLE, J. W. S. (1957). *Insect Flight*. Cambridge. 123.

PUMPHREY, R. J. (1940). Hearing in insects. *Biol. Rev.* **15**, 107–32. 12, 13, 157.

PUMPHREY, R. J. & RAWDON-SMITH, A. F. (1937). Synaptic transmission of nervous impulses through the last abdominal ganglion of the Cockroach. *Proc. Roy. Soc. (Lond.)* B, **122**, 106–18. 11.

QUASTLER, H. (1956). Studies of human channel capacity. In *3rd London Symposium on Information Theory*, ed. C. Cherry, 361–71. London. 263, 339.

RÄBER, H. (1948). Analyse des Balzverhaltens eines domestizierten Truthahns (Meleagris). *Behaviour*, **1**, 237–66. 188, 189.

RÄBER, H. (1950). Das Verhalten gefangener Waldohreulen (*Asio otus otus*) und Waldkäuze (*Strix aluco aluco*) zur Beute. *Behaviour*, **2**, 1–95. 153.

RADEMAKER, G. G. J. & TER BRAAK, J. W. G. (1948). On the central mechanism of some optical reactions. *Brain*, **71**, 48–76. 36, 37, 56.

RAMSAY, A. O. (1951). Familial recognition in domestic birds. *Auk*, **68**, 1–16. 177.

RAMSAY, A. O. & HESS, E. H. (1954). A laboratory approach to the study of imprinting. *Wilson Bull.* **66**, 196–206. 177, 179, 182, 184, 185, 203, 204, 205.

RAZRAN, G. H. S. (1933). Conditioned responses in children. *Arch. Psychol. N.Y.* No. 148. 196.

RAZRAN, G. H. S. (1935). Conditioned responses: an experimental study and a theoretical analysis. *Arch. Psychol. N.Y.* No. 191. 198.

REITAN, R. M. (1955). Certain differential effects of left and right cerebral lesions in human adults. *J. comp. physiol. Psychol.* **48**, 474–77. 82.

REY, A. (1954). Le freinage volontaire du mouvement graphique chez l'enfant. Cahiers de pédagogue et d'orientation professionelle. *L'Institut supérieur de science pédagogiques de l'Université de Liège*, No. 2. 242.

RICHARDSON, C. A. (1933). The growth and variability of intelligence. Pt. I, *Brit. J. Psychol. Mon. Suppl.* No. 18. 194.

RICHTER, C. P. (1922). A behaviouristic study of the activity of the rat. *Comp. Psychol. Monog.* I (2), pp. 55. 235.

RIDDOCH, G. (1917). Dissociation of visual perceptions due to occipital injuries, with especial reference to appreciation of movement. *Brain*, **40**, 15–57. 56.

ROEDER, K. D. (1935). An experimental analysis of the sexual behaviour of the Praying Mantis. *Biol. Bull.* **69**, 203–220. 26.

ROEDER, K. D. (1948). Organisation of the ascending giant fibre system in the Cockroach. *J. exp. Zool.* **108**, 243–362. 11.

ROEDER, K. D. (1953). *Insect Physiology.* New York. 18, 26, 106.

ROEDER, K. D. (1959). A Physiological Approach to the Relation between Prey and Predator. From Studies in Invertebrate Morphology, Smithsonian Misc. Coll. 137. 287–306. 13.

ROEDER, K. D. & TREAT, A. E. (1957). Ultrasonic reception by the tympanic organ of Noctuid moths. *J. Exp. Zool.* **34**, 127–158. 13.

ROSVOLD, H. D. & DELGADO, J. M. R. (1956). The effect on delayed alternation test performance of stimulating or destroying electrically structures within the frontal lobes of the monkey's brain. *J. comp. physiol. Psychol.* **49**, 365–72. 106.

ROTH, L. M. (1948). An experimental laboratory study of the sexual behaviour of *Aedes aegypti* (1). *Amer. Midl. Nat.* **40**, 265—352. 157.

RUCH, T. (1951). Chapter on Motor System in *Handbook of Experimental Psychology*, ed. Stevens. New York. 16.

RUCK, P. (1957). The electrical responses of the dorsal ocelli in Cockroaches and Grasshoppers. *J. Insect. Physiol.* **1**, 109–23. 11.

RUSSELL DAVIS, D. (1957). A note on auditory imperception. (*In press.*) 198.

RUSSELL, E. S. (1946). *The Directiveness of Organic Activities.* Cambridge. 308.

RUSSELL, W. M. S., MEAD, A. P. & HAYES, J. S. (1954). A basis for the quantitative study of the structure of behaviour. *Behaviour*, **6**, 153–206. 371.

SALT, G. (1935). Experimental studies in host parasitism. III. Host selection. *Proc. Roy. Soc.* B. **117**, 413–35. 154.

SALT, G. & HOLLICK, F. S. J. (1946). Studies of wireworm populations. II. Spatial distribution. *J. exp. Biol.* **23**, 1–46. 129.

SANDEEN, M. I., STEPHENS, G. C. & BROWN, F. A., JNR. (1954). Persistent daily and tidal rhythms of oxygen consumption in two species of marine snails. *Physiol. Zool.* **27**, 350–6.

SCHALLER, F. & TIMM, C. (1949). Schallreaktionen bei Nachtfaltern. *Experientia*, **15**, 162. 153.

SCHMIDT, R. S. (1955). The evolution of nest-building behaviour in Apicotermes (Isoptera). *Evolution*, **9**, 157–81. 103, 119.

SCHNEIDER, D. (1957). Elektrophysiologische Untersuchungen von Chemo- und Mechanorezeptoren der Antenne des Seidenspinners *Bombyx mori* L. *Z. vergl. Physiol.* **40**, 8–41. 158.

SCHNEIRLA, T. C. (1946). Ant learning as a problem in comparative psychology. In *Twentieth Century Psychology*, ed. Harriman. New York. 28.

SCHNEIRLA, T. C. (1956). Interrelationships of the 'innate' and the 'acquired' in instinctive behaviour. In *L'Instinct dans le comportement des animaux et de l'homme.* Paris. 155, 160, 161.

REFERENCES

SCHÖNHERR, J. (1955). Über die Abhängigheit der Instinkthandlungen vom Vorderhirn und Zwischenhirn (Epiphyse) beim *Gasterosteus aculeatus*. *Zool. Jahrb. Allg. Zool. Physiol.* **65**, 357–87. 106.

SCHRODER, H. M. & ROTTER, J. B. (1954). Generalisation of expectancy changes as a function of the nature of reinforcement. *J. exp. Psychol.* **48**, 343–8. 258.

SCHWINCK, I. (1953). Über den Sexualduftstoff der Pyraliden. *Z. vergl. Physiol.* **35**, 167–74. 153.

SCHWINCK, I. (1955). Weitere Untersuchungen zum Frage der Geruchsorientierung der Nachtschmetterlinge: partielle Fühleramputation bei Spinnermännchen insbesondere aus Seidenspinner (*Bombyx mori* L.). *Z. vergl. Physiol.* **37**, 439–58. 158.

SCOTT, J. P. (1945). Social behaviour, organisation and leadership in a small flock of domestic sheep. *Comp. Psychol. Monogr.* **18**, 1–29. 209.

SCOTT, J. P., FREDERICKSON, E. & FULLER, J. L. (1951). Experimental exploration of the critical period hypothesis. *Personality*, **1**, 162–83. 192, 208.

SCOVILLE, W. B. & MILNER, B. (1957). Loss of recent memory after bilateral hippocampal lesions. *J. Neurol. Neurosurg. Psychiat.* **20**, 11–21. 83.

SEARS, R. A. & WISE, G. W. (1950). Relation of cup-feeding in infancy to thumb-sucking and the oral drive. *Amer. Journ. Orthopsych.* **20**, 123–38. 192.

SEGAAR, J. (1956). Brain and instinct in *Gasterosteus aculeatus*. *Kon. Ned. Akad. Wet. Proc.* **59**, 738–49. 106.

SEITZ, A. (1940). Die Paarbildung bei einigen Cichliden. I. Die Paarbildung bei *Astatotilapia strigigena*. *Z. Tierpsychol.* **4**, 40–84. 98.

SEITZ, A. (1942). Die Paarbildung bei einigen Cichliden. II. Die Paarbildung bei *Hemichromis bimaculatus*. *Z. Tierpsychol.* **5**, 74–101. 98.

SETTLAGE, P. H. (1939). The effect of occipital lesions on visually guided behaviour in the monkey. *J. comp. Psychol.* **27**, 93–131. 44, 46.

SEWARD, J. & LEVY, N. (1949). Sign learning as a factor in extinction. *J. exp. Psychol.* **39**, 660–8. 253.

SHANNON, C. E. & WEAVER, W. (1949). *The Mathematical Theory of Communication.* Univ. Illinois Press, Urbana. 339, 342.

SHEFFIELD, F. D. & ROBY, T. B. (1950). Reward value of a non-nutritive sweet taste. *J. comp. physiol. Psychol.* **43**, 471–81. 289.

SHEFFIELD, F. D., ROBY, T. B. & CAMPBELL, B. A. (1954). Drive reduction versus consummatory behaviour as determinants of reinforcement. *J. comp. physiol. Psychol.* **47**, 349–54. 290.

SHERMAN, M. (1927). The differentiation of emotional responses in infants. *J. comp. Psychol.* **7**, 335–51. 191.

SHERRINGTON, C. S. (1906). *The Integrative Action of the Nervous System.* New York. 1, 117, 333, 345.

SHOLL, D. A. (1956). *The Organisation of the Cerebral Cortex.* London. 7, 84, 335.

SLONAKER, J. R. (1912). The normal activity of the albino rat from birth to natural death, its rate of growth and the duration of life. *J. Anim. Behav.* **2**, 20–42. 235.

SMITH, K. U. (1937*a*). The effects of removal of the striate cortex upon certain unlearned visually controlled reactions in the cat. *J. genet. Psychol.* **50**, 137–56. 34, 35, 42.

SMITH, K. U. (1937b). Visual discrimination in the cat: V. Postoperative effects of removal of the striate cortex upon intensity discrimination. *J. genet. Psychol.* **51**, 329–69. 32, 33, 34, 40, 42, 43.

SMITH, K. U. (1938). Visual discrimination in the cat. VI. The relation between pattern vision and visual acuity and the optic projection centers of the nervous system. *J. genet. Psychol.* **53**, 251–72. 32, 34, 42.

SMITH, K. U. & BRIDGMAN, MARJORIE (1943). The neural mechanisms of movement vision and optic nystagmus. *J. exp. Psychol.* **33**, 165–87. 34, 38.

SMITH, K. U. & WARKENTIN, J. (1939). The central neural organization of optic functions related to minimum visual acuity. *J. genet. Psychol.* **55**, 177–95. 33, 34.

SMITH, M. P. & CAPRETTA, P. J. (1956). Affect of drive level and experience on the reward value of saccharine solution. *J. comp. physiol. Psychol.* **49**, 553–7. 289.

SMITH, M. and DUFFY, M. (1957a). Consumption of sucrose and saccharine by hungry and satiated rats. *J. comp. physiol. Psychol.* **50**, 65–69. 290.

SMITH, M. & DUFFY, M. (1957b). Evidence for a dual reinforcing effect of sugar. *J. comp. physiol. Psychol.* **50**, 242–7. 290.

SPALDING, D. (1873). Instinct, with original observations on young animals. *MacMillan's Magazine*, reprinted in *Brit. J. Anim. Behav.* 1954, **2**, 2–11. 167. 175, 184.

SPALDING, J. M. K. (1952a). Wounds of the visual pathway. I. The visual radiation. *J. Neurol. Neurosurg. Psychiat.* **15**, 99–109. 51.

SPALDING, J. M. K. (1952b). Wounds of the visual pathway. II. The striate cortex. *J. Neurol. Neurosurg. Psychiat.* **15**, 169–81. 51.

SPEARMAN, C. (1937). *Psychology Down the Ages.* 2 vols. London. 68.

SPENCE, K. W. (1940). Continuous versus non-continuous interpretations of discrimination learning. *Psychol. Rev.* **47**, 271–88. 267.

SPENCE, K. W. (1951). Theoretical implications of learning. (Ch. 18 in Steven's *Handbook of Experimental Psychology*.) 171.

SPENCE, K. W. & FULTON, J. F. (1936). The effects of occipital lobectomy on vision in chimpanzee. *Brain*, **59**, 35–50. 44, 52.

SPIETH, H. T. (1952). Mating behaviour within the genus *Drosophila* (Diptera). *Bull. Amer. Mus. Nat. Hist.* **99**, 395–474. 105, 119.

SPIETH, H. T. & HSU, T. C. (1950). The influence of light on seven species of the *Drosophila melanogaster* group. *Evolution*, **4**, 316–25. 108.

SPITZ, R. A. & WOLFE, K. M. (1946). The smiling response: a contribution to the ontogenesis of social relations. *Genet. Psychol. Monogr.* **34**, 57–125. 190, 191.

STAMM, J. S. (1953). Effects of cortical lesions on established hoarding activity in rats. *J. comp. physiol. Psychol.* **46**, 299–304. 77.

STAMM, J. S. (1954). Control of hoarding activity in rats by the median cerebral cortex. *J. comp. physiol. Psychol.* **47**, 21–27. 76.

STAMM, J. S. (1955). The function of the median cerebral cortex in maternal behaviour of rats. *J. comp. physiol. Psychol.* **48**, 347–56. 76.

STECHE, W. (1957). Beiträge zur Analyse der Bienentänze (Teil I). *Insectes Sociaux*, **4**, 305–18. 121.

STEIN, R. C. (1956). A comparative study of 'advertising song' in the *Hylocichla* thrushes. *Auk*, **73**, 503–12. 108.

STEINBACHER, G. (1939). Zum Problem der Haustierwedung. *Z. Tierpsychol.* **2**, 302–13. 189.

STEIN-BELING (1935). Über das Zeitgedächtnis bei Tieren. *Biol. Rev.*, **10**, 18–41. 214.

STENGEL, E. (1947). A clinical and psychological study of echo-reactions. *J. Ment. Sci.* **93**, 598–612. 192.

STEVEN, D. M. (1955). Transference of 'imprinting' in a wild gosling. *Brit. J. Anim. Behav.* **3**, 14–16. 183.

STEVENS, S. S. & DAVIS, H. (1938). *Hearing: Its Psychology and Physiology*. New York. 13.

STIRNIMAN, F. (1941). Griefversuche mit der Hand Beugeborener. *Ann. Paediatr.* **157**, 17–27. 191.

STONE, C. P. (1922). The congenital sexual behaviour of the young male albino rat. *J. comp. Psychol.* **2**, 95–153. 151.

STONE, C. P. (1929*a*). The age factor in animal learning. I. Rats in the problem box and the maze. *Genet. Psychol. Monog.* **5**, 1–130. 235.

STONE, C. P. (1929*b*). The age factor in animal learning. II. Rats in a multiple light discrimination box and a difficult maze. *Genet. Psychol. Monog.* **6**, 125–202. 235.

STRIDE, G. O. (1956). On the courtship behaviour of *Hypolimnas misippus* L. (Lepidoptera, Nymphalidae) with notes on the mimetic association with *Danaus chrysippus* L. (Lepidoptera, Danaidae). *Brit. J. Anim. Behav.* **4**, 52–68. 108, 163.

STRIDE, G. O. (1957). Investigations into the courtship behaviour of the male of *Hypolimnas misippus* L. (Lepidoptera, Nymphalidae), with special reference to the role of visual stimuli. *Brit. J. Anim. Behav.* **5**, 153–67. 108.

SYMONDS, C. P. & MACKENZIE, I. (1957). Bilateral loss of vision from cerebral infarction. *Brain*, **80**, 415–55. 36.

SZAFRAN, J. & WELFORD, A. T. (1950). On the relation between transfer and difficulty of the initial task. *Quart. J. exp. Psychol.* **2**, 88–94. 386.

TAVOLGA, W. N. (1956). Visual, chemical and sound stimuli as cues in the sex discriminatory behaviour of Gobiid fish *Bathygobius soporator. Zoologica*, **41**, 49–64. 151.

THISTLETHWAITE, D. (1951). A critical review of latent learning and similar experiments. *Psychol. Bull.* **48**, 97–129. 270.

THOMPSON, W. R. & HERON, W. (1954*a*). The effects of early restriction on activity in dogs. *J. comp. physiol. Psychol.* **47**, 77–82. 239, 240.

THOMPSON, W. R. & HERON, W. (1954*b*). The effects of restricting early experience on the problem-solving capacity of dogs. *Canad. J. Psychol.* **8**, 17–31. 239.

THORPE, W. H. (1944). Some problems of animal learning. *Proc. Linn. Soc. Lond.* **156**, 70–83. 154.

THORPE, W. H. (1945). The evolutionary significance of habitat selection. *J. Anim. Ecol.* **14**, 67–70. 175.

THORPE, W. H. (1956). *Learning and Instinct in Animals*. London. 110, 126, 175, 176, 178, 179, 183, 184, 208, 213, 215, 220, 226.

THORPE, W. H. (1958a). The learning of song patterns by birds, with especial reference to the song of the Chaffinch (*Fringilla coelebs*). *Ibis*, **101**, 535–70. 110, 209.

THORPE, W. H. (1958b). Further studies on the process of song learning in the Chaffinch (*Fringilla coelebs gengleri*). *Nature*, **182**, 554–7. 110, 209.

THORPE, W. H. & JONES, E. G. W. (1937). Olfactory conditioning and its relation to the problem of host selection. *Proc. Roy. Soc.* B, **124**, 56–81. 152.

TINBERGEN, N. (1939). The behaviour of the Snow Bunting in spring. *Trans. Linn. Soc. NY.* **5**, 1–95. 147.

TINBERGEN, N. (1942). An objectivistic study of the innate behaviour of animals. *Biblioth. biotheor.* **1**, 39–98. 209.

TINBERGEN, N. (1951). *The Study of Instinct.* Oxford. 102, 115, 116, 125, 178, 332.

TINBERGEN, N. (1952). 'Derived' activities; their causation, biological significance, origin and emancipation during evolution. *Quart. Rev. Bio.* **27**, 1–32. 94, 102, 112, 115.

TINBERGEN, N. (1953). *The Herring Gull's World.* London. 178.

TINBERGEN, N. (1957). The function of territory. *Bird Study*, **4**, 14–27. 127, 129.

TINBERGEN, N. & VAN IERSEL, J. J. A. (1947). 'Displacement reactions' in the Three-spined Stickleback. *Behaviour*, **1**, 56–63.

TINBERGEN, N. & KUENEN, D. J. (1939). Über die auslösenden und richtunggebenden Reizsituationen der Sperrbewegung von jungen Drosseln (*Turdus m. merula* L. and *T. e. ericetorum* Turton). *Z. Tierpsychol.* **3**, 37–60. 160, 177.

TINBERGEN, N. & PERDECK, A. C. (1950). On the stimulus situation releasing the begging response in the newly-hatched Herring Gull chick (*Larus a. argentatus*). *Behaviour*, **3**, 1–38. 98, 152, 156, 160.

TOWE, A. L. & AMASSIAN, V. E. (1958). Patterns of activity in single cortical units following stimulation of the digits in monkeys. *J. Neurophysiol.* **21**, 292–311. 347.

TSANG, Y. C. (1934). The functions of the visual areas of the cerebral cortex in the learning and retention of the maze. I. *Comp. Psychol. Monogr.* **10**, 1–56. 70.

TSANG, Y. C. (1936). The functions of the visual areas of the cerebral cortex in the learning and retention of the maze. II. *Comp. Psychol. Monogr.* **12**, 1–41. 70.

TSANG, Y. C. (1937). Maze learning in rats hemidecorticated in infancy. *J. comp. Psychol.* **24**, 221–54. 72.

UTTLEY, A. M. (1954a). The classification of signals in the nervous system. *E.E.G. Clin. Neurophysiol.* **6**, 479–94. 328, 360.

UTTLEY, A. M. (1954b). Conditional probability machines and conditioned reflexes. *R.R.E. Memo* No. 1045. 328, 360.

UTTLEY, A. M. (1955). The conditional probability of signals in the nervous system. *R.R.E. Memo* No. 1109. 360.

UTTLEY, A. M. (1958). Conditional probability as a principle of learning. *Actes Ire Congrès Cybernétique, Namur*, 1956. Paris. 360.

VERPLANCK, W. S. (1955). An hypothesis on imprinting. *Brit. J. Anim. Behav.* **3**, 123. 184, 189.

VINCE, M. A. (1958). 'String-pulling' in birds. 2. Differences related to age in greenfinches, chaffinches and canaries. *Anim. Behav.* **6**, 53–59. 226.

VINCE, M. A. (1959). Effects of age and experience on the establishment of internal inhibition in finches. *Brit. J. Psychol.* **50**, 136–44. 227.

REFERENCES

VINCE, M. A. (1960). Developmental changes in responsiveness in the great tit (*Parus major*). *Behaviour*, **15**, 219–243. 229, 231, 232, 246, 361.

VOGEL, G. (1957). Verhaltensphysiologische Untersuchungen über die den Weibchenbesprung des Stubenfliegen-Männchens (*Musca domestica*) auslösenden optischen Faktoren. *Z. Tierpsychol.* **14**, 309–23. 159.

VOWLES, D. M. (1954a). The structure and connections of the corpora pedunculata in bees and ants. *Quart. J. micr. Sci.* **96**, 239–55.

VOWLES, D. M. (1954b). The orientation of ants. I. The substitution of stimuli. II. Orientation to light, gravity and polarised light. *J. exp. Biol.* **31**, 341–55 and 356–75. 121.

VOWLES, D. M. (1955). On the foraging of ants. *Brit. J. Anim. Behav.* **3**, 1–3. 27, 153.

VOWLES, D. M. (1958). The perceptual world of ants. *Anim. Behav. (in press).* 28.

WADDINGTON, C. H. (1952). Epigenetics and evolution. *Symp. Soc. exp. Biol.* **7**, 186–99. 104.

WADDINGTON, C. H. (1953). Genetic assimilation of an acquired character. *Evolution*, **7**, 118–26. 104.

WAHL, O. (1932). Neue Untersuchungen über das Zeitgedächtnis der Bienen. *Z. vergl. Physiol.* **16**, 529–89. 218.

WALKER, A. E. & FULTON, J. F. (1938). Hemidecortication in chimpanzee, baboon, macaque, potto, cat and coati: a study in encephalization. *J. nerv. ment. Dis.* **87**, 677–700. 44.

WALKER, E. L., DEMBER, W. N., EARL, R. W. & KAROLY, A. J. (1955). Choice alternation: I. Stimulus versus place versus response. *J. comp. physiol. Psychol.* **48**, 19–23. 281.

WALKER, E. L., DEMBER, W. N., EARL, R. W., FLIEGE, S. E. & KAROLY, A. J. (1955). Choice alternation: II. Exposure to stimulus or stimulus and place without choice. *J. comp. physiol. Psychol.* **48**, 24–28. 281.

WALLACE, B. & DOBZHANSKY, T. H. (1946). Experiments on sexual isolation in *Drosophila*. VIII. Influence of light on the mating behaviour of *Drosophila subobscura, D. persimilis, D. pseudobscura. Proc. Nat. Acad. Sci.* **32** (8), 226–34. 108.

WALLS, G. L. (1940). *The Vertebrate Eye.* Michigan. 161.

WALLS, G. L. (1953). *The Lateral Geniculate Nucleus and Visual Histophysiology.* Berkeley, California. 32, 49, 55.

WALSH, E. G. (1957). *Physiology of the Nervous System.* London. 15.

WARD, A. A. & KENNARD, MARGARET A. (1942). Effect of cholinergic drugs on recovery of function following lesions of the central nervous system in monkeys. *Yale J. Biol. Med.* **15**, 189–229. 53.

WATSON, J. B. (1903). *Animal Education.* Chicago. 225, 234.

WEIDMANN, R. & WEIDMANN, U. (1958). An analysis of the stimulus-situation releasing food-begging in the Black-headed Gull. *Brit. J. Anim. Behav.* **6**, 114. 105, 161.

WEIDMANN, U. (1956). Some experiments on the following and the flocking reaction of Mallard ducklings. *Brit. J. Anim. Behav.* **4**, 78–79. 168, 185, 205.

WEIH, A. S. (1951). Untersuchungen über das Wechselsingen (Anaphonie) und über das angeborene Lautschema einiger Feldheuschrecken. *Z. Tierpsychol.* **8**, 1–41. 163.

WEINSTEIN, S. & TUEBER, H. L. (1957). Effects of penetrating brain injury on intelligence test scores. *Science*, **125**, 1036–7. 82.

WEISENBURG, T. & MACBRIDE, K. E. (1935). *Aphasia: a clinical and psychological study.* New York. 82.

WEIS-FOGH, T. (1956). Notes on sensory mechanisms in locust flight. *Brit. J. Anim. Behav. Phil. Trans.* B, **239**, 553–84. 123.

WEISKRANTZ, L. (1956). Behavioral changes associated with ablation of the amygdaloid complex in monkeys. *J. comp. physiol. Psychol.* **49**, 381–91. 106.

WELFORD, A. T. (1956). Age and learning: theory and needed research. *Experientia Supplementum 4. Symposium on Experimental Gerontology.* 200.

WELLS, M. J. (1958). Factors affecting reaction to *Mysis* by newly hatched *Sepia. Behaviour*, **13**, 96–111. 170.

WELSH, J. H. (1941). The Sinus glands and 24-hour cycles of retinal pigment migration in the Crayfish. *J. exp. Zool.* **86**, 35–59. 214.

WHEELER, W. M. (1928). *The Social Insects.* London.

WIGGLESWORTH, V. B. (1953). *The Principles of Insect Physiology.* Cambridge. 159.

WIGGLESWORTH, V. B. (1959). The histology of the nervous system of an insect. II. *Quart. J. Micr. Sci.* **100**, 299–315. 6, 11.

WIKE, E. L. & CASEY, A. (1954). The secondary reinforcing value of food for thirsty animals. *J. comp. physiol. Psychol.* **47**, 240–3. 283.

WILLIAMS, M. & PENNYBACKER, J. (1954). Memory disturbances in third ventricle tumours. *J. Neurol. Neurosurg. Psychiat.* **17**, 115–23. 83.

WILSON, W. A. & MISHKIN, M. (1955). A survey of the effects of lateral striate and of inferotemporal lesions upon visually-guided behaviour in the monkey. *Amer. Psychol.* **10**, 397. 45.

WOHLGEMUTH, A. (1911). On the after-effects of seen movement. *Brit. J. Psychol. Monograph Suppl.* **1**, 1–117. 352.

WOLF, E. & ZERRAHN-WOLF, G. (1935). The effect of light intensity, area and flicker frequency on the visual reactions of the honeybee. *J. gen. Physiol.* **18**, 853–63. 12.

WOLFLE, H. M. (1949). The importance of the caress in modern child psychology. *Amer. Psychol.* **4**, 249. 191.

WOOD, J. W. (1956). 'Taming' of the wild Norway rat by rhinencephalic lesions. *Nature*, **178**, 869. 106.

WOODWARD, P. M. (1953). *Probability and Information Theory with applications to Radar.* London. 338, 339, 346.

WYATT, S. & LANGDON, J. N. (1932). Inspection processes in industry. *I.H.R.B. Report* No. 63. 257.

YOUNG, J. Z. (1957). *The Life of Mammals.* Oxford. 14.

INDEX